# Automotive Climate Control
# 116 Years of Progress

Gene D. Dickirson

Copyright © 2012 By Gene D. Dickirson

Published By Lulu.com

All rights are reserved. No part or contents of this book may be reproduced or transmitted in any form or by any means, by recording or any information storage or retrieval system without written permission from the author.

First Printing January 2012, Second Printing April 2012

Printed in the United States of America
ISBN 978-1-105-18361-4

This book is dedicated to the climate control engineers that preceded us and to those who will follow in the future.

This page intentionally blank.

# CONTENTS

**PART 1 From Horse and Carriage to Horseless Carriage/1**

**PART 2 From Aftermarket Hang-On Components to Integrated Systems/14**

**PART 3 Air Conditioning Returns to Stay/20**

**PART 4 From Electro Mechanical Systems to Electronic Systems/28**

**BIBLIOGRAPHY/509**

**INDEX/513**

# PREFACE

This book has been compiled for use as a reference for future engineers who will design and build automotive climate control systems. It is also intended for automotive historians and restorers who are interested in the topics covered.

The book is composed primarily of images rather than text because images tell the story very well and minimize misinterpretation.

This book is based upon a booklet that I compiled while working as an engineering supervisor at the Ford Motor Company, Climate Control Division in the early 1990's. The 99-page booklet was titled *Automotive Climate Control the first 100 years*. The booklet was used in conjunction with lectures presented to newly hired engineers to acquaint them with the many aspects of automotive climate control components and systems. Several people have encouraged me to expand the booklet into a book and publish it.

This book includes essentially every topic covered by a paper prepared by Harold V. Joyce in 1960 titled *Progress of Automotive Heating Through The Years*. Mr. Joyce was the manager of the Heating, Ventilating and Air Conditioning department of the Ford Motor Company at the time he prepared the paper. He presented the paper at the Society of Automotive Engineers meeting held at the Sheraton Cadillac Hotel in Detroit, Michigan on March 16, 1960. Mr. Joyce's paper included sketches of many of the various heating components and subsystems discussed in his paper. The sketches were most probably prepared by the graphics arts department of Ford Motor Company.

I have included several items related to windshield wipers in this book because they are a critical automotive system relating to "controlling" the climate.

Antifreeze development is also included because it is an important element in automotive cooling systems. Even though hot water heaters were not introduced for several years on water-cooled engines, it is interesting to track the progress of the coolants used in vehicles.

The book also covers the topics discussed at an SAE technical meeting held on September 18, 1990 at the Hyatt Regency Hotel in Dearborn, Michigan. The title of the session I attended was "50 Years of Climate Control". Paul E. Anglin organized the session. The speakers were Jerold Elson, General Manager of the Harrison Division of General Motors and Dave Bateman of DuPont.

Mr. Elson stated that air conditioning research began in 1930 at General Motors as an advance project and remained a low priority for several years.

Mr. Bateman stated that General Motors engineers, Thomas Midgley Jr., Robert R. McNary and Albert L. Henne, invented R-12 refrigerant. General Motors formed a joint venture with DuPont to manufacture R-12.

Every effort has been made to find the earliest example of the technology being shown. The example of any device or system shown may or may not be the absolute first because I did not search every document available because this book would never have been completed in my lifetime. (There used to be an expression in the office: "There comes a point in every program when it comes time to shoot the engineers and go into production".) In numerous cases when I thought I had found the first application of new technology I subsequently found a prior one. Other researchers will at least have a starting date to work from if they want or need to identify the absolute first.

Note that as this book was completed in December 2011 an effort was underway by Google to copy and digitize every book ever written and make it available on the Internet. At this point in time, the documents and books I used for reference were not digitized and were only available in their original format on paper. Thus it took many visits to libraries and combing thru my personal collection to find and digitize the information contained herein.

This page intentionally blank.

# ACKNOWLEDGEMENTS

I wish to thank the following people for their help in making this book possible:

My wife Barbara for her patience while I devoted nearly 900 hundred hours researching and typing the book and for proofreading the drafts with her comments.

The engineers who looked at the early drafts and their constructive comments: Ralph Emmons, Bob Amano, Gene Smith and Jerry Smutek.

The people at the National Automotive History Collection, Detroit Public Library: Ronald Grantz, Paige Plant, Mark Bowden and others.

The people at the Benson Ford Research Center located at The Henry Ford, Greenfield Village, Dearborn, Michigan.

Barb Dinan and the other Plymouth Library people for their help with MS Word.

Bob Elton for his comments and granting permission to photograph his Hudson.

Mark Kagle at the Parts Department at Hines Park Lincoln, Inc. in Plymouth, Michigan.

The people at the Stahls Classic Car Collection, Chesterfield, Michigan (formerly in St. Clair Shores, Michigan) for hosting the Cadillac LaSalle Club to visit their collection.

Mr. Richard Kughn and his collection executives Debbie Hayes and Robert Ferrand for hosting the Cadillac LaSalle club to visit their collection and the images they provided.

Mr. Ed Meurer for hosting the Cadillac LaSalle Club to visit his collection.

Wayne Schnaidt, Ken Sisk and Dan LaPalm for authorizing me to use their images in the book.

Ron Celano for his comments and suggestions.

Chuck Carlson for his comments and assistance with Photo Shop.

Michael W. R. Davis for his comments and leads to critical contacts.

Thomas J. DeZure for his comments, suggestions and lead to a critical contact.

The CLC Museum and Research Center for access to their historic materials.

Henry Beckwith, Gary Berlin, Jon Churgay and Don Wozniak for their comments regarding in-plant charging and pre-charge.

**The following firms and their people for authorizing the reproduction of their documents included in this book:**

American Bosch- Ambac International, Kenneth W. Smith

Amerigon- Daniel J. Pace, Dan Coker

Anderson Company, Federal –Mogul Corp.- Tracy Neil

The Auto Collections- Rob Williams

Automobile Quarterly Magazine-Rod Hottle, Jeffrey K. Leestma

Robert Bosch LLC- Cheryl Kilborn, Chandra Lewis

Chrysler Group, LLC- Donna L. Berry, Danielle M. McGurk

Corvair Underground- Lon Wall

Denso Corporation- Bridgette Gollinger, Tomohiro Miura, Goro Kanemasu

DuPont- Helena Bradley

Ford Motor Company-Raymond L. Coppiellie, Louis J. Ghilardi

Fountainhead Antique Auto Museum- Candy Waugaman, Nancy DeWitt

General Motors Company- Janice Mondary, Scott Farrar, Mark L. Reuss, Thomas Arnold

Richard & Regina Jandrey

Nissan- Rusty Reid

Mike and Debbie Rowand

Saab Cars North America- Nicole Jankowski

Stewart Warner, Meggitt (Troy), Inc.- James W. Tarter, Jr.

Toyota- Jennifer Mathis

Trico Products- James Crostin, James E. Finley

VDO- Caitalin Smith, Diane Esch, Esq

# PART 1
# From Horse and Carriage to Horseless Carriage

The focus of automotive engineers in the late 1800's and early 1900's was on getting the vehicles to go and stop. Ventilation was abundant because the vehicles were open and had no windshields. Keeping drivers and passengers warm in the early vehicles was achieved using the same techniques as with horse and carriage- charcoal foot heaters, carriage robes and gloves, coats and hats.

The Lehman Brothers and Chicago Flexible Shaft charcoal heaters were initially used for horse-drawn carriages. The 1901 Lehman ad states "150,000 are in actual use-an average yearly sale of 15,000- with a special coal they sell a million pieces a year." The special coal briquettes would be ignited in a wood or coal burning stove and transferred to the cloth-covered heater. The Lehman ad claimed the heaters would give off heat for 15 hours at a cost of ¼ cent per hour. The heaters were intended to be used in conjunction with lap robes.

On June 20, 1903, a Packard Overland vehicle departed from San Francisco on a cross-country journey to demonstrate the reliability of the brand. The journey included crossing the desert and the weather was naturally hot. The driver, E.T. (Tom) Fetch and photographer Marius C. Krarup, fitted an umbrella to shade them from the blistering sun. The concept of air conditioning an automobile was unimaginable because it would be many years before the required technology would be invented. The journey was successfully completed when the "Old Pac" [for Old Pacific] arrived in New York on August 21, 1903.

An ad in the February 1907, issue of Motor magazine listed a heater for a motorcar. This is the earliest mention of an automotive heater that came up in my research. The ad states, "The new Motor Car heater is the neatest thing yet invented". The illustration shows a jacket surrounding the vehicle's exhaust muffler and a duct from the jacket to a box. The box was intended to be attached to the vehicle's floor or toe board. A door and register in the box permitted the hot air to be directed into the passenger compartment or into the outside air. This represented rather sophisticated technology for the era.

Lap robes in the era were elaborate and were offered in many styles and prices. The 1908 Sears, Roebuck & Co. catalog had two full pages of lap robes. Prices ranged from $1.10 to $5.87. Lap robes would continue to be offered for many years by many different firms.

The cover offered by the Mandeville Steering Wheel Muff Co. was an example of how the automotive specific technology evolved. The device shown in the January 1908 ad in The Automobile magazine was made of either "leather covered Buffalo robe, cloth lining" or "leather covered lined with fur" and both types were offered with a window. The windows would have permitted the driver to see the steering column mounted spark and throttle controls. The devices were used to keep the drivers' hands warm and sold for $10.00 to $14.33.

The early vehicles were fitted with canvas covers to keep out the rain and snow. Some used small window openings made of transparent sheet mica that was commonly called "isinglass". Small pieces of glass were also used. Curtains using rigid sheet mica or glass would have been folded for storage and not rolled. Flexible side curtains as well as some early windshields were made of transparent celluloid and could be rolled for storage.

Glass windshields were introduced circa 1904. The Vasco windshield ad shown in the images section features "French plate glass". The website glassonline dot com described French plate glass as poured, ground and polished with good optical transmission qualities. This type of glass was dangerous when it was struck or bent because it broke into razor like shards that could severely injure people. The website glasslinks dot com states that John C. Wood invented safety plate glass in 1905 in England and it was called Triplex. It was brought to the U.S. in 1926 and Ford began using it in 1927.

Antifreeze was a major factor in automotive technology for vehicles equipped with water-cooled engines. It permitted the vehicles to be driven in cold climates by preventing the cooling water from freezing, expanding and breaking the engine block or cylinder head. The early antifreezes were calcium chloride, denatured alcohol and wood alcohol. An article in the September 1909 issue of The Automobile magazine states that light oil might function as antifreeze if mixed with the water in the cooling system if the engine had a water pump.

It is interesting to note that Mary Anderson received her patent for a "window cleaning device" in 1903 at the very beginning of glass windshields. An ad for a windshield-cleaning device appeared in the January 1912 issue of The Automobile magazine. It states that the hand-operated wiper operates from the niside [inside] and asks, "Which is cheaper and better- the old-fashioned way of soap and water or the up-to-date SECURITY method?"

Electrically heated gloves and electrically heated steering wheels were offered in 1912 and 1914. They were powered by the vehicle's battery, magneto or a portable dry cell battery.

In the 1915 issue of Dykes Automobile and Engine Encyclopedia the following article described a method of keeping a windshield clear. It appears to be similar to a windshield coating product being sold in 2011 called Rain- X. "A very practical method of keeping the windshield from clouding in wet or cold weather is to rub a half-and-half mixture of kerosene and glycerin on the clouded surface of the glass. A small bottle of the mixture tightly corked, may be wrapped in a little waste, put in a half pound baking powder can and stored away in the tool box or some other convenient place in the car. When it begins to rain, pour a small portion of the mixture on the waste, rub over the damp surface of the glass and all the water will spread out in a thin sheet where ever it comes in contact with the glycerin, instead of forming in little globules, which are so detrimental to the vision."

The 1915 Chevrolet Baby Grand on display at the Buick Gallery & Research Center in Flint, MI in 2011 featured two hand-operated windshield wipers. It is not known if they were factory installed or aftermarket units.

The 1916 Detroit Electric, owned by Jack Beatty, is interesting because it features a ventilation register above the windshield and an electric windshield wiper motor.

An early fresh air heater was offered in 1916 and it was called the Brickey Auto Heater. Air was ducted from the vehicle radiator and fan, thru a chamber surrounding the exhaust manifold then into the passenger compartment. This would have been a combination exhaust heater and fresh air heater.

Hudson offered a radiator shutter system as an option in 1916. The system was operated with a push-pull rod/ knob on the instrument panel. It permitted the driver to control the temperature of the coolant in the radiator by opening or closing the shutters. The device resulted in faster winter warm-up of the engine and passenger compartment if fitted with a heater. The vehicle needed an optional Motor Meter radiator cap that contained a thermometer visible from the driver's seat to determine when the desired temperature of 130 degrees F was reached. The shutter system sold for $15 and the Motor Meter sold for $10.

The Autocraft-Bovey ad in the October 1916 issue of Motor Age magazine stated the device was the first to incorporate an air-to-air heat exchanger. The ad focused on the safety aspects of the device claiming that it "WONT BURN YOUR SHOES" or set fire to the car. The heaters were priced at $25 to $35.

A factory-installed heater was featured on the 1917 Jordan. Up until the Jordan, most- if not all- heaters were dealer/owner installed aftermarket units.

The Motor World magazine in 1917 listed several manufacturers of antifreeze: S.C. Johnson and Sons, Racine, WI; Hall Thompson Co., Hartford, CN; Durkee Atwood Co, Minneapolis, MN; Columbus Varnish Co., Columbus, OH; Frezalene Mfg. Co. Inc, Chicago, IL; Orez Mfg. Co., Middletown, OH; Northwest Chemical Co., Marietta, OH. The list did not describe the active ingredients used in the antifreeze.

One of the earliest hot water heaters was described in the January 1917 issue of Motor Age magazine. The Peerless heater used two water- to- air heat exchangers. This was the forerunner of the method still in place today for heating vehicles with water-cooled engines.

The same issue of Motor Age also listed an electric heater manufactured by the Konserve Electric Co., Cleveland, OH. The article stated the heater gave "steady heat, but does not get hot enough to burn or scorch." This is the only electric heater mentioned in all of the documents I reviewed for this era.

The Motor Age article also described a hybrid water heater by the J.M. Shock Absorber Co., Philadelphia, PA. The heater used a water jacket around the exhaust "passage" and a footrest with a water jacket built in. Heater temperature could be controlled with a valve. The Mason Thermo heater featured a thermostat to control the temperature from the exhaust type heater. This is the first mention of a thermostat that I came across in my research.

The 1917 windshield "cleaner" ad by John W. Jepson featured two blades that were attached to a flat handle that slid between the upper and lower windshield glass plates. The driver hand-operated the device.

An article by Jim Donnelly in the January, 2010 issue of Hemming Classic Car magazine stated that John R. Oishei invested in the wiper system developed and sold by John Jepson beginning in 1916. Mr. Oishei purchased the firm in 1919 and later named it Tri-Continental Corporation or Trico. The article listed the following milestone events in the development of wiper/washer technology: refillable blades 1934, windshield washers 1936, and wipers standard on all U.S. cars 1937.

A cowl-top ventilation scoop was offered for the Ford Model T by the N.A. Petry Company in an ad in the Automotive Industries magazine dated October 1918. This was offered to fill a need for improved ventilation in a closed vehicle.

The windshield cleaner described in the 1918 Burke Clearsight Centrifugal brochure is interesting for two reasons. It is an early use of an electric motor to clear a windshield and it used a rotating disc of glass that was 7 inches in diameter. The motor and glass disc were attached to the outside of the windshield, perhaps by suction cups. The disc spun at 2000 RPM and slung rain, snow and dust off. An on-off switch mounted on the instrument panel controlled the device that was powered by the vehicle's battery. The motor was advertised as water and rustproof and was guaranteed for 5 years. The manufacturer was the Automobile Devices Co., Philadelphia, PA.

The 1919 Stadco See-Safe windshield wiper was hand-operated. It could be moved side-to-side or rotated. It was made by the Stadeker Metal Specialty Co., Chicago, IL and sold for 50 cents.

The 1920 Temptrol was a hand-operated thermostat to control the temperature of the engine coolant. It was operated by a push-pull rod/knob located on the instrument panel. It controlled a butterfly valve in the "circulation pipe". It was manufactured by the Johnson Co, Detroit, MI and sold for $7.50.

The 1920 Perfection Heater & Mfg. Company ad is interesting because it lists the forty various automobile manufacturers that used the heaters as standard equipment. The company offered 6 different heater models in the ad.

The Trico windshield wiper company listed an "Automatic Vacuum Wiper motor" introduced in 1921 on its history website tricoproducts dot com/About/Innovations. This was an early use of a vacuum motor for windshield wipers.

The 1922 issue of Dykes Automobile and Gasoline and Engine Encyclopedia carried an article on top, window and curtain repairs. Both glass and celluloid were mentioned as materials used for the transparent coverings. The same book described a method of making a windshield wiper by bending a quarter inch diameter steel or brass rod into two loops and attaching rubber hoses to the open ends using rubber cement.

The 1923 Buick brochure image shows a hand-operated windshield wiper and a cowl-top ventilation air scoop.

The Helzen Pedal Slot closures described in the Motor Age magazine ad in January 1924 were intended to prevent outside air from entering the passenger compartment. Vehicles in this era had large openings in the floor for the foot pedal arms to slide through. The aftermarket firm offered a product to fulfill a driver's desire to keep out cold/hot/dirty engine compartment air.

The Presto-Felt windshield cleaner in the 1924 Motor Age ad features a hand-operated wiper with a felt tipped blade that cleaned the inside of the glass as the rubber blade cleaned the outside.

The 1925 Kissel was fitted with a single vacuum windshield wiper motor. Note the shaft passing thru the glass.

Another early vacuum motor-powered windshield wiper system is described in the 1924 American Automatic Windshield Wiper brochure. The retail price was listed as $5.00. The Stromberg Motor Devices Co. offered an electric motor windshield wiper system for $9.00 in its brochure. The same year, the Folberth Auto Specialty Co. offered its vacuum motor wiper system for $7.00.

The Auto Shutter offered by the Haines Mfg. Corp. in 1925 was designed to fit Ford vehicles. The radiator shutter was operated with a hand control from the dash. The attachment to the radiator core was interesting because it used wires passing thru the core and held in place with push nuts. That type of push nut is still in use today.

The Kunkle heater described in the 1926 ad in the Ford Dealer and Service magazine featured a funnel and duct to push heated air from the radiator using the vehicle's fan. The air passed thru a 97-inch long exhaust-to-air heat exchanger then thru a register in the floor into the passenger compartment. The ad stated "Keeps fatal fumes from car." Prices were listed at $6.00 single- $9.00 double.

The 1925 Chevrolet featured a rollup windshield for ventilation. The system was called the Fisher VV windshield. The 1927 Chrysler, and perhaps other vehicles, also had rollup windshields. Note that Chevrolet also fitted pedal slot closures and robe rails as standard equipment on its 1926 models. The windshield wiper was mounted forward of the windshield.

Pierce Arrow's were fitted with parallel vertical wipers in the 1927 model year. The unique system appears to have used vacuum to power the arms horizontally. Two hoses are attached to the left side of the wiper drive mechanism on the vehicle in Richard Kughn's collection.

The Arvin ad in the December 1927 issue of Ford Dealer and Service listed six different heater models for the Ford Model T. They were all the exhaust type and prices ranged from $1.75 to $9.00.

In 1928, Trico offered a "Sleet Wand" system that used a fabric bag filled with rock salt attached to the windshield wiper arm. The product was intended to fight windshield icing. The source of this information was tricoproducts dot com/About/Innovations.

The patent issued on May 1, 1928 to O.S. Caesar, described the type of automotive heater system that is still in use today. The system consisted of a water-to-air heat exchanger and an electric fan to blow the hot air into the passenger compartment. This was a major change from the exhaust heaters then in use. The patent was assigned to Tropic-Aire.

The 1928 Chevrolet, owned by Jim and Barb Morningstar, has a chrome-plated floor register for the heater outlet that includes the Chevrolet bowtie logo. Note the molded rubber floor mat also has the logo pattern molded in.

The 1928 Ford Model A brochure listed a swing-out windshield for ventilation. The windshield was made of Triplex safety glass. Both were standard equipment. Even though the windshield would swing out for ventilation, two aftermarket devices were offered to improve the Model A's ventilation. One was the Ventilating Eave by Hubbard Products in 1928. The Hubbard visor was protected by U.S. Patent 1,609,409, which stated the purpose of the device was to shed rain or snow while the door glass was lowered for ventilation. It also claimed to improve ventilation and serve as a sunshade or visor. The 1929 Indian Sales Corporation's ventilator ad for the Model A claimed it was "Scientifically designed by Famous Automotive engineers to draw fresh clean *in* under pressure and to take old, stale air, and tobacco smoke and gas *out* by suction." The retail price for the Indian device was listed as $3.00 for a set of two.

The 1928 Whippet featured a windshield wiper motor mounted above and forward of the windshield. The wiper arm was a wire rod and the blade was attached to the arm using a cotter pin.

The 1929 Kissel had its windshield wiper motor mounted low and on the inside surface of the windshield. This was similar to the 1925 Kissel system discussed above.

The image of the 1929 Chevrolet being restored illustrates the mechanism used to raise and lower the Fisher VV windshield. The two circular spur gears with pins projecting forward were tied together with a shaft that was driven with the hand crank. As the gears rotated, the pins would move thru slots in the windshield glass causing it to move up or down.

The Greb "SLEETOFF" device advertised in the January 1929 issue of Ford Dealer and Service Field was a blade that swept the inside of the windshield. The same electrical circuit that powered the driving motor heated the blade. It was intended to defog the windshield and was "Your windshield wipers twin brother."

The migration to hot water heaters from the exhaust type began in 1929. An early ad for a Tropic-Aire brand hot water heater with an electrically driven fan appeared in the June 1929 issue of Ford Dealer and Service Field. The ad explained the new heaters had been used previously only on higher priced cars and advised the auto dealers that the heaters would be supported with a national advertising campaign that would reach 8,260,000 readers. An accompanying article elsewhere in the same magazine emphasized the fact that the system had no connection with the vehicle's exhaust. The Ford Model A heater kit included a new cast housing and thermostat that replaced the standard engine water outlet. The HaDees brand hot water heater was similar to the Tropic-Aire.

Some alternative heaters were offered in 1929. An example of a hot air heater was the Donaldson HOTTENTOT heater that was described in an ad in the September, 1929 issue of Ford Dealer and Service Field. It used a duct behind the Ford Model A's radiator fan to direct air into the passenger compartment. The kit included a radiator shutter and a push-pull rod/knob to control the shutter from the passenger compartment. A separate knob operated a flapper valve to control the air into the passenger compartment. The ad exclaimed that the heater had been used on a 3,000-mile test trip and the driver wore only short sleeves in 20-degree weather and "On the coldest nights he slept in his car with the engine idling." The list price for the kit was $13.75. The same issue of the magazine carried an ad for the DONALDSON Heat Blaster. This heater kit featured an electric fan to be mounted on the existing exhaust manifold heater plenum and a deflector in the passenger compartment. The kit sold for $11.00.

The Cooper Custom Built ad in the same issue described a complete heater for only $3.00. The Cooper heater was a box fabricated from steel sheet that enveloped the exhaust manifold and used ram air from the engine's fan to force air into the passenger compartment.

The Peters and Russell PAR heater ad was of interest because it included a heater performance curve showing the PAR exhaust heater produced more BTU's than the three others tested in 40 minutes at 30 and 45 miles per hour. Test performance curves were seldom published in trade journals.

The October 1929 issue of Ford Dealer and Service Field included an ad for the GOPHER HEATERS by the ASHCO Corp. One of the heaters was called the TORRID-HETE and the illustration showed an exhaust-to-air heat exchanger with an electric blower on a unit that was intended to be installed in the passenger compartment. The ad stated, "Cannot leak or allow motor fumes to enter car because of special construction and installation."

Robert Bosh A. -G. published a brochure in 1929 on its electric motor windshield wiper systems. It listed dual wipers as one option. 1929 may have been the first year for powered dual wipers. The brochure showed several alternative mounting configurations including fixing directly to the glass by drilling three holes 10mm in diameter. The motor was 6 ¼ inches long and 2 ¼ inches in diameter. It weighed 2 pounds and was offered in both 6 volts and 12 volts.

The 1929 Smith Silent Wiper made by S. Smith & Sons, London, England used a rotating cable to drive the wiper. The wiper cable was driven off the vehicle's speedometer cable. The cable and motor and other components were to be disassembled and lubricated "about every three thousand miles" or ten thousand miles with thick grease, thin grease or thick oil depending on the component. The wiper provided a hand-operated override lever for use in an emergency. The instructions include a method of drilling the glass windscreen to fit the motor if necessary on certain model vehicles. Special drills were offered by the firm and it was recommended to use "turpentine mixed with a little camphor" when drilling the glass.

The APCO thermostat article in the 1929 Ford Dealer and Service Field stated that most vehicles were fitted with engine cooling system thermostats however Ford Model A's were not. The thermostat was an important component in developing hot water heaters. The aftermarket was again filling a need by tooling, manufacturing and selling a product not being offered by the automobile manufacturers.

The 1929 Cadillac Accessories brochures illustrate several climate control related features including the De Luxe "Visionall" parallel vertical wiper system that was powered by "surplus suction from your motor." The Breeze Filter for the windshield screen was needed to prevent insects from entering the vehicle thru the open Fisher VV windshield. A small fan was offered in Green Zapon Lacquer or Satin Silver. The fan was intended to circulate air in winter and summer. Ventilating eaves and ventilating wings were offered in various models and were made of glass. These were the forerunners of the Fisher Body ventipanes fitted to all sedans beginning with the 1933 models.

The 1930 Cadillac accessory brochure featured monogrammed lap robes in fabric to match the vehicle upholstery for prices ranging from $37.50 to $85.00. Matching pillows were offered at $12.50 to $15.00 and foot muffs sold for $22.50 to $27.50 depending on the fabric selected. Other items included a Kool Kooshion for $3.50 and tire chains for $8.50 to $10.25. Cadillac also offered a hot water heater and a Kelch recirculating exhaust heater. The hot water version was priced at $42.50; the exhaust type was $42.50 plus $12.50 for the optional front register to warm the driver.

The Auto-Lite brand of exhaust heaters shown on the two 1930 Ford Model A's had different passenger compartment inlets. The vehicle owned by D. Gray had a simple rotating disc to cover the inlet when the heater was in the OFF position. The device fitted to the Model A, owned by Greg Gardner, was very sophisticated for the era. It is a cast metal distribution chamber that permitted control of the heated air to be shut off completely, directed to the left and/or the right side of the floor and to the left and/or the right defroster nozzles. Note the defroster ducts and nozzles were not fitted to the device in the photo.

The 1930 Monogram heater by the Kingsley-Miller Co. was similar to the HOTTENTOT Forced-hot air heater discussed previously, except the Monogram unit had two ducts versus one on the HOTTENTOT.

The 1930 model Otwell exhaust heater was unique because it consisted of a cast iron exhaust manifold with an integral passage for air from the radiator to pass thru the device then into the passenger compartment. A simple rotating disc was used to close the outlet for driving in warm weather.

The 1930 Ford Model A, owned by Ryan Johnson, was fitted with an electric windshield wiper motor forward of the windshield glass. The Ford Model A began fitting vacuum-windshield wiper motors later in 1930 as a cost reduction according to Jim Schild in his book *Original Ford Model A*. He also stated that hand-operated wipers were fitted to Model A commercial vehicles until April 1930.

Arvin ran an ad in the July 1930 issue of Ford Field and Service describing its hot water heater with a price of $27.50. This may have been one of Arvin's earliest offerings of a hot water heater. It had offered exhaust heaters for several years. The same issue carried an ad for the Liberty Car Heater manufactured by the Liberty Foundries Company, which also made the HaDees hot water heater. The Liberty heater used fresh air from the vehicle's radiator and sold for $10.00. The ad mentioned the importance of outside air, which was "humid and not dried out."

In the August 1930 issue of Ford Field and Service, HaDees announced record-setting sales and improvements to its hot water heater selling for $27.50. It also announced its new Concentrated Non-Freezing Fluid made under special formula. The ad listed the features of the antifreeze as "Non-corrosive, non-explosive, non-evaporating. Positively will not foam. Will not clog or injure cooling system; does not harm paint or Duco. Super-concentrated- more safety and freeze protection per dollar for Ford owners and more profit for you."

An early windshield defroster was offered by the Fulton Company in the form of an electric heating element affixed to a glass frame that attached to the windshield with suction cups. The ad, in the September 1930 issue of Ford Dealer and Service Field, stated that the device was "Guaranteed to keep its space on the windshield clear of frost, ice, snow and sleet at any outside temperature." It was priced at $6.50. Casco Products offered a similar device for $3.50 in its ad in the November 1930 issue.

The Francisco heater described in the same September 1930 issue was interesting because it stated the improvements made to the new model will "make the 1930

Francisco the greatest car heater ever built." The exhaust heater offered a front blower to draw heated air from the manifold box and push it into the front passenger compartment. A rear heater blower was offered to force heated air into the rear passenger compartment. The ad further stated the system was "FUME-PROOF" due to a tight fit around the manifold and internal ducting. The selling prices were $3.75 for the basic system, $12.50 for the front blower and $15.00 for the rear blower.

An ad for the Red Cat Instant Heater was also in the same issue. It was an exhaust heater that used an electric blower to circulate air from the interior thru the heat chamber then back into the passenger compartment. The ad included a test result: "Zero to 76 degrees in 10 minutes." The retail price was listed as $22.50.

The Metal Stamping Company offered its Weather King multiple action shutter in its ad in the October, 1930 issue of Ford Dealer and Service Field magazine. The radiator shutter could be fully open for hot weather, fully closed for cold weather driving or partially open for moderate weather. The louvers were controlled with a push-pull rod/knob attached to the steering column. The price was listed at $6.75.

HaDees offered a hybrid heater in its ad in the November 1930 issue of Ford Dealer and Service Field. The "flash" system used the exhaust manifold to heat the water for the hot water heater. The ad stated the system delivered heat to the vehicle interior "almost instantaneously after starting the motor." The ad stated that the $27.50 system eliminated the need for "special thermostats or winter fronts."

The same issue of the magazine carried an article on a horizontal parallel vertical windshield wiper system made by the Handy Governor Corp, Detroit, MI. This design was the only electrically powered one that I came across in my research. The two others were vacuum-powered.

The Zenith exhaust manifold heater was unique because it had an integral blower driven by the vehicle's engine fan belt. The system had a list price of $12.50. It was shown in an ad in the December 1930 issue of Ford Dealer and Service Field.

The brochure for the 1930 Chevrolet listed a Genuine Chevrolet engine thermostat for $2.00 installed and an exhaust heater for $9.85. The brochure stated there had been over 100,000 heaters sold by Chevrolet dealers since it was introduced the previous year.

Packard, in its 1931 brochure, listed "Rugged windshield cleaners clear the way for good vision." It also mentioned the cowl-top ventilators were fitted with screens to "keep disturbing insects out of the car." The picture showed the windshield opened at a wide angle for ventilation.

U.S. Patent 1,833,847, issued November 24, 1931, was for a refrigerant that would come to be known as R-12. This refrigerant invention would lead to air-conditioned vehicles.

The 1931 DeVaux, owned by Howard E. Reinke, has ventilators fitted that would come to be known as ventipanes. It is not known if DeVaux or the aftermarket made the units.

The 1931 models, and earlier American Austin roadsters were fitted with a single windshield wiper motor mounted in front of the windshield. The motors were not chrome-plated and they were mounted at the lower edge of the windshield. Typical mountings in the era were on the inside of the windshield and many were chrome-plated.

The 1931 Cadillac accessory brochure listed an exhaust heater. It sold for $41.00 for the base model. Double register models sold for $55.00.

The 1932 Oldsmobile, owned by Doug Width, was fitted with a chrome plated vacuum motor and vacuum tube that were finished like decorative trim pieces.

At the SAE Technical meeting held at the Hyatt Regency hotel in Dearborn Michigan on September 18, 1990 Mr. Jerold Elson, General Manager of Harrison Division of General Motors, spoke on the evolution of automotive air conditioning. He stated that in 1933 a Cadillac Town Sedan was modified to determine the amount of refrigeration that would be required to cool a vehicle. The rear seat was removed and an "ice bunker" was suspended with straps in the rear passenger compartment. The vehicle was driven at 65 mph in an ambient temperature of 95 degrees F, 40% RH with the windshield open one half inch. The ice bunker was weighed every 10 minutes. The engineers calculated that 12,000 BTU/hr would be required to maintain the interior at 85 degrees F. It was believed that 10 degrees F was the maximum temperature differential people could withstand without fainting when leaving a cold vehicle.

The 1933 Fisher Body No-Draft Ventilation brochure tells the story of the new ventilation system offered for the first time on General Motors vehicles. The brochure states, "This new Fisher ventilation system is the most important contribution to closed car comfort and safety since closed bodies were first offered (by Fisher) in 1910." The system was comprised of quarter glass windows at each corner of the vehicle. The small windows would later be called ventipanes. They were operated by hand cranks and could be rotated from fully closed to fully open, which was approximately 150 degrees of travel. When partially open, they created a low-pressure area behind the glass that drew air from the vehicle. In the fully open position, the glass acted as an air scoop and forced outside air into the vehicle. A separate Cadillac brochure describes the cowl-top air inlet scoops as rain proof.

The 1933 Packard ventilation system was similar to the Fisher Body system. The Packard system also utilized ventipanes at each corner of the body. The front door glass was unique because it was divided into two portions. Both portions rotated outward rather than the traditional rear portion lowering.

The 1933 Continental Flyer was among the first production vehicles to have the windshield wiper motor and linkage mounted below the cowl. Prior to this the motors had been mounted above the windshield. The next early adopter of the cowl-mounted wipers was the 1934 Studebaker.

The 1934 Ford ventilation system consisted of a conventional swing-out windshield, a cowl-top ventilator scoop and side door glass that moved rearward approximately 2 inches.

The 1934 Ford exhaust heater brochure had the headline "AIR CONDITION your FORD V8 for winter driving". This was an early use for the term that came to mean refrigerated air. The system sold for $14.00 and could "be installed while you wait."

The 1934 Oldsmobile brochure describes a significant revision to the Fisher Body No-Draft Ventilation system. The rear facing cowl-top air intake scoop now opened toward the windshield. This departure from the traditional front facing scoop was based upon tests that showed that a high-pressure area existed in front of the windshield that would push more air into the car than the previous front facing design. The brochure also lists an optional "dual windshield wiper and vacuum booster fuel pump." The companion 1934 Buick brochure highlights its "two windshield wipers that won't slow down on a hill or while accelerating…assure clear vision under all driving conditions." The vehicles were equipped with vacuum-powered windshield wiper motors that were common in the era. Vacuum motors were vulnerable to changes in engine vacuum that resulted in slow moving or stalled wipers, often at critical times. The vacuum booster fuel pump option was intended to minimize the slow down in wiper action.

The 1934 Packard ventilation brochure shows that the rear portion of the front door glass now dropped down rather than rotating out like it did in 1933.

The 1934 Chrysler Air Flow featured a liberal ventilation system with dual cowl-top air scoops, two swing-out windshields and ventipanes.

According to the website classicaccessories dot org, the Delco steam heater may have been used on the following 1934-1935 vehicles: Cadillac, DuPont, Duesenberg, Marmon and Pierce. The February 1934 issue of Popular Mechanics carried an article on the heater and said it was just introduced. The 1940 Dykes Encyclopedia contains a description of the Delco steam heater. This was a rare heater system that heated water with the vehicle's exhaust-to-steam and circulated it thru a steam-to-air heat exchanger. An electric blower forced the hot air into the passenger compartment.

The image of the 1935 Chrysler Air Flow, taken at the Stahls Collection, shows a fan mounted to the steering column that blew air toward the driver for cooling rather than at the windshield for defogging.

The 1935 Oldsmobile was one of the earliest Fisher Body car lines to adopt cowl-mounted windshield wipers.

The Smith Heater Booster water-jacketed exhaust manifold, described in its brochure circa 1935, was similar to the 1930 HaDees FLASH hybrid heater. The brochure stated that the water temperature from the manifold would be twenty-five to fifty degrees above engine temperature.

The 1935 Ford coupe was one of the first vehicles to offer a lowering rear window. A handle inside the vehicle operated it.

In 1935, Arvin ran an advertisement offering both hot water and steam heaters. This was the second steam heater that I came across in my research. The other was the 1934 Delco unit discussed above.

The 1935 Hudson accessories bulletin listed a set of lap robes and pillows made "in our own Trim Department of the same cloth and as is used in car production, assuring you of an accurate match of color and material with the car in which they are to be used." The dealer cost for the items was: pillows $2.25, robes $9.76.

Another interesting accessory offered by Hudson in 1935 and 1936 was Windshield Wiper Auxiliary Arms. The linkages were probably intended to prevent the blades from wobbling. They are nicknamed "Duckbill" by some auto collectors. The 1936 Plymouth DeLuxe convertible coupe was also fitted with the unique double arm system.

Windshield washers were introduced in 1936 according to the Trico website tricoproducts dot com/About/Innovations. The reservoir was a glass jar as shown in the November 1936 issue of Popular Mechanics.

The 1936 Hudson ventilation system featured a large air intake located under the rear seat of the vehicle. The intake was covered with a fabric filter similar to the material used in vacuum cleaners. This is an example of the extensive engineering effort that was devoted to ventilation systems in that era.

In 1936 Ford offered an accessory passenger side wiper. The passenger wiper was linked to the driver side arm with an external link. The wiper system was chrome-plated.

The 1936 Studebaker President coupe featured a rear window shaped like an inverted triangle that could be opened.

The 1937 Studebaker featured integrated defroster nozzles and body insulation.

The images of the 1937 Ford show dead air devices on both sides of the windshield and on one of the rear windows. These were probably aftermarket items intended to minimize fogging.

# PART 2
# From Aftermarket Hang-On Components to Integrated Systems

The 1938 Nash represented a major change in automotive climate control design strategy. The vehicle had a fully integrated heater and ventilation system. The vehicle had included the climate control system from the beginning of the design. Prior to this milestone, the heating systems were basically designed by aftermarket manufacturers and added to the vehicle primarily at the dealerships. The 1938 Nash brochure headline read: "CONDITIONED AIR FOR WINTER DRIVING". The 1938 Nash system promised 70-degree interior temperatures "(80 if you like)" at outside temperatures of zero at the "price of a regular heater." The system featured an air filter on the intake system, warmed the air then distributed it under positive pressure to all parts of the interior of the vehicle. Positive pressure inside the vehicle prevented cold air from leaking in thru small openings in the body. The 1939 Nash added mechanical automatic temperature control to the heater system. Integrated defroster ducts and nozzles were also added to the 1939 system.

In his 1960 paper, *Progress of Automotive Heating Through the Years*, Harold V. Joyce, Manager Heating, Ventilating and Air Conditioning Department, Ford Motor Company explained why the popular exhaust heaters in use in the late 1930's and early 1940's would soon be replaced with fresh air hot water heaters. He mentioned that the exhaust heaters had desirable high discharge air temperatures of 200 degrees F. He wrote "The writer remembers on one occasion working on a manifold heater so hot that it actually melted the vinyl plastic between two layers of glass in the windshield." However, hot water heaters would displace manifold heaters because "the constant whittling down in heat exchanger size due to less and less engine compartment space; a different unit for six and eight cylinder engines; and the inability to provide satisfactory and economical temperature control doomed this type of heater".

The 1938 Oldsmobile offered an optional hot water rear seat heater. This was an early application of this type heater for the rear seat.

Studebaker, in 1939, announced its new Climatizer heater system. This was an integrated system that was mounted under the front seat rather than on the dash panel.

The image showing the cover of the 1939 Ford Service Bulletin shows a technician conducting a wind tunnel test. The test is being run at low temperatures, as shown by the ice caked on the side cowl and door.

The 1939 Ford was the last model to use a swing out windshield. The windshield control knob and cowl-top ventilator controls are shown in the image from the owner's manual.

The 1939 Arvin hot water heater ad featured models for several U.S. automobiles including Ford, Chevrolet and Chrysler. The "special adjustable thermostat for the cars that need it" was interesting because it implied thermostats were not yet standard equipment on all vehicles. Another interesting feature was "7/8 inch ready-formed metal water pipes" to make installation easier. Most hot water heater installations used rubber hoses rather than metal tubes. Double defrosters were also mentioned in the ad.

The Ford heater advertisement in the September 1939 issue of Ford Field magazine lists both hot water and hot air heaters. The ad also lists the two manufacturers of the Ford branded heaters: hot water type by Eaton Manufacturing Co., Cleveland, OH and hot air type by Novi Equipment Co., Novi, MI.

The Tropic-Aire heater ad in the same issue uses the phrase "the finest winter air conditioner unit available" in the text. The unit featured a reversible motor, switch and fan. The air could be sent directly toward the passengers or indirectly toward the dash panel. The list price for the kit was $16.95 installed.

The rubber bladed fan shown in the 1939 Fulton ad was produced under license of U.S. Patent 2,095,223, which was for a safety feature to prevent/minimize injury to anything getting in the path of the blades. The fans were used to defog the windshield. The ad also listed the firm's electric sleet frost shield.

The brochure for the 1939 Oldsmobile lists the ventilation and demisting benefits of the Fisher Body ventipanes and states that the cowl-top ventilation scoop can be locked from the inside "against intruders". The ad also states that the vehicle is fitted with defroster outlets under the windshield for optional defrosters. "These are inconspicuous slits through which warm air from the heater can be distributed over the windshield panels to permit clear vision. This arrangement completely eliminates the need for unsightly pipes and connections." The other brochure image states that the windshield wiper mechanism is concealed under the cowl and windshield washers are offered as an accessory. Standard models were fitted with one wiper; deluxe models had two.

Packard introduced the world's first air-conditioned car in the 1940 model year. The system was packaged in the trunk of the vehicle. The system both heated and cooled the interior of the vehicle. It had no clutch on the compressor and Packard recommended the drive belt be removed for winter driving. The Bishop and Babcock Company built and installed the system in their plant in Cleveland, Ohio, as told by Beverly Rae Kimes in her book, *Packard: A History of the Motor Car and the Company*. Ms. Kimes reported in her book the price for the system was $275. She stated that Packard engineer W.H. Graves estimated approximately 2,000 units were sold before production ended in 1942.

The image of the 1940 Cadillac shows the vehicle fitted with a sunshade. These devices were popular for several years. They were used to reduce the glare of the sun and the heat load going into the interior of the vehicle.

The 1940 Buick offered a major new set of optional hot water heater systems. The prominent features were an underseat heater for both the front and rear seat occupants and a fresh air defroster system that permitted cool air at breath level to minimize driver drowsiness.

The Stewart Warner South Wind heater was fueled by the vehicle's gasoline system. It featured a small combustion chamber and an ignition device. Heated air was circulated in the passenger compartment with an integral electric fan. The heater was introduced circa 1937 and by 1941 more than 1,250,000 units had been sold.

The 1940 Chrysler heater and ventilation system was an all-new, fully integrated unit. An interesting new feature was high velocity sidewall air vents located in the side cowls. Like the 1938 Nash, the Chrysler unit had an integral air filter and a rain separator. Another interesting accessory listed in the Chrysler brochure was a Kool Kooshion with an installed price of $2.95. The device is similar to the ones listed in the 1930 Cadillac and 1942 Pontiac brochures and the one pictured in the August 1952 issue of Motor Trend. These devices were made with several sets of wire coils covered with an open weave fabric. The devices allowed air to circulate between the occupant and the vehicle seat and minimized the uncomfortable situation of "sticking to the seat" in hot weather.

Cadillac fitted air conditioners to approximately 200 vehicles in the 1941 model year according to Paul Ayres, President of the Cadillac LaSalle Museum & Research Center. Mr. Ayres said only three air-conditioned vehicles are known to exist in 2011. Mr. Richard Kughn owns one of the existing Cadillacs with factory air conditioning. It has been fully restored so it represents an extremely rare vehicle in 2011. Another 1941 Cadillac fitted with factory air conditioning is owned by Mr. C.D. Houston, Jr. Mr. Houston owns a coupe model and he reports the A/C system is fully functional and cools the vehicle well. Mr. Houston stated the Bishop and Babcock Company also installed the Cadillac system like the 1940 Packard system.

The 1941 Buick is shown fitted with an aftermarket window cooler. These devices were used to cool the interior of the vehicle by forcing ram air across a jute pad saturated with water. The moving air was cooled by the evaporation of the water inside the unit. These units would only function in dry climates. This type of cooler was still being manufactured in December 2010 by J&M Engineering, Inc., Camas, WA. They were intended for use on older collector vehicles to maintain authenticity. Their website classicaire dot com included a chart showing air temperature reductions of 8 degrees F at 50% RH to 25 degrees F at 10% RH.

Studebaker's Climatizer system in 1941 was comprised of two hot water heat exchangers and two blower motors. One set of components was for the underseat heater and the other was for the defroster system. The underseat unit featured an air intake filter that was intended to be removed and cleaned three or four times per year if the vehicle was operated on paved roads, more frequently if operated on dusty roads.

Pontiac's 1942 optional Venti Heat heater system was described as delivering four times more air than the previous year's system. The new system maintained a positive pressure in the passenger compartment by forcing air in faster than it could leak out. It also offered the feature of a separate defroster subsystem so the driver could select cool fresh air at breath level to minimize drowsiness. A thermostat was also used to maintain constant air temperature in the passenger compartment. Pontiac also offered a traditional hot water heater called the Weatherchief Dash heater for coupes and when the ventilating feature was not desired. The accessory list also included a Kool Kooshion as discussed earlier, a rear window sun shade venetian blind device, a rear window wiper, a front windshield washer and a "Booster Pump for uniform continuous windshield wiper operation at all speeds".

Oldsmobile's optional premium heater system in 1942 was called "Condition-air" and it featured a unit integrated into the dash panel with a filter, heater core, blower and defrosters. Two other heaters were offered: DeLuxe Dual Flow with a variable speed reversible motor and the Underseat Heater and Defroster. The defroster was a separate installation according to the brochure.

The 1946 Studebaker had an improved Climatizer premium heater option. The changes from the prior model included a larger blower motor, larger heater core, air intake thru the right hand cowl side and a new in-line cactus-type air filter. A separate blower and heating coil was used for defrosting the windshield. A low cost hot water heater was available and it was called the Quad Duty Heating System. Studebaker also offered Thermo Royal brand antifreeze that emitted no poisonous fumes or unpleasant odors and it was non-corroding.

Motorola offered two models of a gasoline heater in the late 1940's: underseat and cowl-mounted. The systems reportedly started in 33 seconds and produced heat with the vehicle's engine on or off.

The 1947 Ford accessory brochure lists windshield washers that use two special fluids: one to prevent winter freezing and the other to reduce surface tension in the summer. The other item in the brochure was a vacuum storage tank that mounted under the driver side front fender. The device was intended to "keep windshield wipers or fans operating at any speed you desire, regardless of engine speed. This eliminates wipers slowing down, or stopping, when engine is accelerating."

The 1948 Ford truck hot water heater featured a fresh air intake and a thermostat-controlled water valve to modulate the discharge air temperature. This was an early Ford hot water heater system that pressurized the passenger compartment. The blower assembly was mounted in the engine compartment as this location minimized noise. The air intake was above the headlamp and was designed to prevent rain from entering the system. The system could be used for forced ventilation in the summer by turning on the blower and manually turning off the water shutoff cock at the engine. A dealer sales brochure for the system stated the system delivered "nearly two complete air changes per minute". Window fogging was minimized or eliminated. The system controlled temperature accurately and thereby controlled humidity. "This, in combination with the fresh air supplied to the interior, reduces window fogging. Before the air becomes saturated, it is replaced by more fresh air." Another feature was the system could be used as a recirculation system by closing off the outside air intake. This prevented objectionable outside odors from entering the cab and could also be used for faster warmup, if desired.

The 1948 South Wind gasoline heater was available in kit form for numerous U.S. cars and trucks. Model 977 was a fresh air system unlike the late 1930's model that was only a recirculation type.

The 1948 Anderson Company brochure shows windshield wiper blades for both flat and curved windshields. It also shows the myriad of arms and blade connectors that were available in the era.

American Bosch in 1948 offered an electric rear window wiper system for numerous vehicles from 1942 to 1948. This was an early rear window wiper system.

The 1948 Ford heater was interesting for several reasons. It had a fresh air duct that was fitted to the inside of the hood. The duct opened with the hood and sealed against a cylindrical gasket on the blower module when the hood was closed. The heater core was relatively long and narrow when compared to other cores in the era. The air inlet duct had a damper door that was controlled by a Bowden cable with a knob in the passenger compartment.

The 1948 Packard had a machined and chrome-plated windshield washer nozzle that was as attractive as it was functional.

The 1949 Ford premium optional hot water heater system was totally new and featured two fresh air intakes. It was called "Magic Air" and it had a thermostat-controlled water valve. It had all of the features of the 1948 Ford truck heater system discussed above plus a second air inlet ventilation duct on the driver side. Both air inlet ducts were fitted with damper doors controlled by Bowden cables with control knobs fitted to each side of the steering column. Ford also offered a dealer installed recirculation heater for vehicles not equipped with a factory fresh air system.

Cadillac, in 1949, offered two noteworthy features: dual underseat heaters and side window demisters. The system also included a third heater/defroster unit for a total of three heater cores and three blowers. This was leading edge technology for the era.

Ford engineer Ted Daykin, now retired, designed the 1949 Mercury wiper system. The system was unique because it used an electric motor in the era of vacuum motors. An article on the website autopartsway dot com described the two speed system as using an electric motor to drive a worm gear. A circuit within the system sensed when the wipers were in the down position and the power would be stopped to park the wipers. Mr. Dayken said the system was controlled with a Bowden cable rather than an electric switch to minimize complexity. He said the system was referred to as "Depressed Park".

The 1949 Mercury featured monogrammed lap robes in various colors in its accessory catalog. This was another example of lap robes offered as accessories by vehicle manufacturers. Cadillac and Hudson were the others discussed above. Also, the 1949 Mercury offered an optional rear window wiper.

The windshield wiper system fitted to the 1950 MG was interesting because it had a link between the two arms and an idler arm. The idler arm was probably used to minimize bending or bowing of the link during operation of the wipers.

The 1950 Studebaker, owned by Harvey and Julie Snitzer, is fitted with venetian blinds. The blinds were sold as genuine Studebaker accessories.

The letter dated June 21, 1950 from the Chevrolet Engineering Department to the Chrysler Engineering Department lists the windshield wiper motor types and booster equipment used with vacuum systems. Only four out of 29 systems used electric motors. The list also shows that all the vehicles listed had two windshield wipers as standard equipment by 1950.

The 1950 MoPar windshield washer was operated with a foot switch. Previously most washers were operated with a switch on the instrument panel.

The windshield wiper drive system used on the 1950 Hudson was via a cable system and a vacuum motor.

The 1951 Ford truck fresh air hot water heater was an early application of the "blend air" system. The thermostat-controlled water valve was replaced with a Bowden cable-controlled door in the heater plenum. Outlet heater air temperature was controlled by the amount of air passing thru the heater core. The heater core was receiving engine coolant at all times. In the max heat position, all incoming air passed thru the core and into the passenger compartment. In the max cool/off position, the door prevented any air from passing thru the heater core. Any control lever position between max heat and max cool moved the door and blended the air in proportion to achieve the desired temperature.
In Harold Joyce's paper, *Progress of Automotive Heating Through the Years*, he listed the merits of the blend air system compared to a water valve system: 1. Reduced fluctuations in discharge air temperatures due to the continual changes in car speeds; 2. No cycling of discharge air temperatures due to changes in the temperature control lever setting; 3. Immediate response to changes in temperature control lever setting; 4. Simple, more trouble-free design; 5. Elimination of the water shut-off valve on some installations has resulted in the increase in maximum discharge air temperature due to increased water flow.

Mr. Joyce also stated that average interior temperatures, due to increased heater performance, had improved considerably between 1930 and the 1950's. They were about 40 degrees F in the 1930's and had risen to the 80 degree F range in the 1950's.

The 1951 Buick Le Sabre concept vehicle was fitted with heated seats. The system was described in the brochure titled, *General Motors LE SABRE AN EXPERIMENTAL LABORATORY ON WHEELS* as "The seats have thermostatically controlled warmers similar in principle to electrically heated flying suits."

The sister concept car, the 1951 Buick XP 300, was fitted with four windshield wipers. The outboard set was needed to clear the corners of the wraparound windshield.

The 1952 Ford featured a rear window defroster. It was an electric motor and fan system that blew ambient air onto the backlight. This was an early application of the feature. Also the heater control used on the 1952 Ford had two slide levers that Ford had introduced in 1950. Prior to 1950, the controls had been push–pull or rotary knobs. The 1950 Ford heater control had its own illumination lamp and switch.

The 1952 Bosch brochure for its electric windshield wiper motor stated that the unit had several good features including an "Improved, simplified design" with "Automatic adjustable blade parking". The 6-volt system was being offered as an alternative to the vacuum motors that were prevalent during the era.

The two images in the August 1952 issue of Motor Trend show a window mounted evaporative cooler discussed above and the "Kool Kooshion" that was offered as an option by Cadillac, Chrysler, Pontiac and possibly other vehicle manufacturers. The two devices were intended to reduce the discomfort of driving in hot weather.

# PART 3
# Air Conditioning Returns to Stay

In the 1953 model year, air conditiong was offered as an option on several vehicles including Buick, Oldsmobile, Cadillac, Chrysler and Lincoln. Notably absent was Packard, which had been the first to offer air conditioning in the 1940 model year. All of the units offered in 1953 were similar to the Packard: the evaporators were mounted in the trunk compartments, cooler ducts ran thru the package shelf and their compressors were direct drive- they had no clutches. Most of the systems used R-12 refrigerant except Chrysler that used R-22. Chrysler was also unique with two condensers.

The letter from Oldsmobile's General Service Manger, E.E. Kohl, in the 1953 Oldsmobile air conditioner service manual to ALL OLDSMOBILE SERVICE MEN, is interesting because it tells the mechanics that the systems are so new they will be "faced with a new experience in learning how to service it."

Improvements were released on several of the 1954 air conditioners notably compressor clutches that permitted winter driving without the need to remove the drive belt. Some revisions were made as running changes. For example, the 1954 Dodge service manual states that air conditioner air filters were added later in the model year.

The Dodge manual also includes a specification list that describes the major components in the air conditioner system. Note the use of 5 pounds of R-22 refrigerant and the materials and weight of the two condensers: steel tubing and steel fins weighing 10 pounds in the upper condenser, copper tubes and aluminum fins weighing 20 pounds in the lower condenser. The weight of the compressor at 75 pounds is enormous by today's standards. Tube fittings were the flare type in the Chrysler vehicles.

Packard again offered air conditioners on its 1954 models. The new Packard heater system featured two heaters.

New air conditiong packaging architecture was introduced on the 1954 Pontiac. The evaporator and blower assembly was mounted in the engine compartment rather than in the trunk. This architecture is still in place today.

The 1954 Nash also mounted the air conditioning evaporator and blower in the engine compartment rather than in the trunk. The Nash system sold for "around $330" according to an article in the August 1954 issue of Motor Trend. The article stated the air conditioner system "weighed 115 pounds (up to 125 pounds less than other systems)". The system also "cooled down 19 degrees in five minutes" during a desert trip.

The July 1954 issue of Motor Trend described THE IDLER blower type car cooler that mounted to the vehicle transmission tunnel as "just the thing for hot weather driving". The unit sold for $89.50. It was not a refrigeration device. It was simply a new configuration of the window-mount evaporative cooler discussed earlier.

An air conditioner for the 1954 Ford was built by Novi Equipment Company, Novi, MI for dealer installation, according to an article in the August 1954 issue of Motor Trend. The article also stated that factory-installed air conditioning had been planned for the 1954 Ford and Mercury "but for some reason was postponed indefinitely". The Novi system was priced at $345.00 plus tax and labor.

In 1954, Newhouse Automotive Industries offered a heated seat cushion. It was similar to the Kool Kooshion discussed above with the added feature of a heating device powered by the cigarette lighter and it drew 4 amps. The price was listed at $5.95.

A Trico brand vacuum-powered windshield fan was advertised in the January 1954 issue of Hot Rod. The unit was priced at $1.95, shipping included. The ad stated, "more than one million sold".

The JC Whitney Company offered venetian blind kits for all cars from 1939 to 1954. They were intended to keep out the sun without impairing vision. They were priced at $3.25.

The 1954 Buick, owned by Philip Fischer, is shown with a desert water bag hanging from the front bumper. The bags were popular for driving thru the desert because they cooled the water for drinking or for replenishing the radiator if it boiled over. They cooled the water by evaporation. A small amount of water seeped thru the jute cover and the rushing air caused it to evaporate rapidly. This is the same principal used by the window cooling devices.

Mr. Fischer's 1954 Buick was fitted with the GM windshield washer system that used a round glass jar for its reservoir. GM used the round jars for most of the 1950's.

The 1954 Cadillac air conditioner compressor was fitted with a clutch. The clutch was actuated by an electrical solenoid that pulled an arm against the face of the clutch. The face forced the clutch plate against the pulley that engaged the compressor shaft and caused it to turn. This unique clutch system was used in 1954 only.

In 1954, Chrysler offered two air conditioner systems. The standard system used R-22 refrigerant without a clutch. The premium system used R-12 refrigerant and used an electromagnet clutch similar to those still in use today.

An article in the November 1954 issue of Motor Trend listed a comparison of antifreeze types then available and the merits of each. Note that methanol-based antifreezes were poisonous.

The 1954 Trico brochure for its CAM-O-MATIC wiper system was introduced to clear the corners of the wrap-around windshields. It eliminated the need for the Buick XP 300 four-wiper system.

The brochure for the 1954 Trico SUPER-SPEED wiper system describes a vacuum device to offer the driver two speeds and two angles of arm travel. The opening sentence in the text of the brochure explains that the device is needed to increase safety while driving in the rain or snow at the "new legalized higher speeds on thruways and express highways where legal limit of travel is from 60 to 70 miles per hour…" For high vehicle speeds, the driver could select the fast blade speed and the narrow angle of travel. For lower vehicle speeds the driver could select the normal blade speed and angle. Researchers on the subject of wrap-around wipers are advised to review a paper titled *Wiping Wrap-Around Windshields Was a Problem*, by John W. Anderson, President, The Anderson Company. It was delivered at the American Society of Body Engineers, Rackham Memorial building, Detroit, MI, October 28, 1954.

Chrysler MoPar offered three heaters in 1954. Two fresh air units and one recirculation unit. The premium model of the fresh air unit featured automatic temperature control.

The following text is excerpted from an article in the August 1955 issue of Motor Life titled LOW COST AIR CONDITIONING. Ken Fermoyle wrote the article.

**The prosperity of this country is reflected in the comforts we demand; the efficiency is reflected by the way those demands are met-and at increasingly lower prices.**

There is no better recent example of this than air conditioning for automobiles. Only a few years ago such a luxury was almost unheard of; attention was focused on such other accessories as power brakes, automatic transmissions, power steering and a variety of others. Now most of these are commonplace. They don't show up on every car but they're out of the novelty class.

Air conditioning however is still something quite new under the automotive sun, though indications are it will become as widely used and demanded as some of these other accessories. In 1953 production of auto air conditioning units totaled 41,000. Last year the figure jumped to above 67,000. This year estimates are that from 150,000 to 200,000 or more will be sold. The first units were priced so high that they were out of reach of all but the best–heeled auto buyers. Prices are still high compared to other accessories but there are now several units tagged at under $400.

That isn't cheap, but it's well under prices of original units. Understand, not all air condition systems can be had for under $400, in fact, most systems are still priced well above that figure. The significant thing is that prices have tended downward and will, no doubt, continue to do so.

The lowest priced optional equipment air conditioning systems now being offered is that for American Motors' Rambler at $345. Highest-priced is the Lincoln unit which goes for nearly $750.

Between these are a variety of others being sold for a variety of prices. In addition to the systems being sold and installed by manufacturers or their dealers, several units are being built and sold by firms without official factory connections. They offer air conditioning systems on the same basis that independent accessory manufacturers offer other accessories for a car.

One of the foremost of these independent manufacturers is the Novi Equipment Company, which manufacturers a unit for many late model cars. The Novi unit sells for $395 installed. The company has systems for late models of Ford, Chevrolet, Buick, Oldsmobile, Lincoln, Mercury and Cadillac. Installations for Plymouth and several others were being studied by Novi engineers at press time.

(Incidentally, the Novi firm is headed by Lou Welch, well-known to auto enthusiasts for his sponsorship of the fast but unfortunate Novi cars which have appeared annually at Indianapolis for a number of years.)

Although the refrigeration circuit on all automobile air conditioners is basically the same, there are two different methods of mounting the equipment in the car. Most of the earlier systems and many in current production have part of the system in the front part of the car; part goes in the trunk. In some systems all components are installed in the front portion of the car.

*[A portion of the article describing the refrigeration cycle is omitted here for brevity.]*

Freon 12 incidentally is not especially expensive. It costs from $4 to $6 or so for enough to fill a unit depending on the size of the system and the cost of the gas in that locality. It takes from about 4 to 6 and one-half pounds of Freon to fill the average system.

What does the future hold for automobile air conditioning? One of the foremost experts in the field G.T. Etheridge, manager of automotive air conditioning, Kelvinator division of American Motors, discussed this in a recent talk before refrigeration engineers. "We have the twin problems of lowering the price to the ultimate consumer and the ever-present problem of improving performance," he stated.

Etheridge also points out that equipment will have to be made more compact, controls simplified to eliminate unnecessary components and the cost of wiring and installation. He feels that systems should be set up so they will circulate either cool air in summer or warm air in winter. (The All-Season system used in American Motors products already does this, by the way. This means a big savings because it eliminates the extra expense of a heater.)

Etheridge predicts that "it will be only several years before automobile air conditioner volume reaches the 1,000.000 per year sales rate". This indicates that he feels many of the problems mentioned above will be at least partially solved in that time and that prices will come down to a level which more buyers can afford. This is good news for car buyers particularly in very hot climates where air conditioning is a real boon.

The market for automobile air conditioning is there, certainly. A survey covering the desires and habits of passenger car buyers taken last year revealed that 25 per cent of them would like to have air conditioning in their cars. This is only 12 per cent less than the number who want power steering and just slightly less than half the total who want automatic transmissions.

This is significant because the auto industry has always been alert to the desires of its customers and has made great efforts to meet and satisfy them. It is logical to assume that it will do so in the field of air conditioning also.

Thus it might not be too long before most of us will be spared the feeling of entering a Turkish bath when we get into our cars on a hot summer day. With the flick of a switch we will be able to reduce the temperature to a comfortable level, just as we now turn on our car heaters in the winter.

The road test of the 1955 Nash Ambassador V8 in the July 1955 issue of Motor Life stated that the Weather Eye system was considered by many people to be " the finest automobile air conditioning system available. The heater still is one of the best, has an excellent defrosting system also which is a terrific asset in cold climates." The accessory was priced at $395.00.

The 1955 Trico brochure, listing its British products, shows the numerous blade types and arm attachments available in the era. Note the lower left hand corner box listing of "Rubber pegs, stops, pads, etc."

The 1955 American Bosch brochure described the features of its electric wiper motor and stated that the unit was available to fit "many late model cars and trucks".

The 1956 Frigiking brand air conditioner was an aftermarket hang-on unit that listed at $298 plus installation. This was an early advertisement for a segment of the air conditioner market that would grow enormously then fade away as the factory installation rates increased.

The construction of the 1956 Ford air conditioning components is noteworthy when compared to today's technology. The evaporator housing was fabricated from sheet metal and held together with screws. The instrument panel control was fabricated from metal stampings and castings. The Bowden cables were attached to the control with steel clips and screws. Note the air outlet registers were located on top of the instrument panel. For defrosting, the instrument panel registers were rotated toward the windshield. The windshield wiper system consisted of a vacuum motor driving two metal rods attached to the pivot assemblies.

The 1956 Anco catalog for vacuum motors for windshield wiper systems states that 10 models of the service replacement motors would fit most vehicles from "1935 to present". The retail prices of the components were: motors $7.45 to $9.55, shafts of various lengths $0.50 and the instrument panel control switches, Short 15 inches, Medium 18 inches $0.95 each. An interesting feature was the motors could be adjusted for arc angle to ensure "NO BLADE SLAP". Another accessory was a Primer Valve that permitted the washer and the wiper to be operated simultaneously with the same instrument panel control switch. Without the valve, the switch would operate the wiper and then had to be rotated in the other direction to operate the washer.

JC Whitney was offering electric motors to replace vacuum wiper motors in 1955 for $13.69 postpaid. Their ad offered models to fit all vehicles except 1948-52 Hudson and Nash. Wipe angle was approximately 110 degrees. The same firm offered an electric wiper motor in 1956 to fit a Ford Model A for $7.95 postpaid. The ad claimed the motor was powerful enough to operate two blades.

It is interesting to note that the aftermarket in 2011 is selling electric windshield wiper motors to replace vacuum motors for use on older collector vehicles. One source found on the Internet is at forwhatyouneed dot com. The electric motors are priced at $197.00.

The 1956 Buick air conditioner had an early example of a cam-operated door. The cam would be actuated with a Bowden cable and the cam would cause the door to dwell thru part of the cable/cam travel. This was leading edge technology for the era.

Chrysler offered a gasoline heater system in some models in 1956. Chrysler called it the "Instant Heater". It also offered a Custom Conditioner Air Heater and Air Temp air conditioning. The Crown Imperial Limousine was fitted with a rear seat heater with a switch controlled by the rear seat passenger.

The 1956 DeSoto featured a flexible rubber bag for its windshield washer reservoir. Some other vehicles would also use the soft bags for several years.

In 1957, Ford used a cable wiper system similar to the Hudson system discussed above. This was a more complex system than the one used on the 1956 Ford. Ted Daykin, a Ford engineer who worked on the 1957 Ford wiper *motor* program said the more complex design was probably implemented to resolve a packaging problem that could not be worked out using the rigid arm design.

The 1957 JC Whitney ad for the Multi Spray Windshield Wiper attachment was interesting because it was a forerunner to the 1971 "wet arm" by Trico.

An article was published in the July 1957 issue of Motor Trend. It was titled ACCESORY TYPE AIR-CONDITIONING UNITS. The article listed the following aftermarket makers of air conditioners: AIRTEMP/ MOPAR, ALLSTATE, A.R.A, ARCTIC, ARTIC-KAR, CLIMATIC-AIR, COOL QUEEN, FORSTON, FRIGETTE, FRIGIKING, LO-MERC, MARK IV, MOBILETTE WEATHER MATIC, NOVI, PARKOMAT, TOWNE AND COUNTRY, VORNADO and WIZARD. Most of the makers were located in Texas. The article included the prices for the units installed that ranged from $295 to $565. The article also listed the prices of factory-installed air conditioners for 18 different vehicle lines. They ranged from $362 for Rambler to $590 for Imperial.

The 1958 Mercury featured two instrument panel heater controls and one air conditioner control. The standard heater control was the mechanical slide lever type. The premium heater control and the air conditioner both were rotary knob type. The premium system was an early example of using vacuum motors rather than Bowden cables to open and close doors and control the coolant water valve. Even though vacuum motors were more expensive than Bowden cables, they were superior in providing a smooth feeling and minimal operating forces to the instrument panel control.

Early Chevrolet Corvettes used cable wiper systems as shown in the 1958 Corvette parts book. Other vehicles with cable wiper systems were 1953-57 Chevrolet cars and 1946-1954 Buicks, 1948-49 Cadillacs, 1958 Edsel, 1947-59 Chevrolet truck. This list was compiled from the website chevychevy dot com. The 1949–50 Nash also had cable wipers.

The 1958 Lincoln air conditioner system had two units on each side of the dash panel in the engine compartment. The units housed the air conditioner evaporator cores, blower assemblies and two heater cores. The control system was via eight Bowden cables driven by an electric servo. The instrument panel control consisted of one rotary knob and a rotating dial face that matched the clock face for shape and style. Rear seat passengers could control the direction of the flow of air to the rear compartment via knobs located in registers on the vertical rear faces of the front door arm rests.

The 1959 Ford, Mercury and Edsel heaters were the blend air type. Their operation was similar to the 1951 Ford truck heater discussed above. The housings for the heaters were molded thermoplastic. Prior housings were fabricated from sheet metal.

The 1960 Ford offered two air conditioners; the low cost PolarAire for 6 and 8 cylinder engines and SelectAire for V8's only. Both were the hang-on type. The heater choices were a low price recirculation unit or the deluxe MagicAire fresh air unit. Other options available were: windshield washers and electric motor-driven wiper systems with two speeds rather than the standard vacuum motor. This was the last year Ford offered a recirculation heater.

The 1960 Chevrolet Corvair offered an optional gasoline heater to warm the occupants of the new vehicle with an air-cooled engine. An alternate hot air system would be offered as an additional option in 1961. The website corvairunderground dot com stated that the fuel-fired heater was not popular with Corvair owners because "it met with customer resistance" and about 98 % of all Corvairs were fitted with hot air heaters. The gasoline heater was discontinued in 1964.

Several pages from the 1960 Ford Thunderbird service manual are included because they illustrate the state-of-the-art for the era. Examples: a two-cylinder A/C compressor, a tube fin evaporator core, a three-slide lever instrument panel control made of stampings and a casting that operated three Bowden cables, a vacuum motor/cable windshield wiper system and a rubberized bag windshield washer reservoir.

The 1961 Lincoln used a hydraulic motor to drive its windshield wiper system rather than vacuum or electric. The hydraulic system was required to move the large arms and blades, according to an article by Nick Cole in the Forth Quarter 1985 issue of Automobile Quarterly.

At least two British cars were fitted with three windshield wipers. Besides the 1963 MG shown below, some models of the Morgan, such as the 1986, had them.

Chrysler vehicles in 1963 featured airfoil wiper blades. They were intended to increase the force of the blades against the windshield at high vehicle speeds.

Ford made heaters standard equipment with the option to delete in the 1963 model year.

General Motors introduced the air conditioner and heater reheat system in 1963. All of the time this system was ON it pumped refrigerant thru the evaporator core and also pumped engine coolant thru the heater core. Discharge air temperature was maintained at the desired level by directing refrigerated, dry, clean air thru the heater core in proportion to the control setting. Other vehicle manufacturers adopted this method within the next few years.

The 1964 Lincoln utilized a molded plastic floor duct to direct air to the rear seat passengers.

An electro-pneumatic module and a vernier switch mounted on the instrument panel were the two major components of the 1964 Trico intermittent windshield wiper control system. Dwell time could be as long as 50 seconds out of every minute when 10 cycles per minute was selected. It was for use with an electric motor.

The 1964 Chrysler windshield washer system used an electric pump to force the cleaning fluid to the windshield.

# PART 4
# From Electro Mechanical Systems to Electronic Systems

The 1964 Cadillac was fitted with automatic temperature control. This unit used a transistorized amplifier and thermistors that were very early applications of electronics to climate control systems. The system instrument panel control had a thumbwheel temperature selector and one slide lever to select OFF, AUTOMATIC or DEFROST.

By 1965, the majority of the windshield wiper motors used on Ford Motor Company vehicles were electric. The chart shows only one vacuum motor was still in use on the Bronco and the Thunderbird was using a hydraulic motor.

The images of the 1965 Mustang show the hang-on air conditioner evaporator in the passenger compartment and the air conditioner compressor. The compressor had service valves that could be opened and closed in the event the compressor needed service. The valves would be closed and the refrigerant would not escape to the atmosphere and have to be replaced. The valves were also used to pre charge the heat exchangers at the plant that built the heat exchanger and attached the refrigerant hoses. The heat exchanger and hose assembly would then be filled with the appropriate amount of refrigerant, leak tested, then shipped to the vehicle assembly plant. Because the air conditioner installation rate was still relatively low, few vehicle assembly plants were facilitized to fill the air conditioner system with refrigerant on the assembly line. In-plant refrigerant charging began in the mid 1970's at some Ford car assembly plants and the late 1970's to early 1980's in Ford truck plants.

The German 1965 VDO brochure illustrates the popularity of the small aftermarket fans worldwide. They were used to augment the vehicles' defroster and rear window defogger systems.

The 1965 Pontiac GTO windshield washer nozzle was a metal tube with the outlet end flattened and drilled. The tube and bracket assembly was fitted between the louvers in the cowl-top air inlet grille.

The 1965 Chevrolet Corvair utilized the GM windshield washer fluid triangular shaped glass bottle. It also used the molded plastic washer reservoir that replaced the previously used round glass jar.

The newly redesigned 1965 Chevrolet Corvair also included a new hot air heater system. The system utilized radiant and convected heated air from the engine and forced it into the passenger compartment thru a duct. The engine air was captured in a metal container that surrounded the lower portion of the engine.

Car Life's August 1966 issue showed a graph of the rapid increase in air conditioner installations for both factory and aftermarket. Aftermarket units had increased from 142,656 in 1958 to 379,056 in 1965. The factory units had increased from 198,254 in 1958 to 2,042,423 in 1965.

In the 2011 model year, only the following vehicles sold in the U.S. were offered with out air conditioning as standard equipment: Chevrolet Aveo, Honda Civic DX, Hyundai Accent GL, Kia Rio, Mitsubishi Lancer, Nissan Versa. This is based upon a search of the manufacturers websites.

Lincoln, in 1966, offered automatic temperature control that utilized electro mechanical components. Note the instrument panel control used pushbuttons and a rotary knob. The rotary knob controlled a moving pointer and a potentiometer for selecting the desired temperature. The control head was constructed with a die cast housing and the sliding pointer was illuminated.

Ventilation was still an important feature as shown on the 1966 Buick Riviera. Even though air conditioning was readily available, not all car buyers ordered it as they were satisfied with a well performing ventilation system. This vehicle was a leader in the early trend to eliminate ventipanes that were introduced in 1933.

The 1966 Cadillac air conditioner evaporator core was the plate fin type. This new design had performance much higher than the tube fin type that had been used from the 1940 beginning. This was an early application of the high technology heat exchanger.

The patent for intermittent windshield wipers was awarded to R.K. Kearns in 1967. This patent would be the basis for several multi million-dollar lawsuits that Mr. Kearns won. The story of the patent and lawsuits was told in the 2008 movie *Flash of Genius*. The patent was for an electronic circuit to control electric wiper motors. Details of the first lawsuit are described in an extensive article in the Detroit Free Press, November 15, 1990, pages 1E and 3E. The second lawsuit is covered in the same newspaper on June 12, 1992 pages 1E and 2E.

The Bacho 3000 preheater was shown in the May 1967 issue of Car Life. It was a device that sold for $200. The unit was mounted in the engine compartment and had ducts running to and from the passenger compartment. Several electrical wires were attached to the unit. The Stanley Works, New Britain, CT, offered it.

The 1968 Buick shop manual image shows how Bowden control cables had evolved. They were fitted with molded plastic end fittings that expedited assembly by eliminating the cumbersome separate steel clip. Some of the cables that needed to be adjusted after assembly were fitted with turnbuckles that expedited the delicate procedure needed to align the door with the control lever position.

The 1968 Chevrolet Corvette utilized a heater and air conditioner control that was operated with two thumb wheels and a lever. The control was fabricated from metal stampings and a casting and held together with screws. The thumb wheels were molded plastic. The Corvette also featured a movable panel that raised and lowered to conceal the windshield wipers. The panel was powered with vacuum motors.

A significant improvement in rear window defogging, defrosting and de-icing was achieved with the electric rear window defroster introduced on the 1969 Lincoln Continental Mark III. The system utilized metal resistive wires bonded to the inside of the glass and electric current to heat the glass.

The 1969 Pontiac GTO and some other vehicles in the era had windshield wipers hidden below the hood that actually traveled across and parked on metal trim moldings.

Saab invented headlamp wipers and introduced them in the 1970 model year.

The hang-on air conditioner used on the 1971 Ford was thin and spanned the entire lower edge of the instrument panel. This design provided more legroom for the center passenger.

The 1971 Chevrolet Corvette introduced windshield wiper arms with the washer nozzles attached. Trico referred to these as "wet arms".

In 1972, Saab introduced heated seats, which was probably the first for a high volume production vehicle manufacturer.

The 1972 Ford Torino utilized the instrument panel control that would be called the "Universal Control" because it was intended to be used on several car lines for several years. It featured a zinc die cast bracket and stamped steel slide levers.

A major defrosting feature was offered on the 1974 Ford Thunderbird and Lincoln Continental Mark IV. The system was called Quick Defrost Windshield and Rear Window. The system employed metallic gold deposited on the glass before lamination. The gold served as a conductor/resistive-heating element when current passed thru it. The gold film was barely visible to the eye. A separate high voltage alternator was used to power the system.

Chevrolet introduced the new compact R-4 compressor in 1975. It was a 10 cubic inch displacement unit.

A tube-o type refrigerant fitting is shown in the illustration from the 1975 Ford service manual. The tube-o fitting was used in place of the flare type because it was able to achieve a robust seal. The o-ring is in controlled compression compared to flare fittings that are dependent on sensitive tightening torque.

Several pages from the 1975 Ford service manual are shown to illustrate how the heater, air conditioner and ventilation systems were evolving.

The 1975 Saab windshield wiper drive system is interesting because it used a rotary cable. A similar system is being offered in 2011 for aftermarket fittings that use the Lucas 14W two-speed wiper motor with a "park" position.

The 1975 Ferrari 512 BB wiper system utilized two blades attached to the same articulated two bar linkage. One of the blades was long and the other was short. This could be referred to as a one and a half blade system.

The 1976 Ford Mustang was the first application of a plastic instrument panel control bracket on a U.S. Ford vehicle. The zinc die cast bracket on the Universal Control was converted to DuPont's mineral filled nylon called Minlon. The change was made to reduce weight and cost.

The 1977 patent for the spring lock coupling was for a design that would be used on millions of applications for both refrigerant and fuel systems. It was intended to achieve robust sealing equal to tube-o fittings without the use of hand tools. It is still in use in 2011.

An early application of a heated rear view mirror was on the 1979 Lincoln Mark V.

Several pages of information are included that show the transition of refrigerant control systems and compressors at Ford Motor Company from 1979 thru 1981.

The 1981 Ford utilized Power Ventilation whereby the outside air was forced into the passenger compartment utilizing the heater blower motor.

The 1984 Lincoln Continental featured fully electronic automatic temperature control.

The booklet titled *The ABC's of Air Conditioning* was published by Ford Motor Company circa 1985 and it explains the need to transition from R-12 refrigerant to R-134a. The new refrigerant, R-134a was found to have a smaller impact on the ozone layer.

1986 was the model year the Ford Taurus began using a three knob rotary control. This was the first application of this type of control on Ford's U.S. vehicles. This control was also noteworthy because it was made almost entirely of plastic. Note the rack and pinion to drive the temperature Bowden cable. The cable snapped into place and no screw was needed to hold it. This control was the genesis for the millions of three knob rotary controls that would be used on Ford and other vehicles. Andy Pastoria engineered it. The Taurus also offered fully electronic automatic temperature control as an option.

The 1986 Daimler used a single wiper blade. Other vehicles that have and/or currently use this configuration include: 1967 Ferrari Dino, 1996 Jaguar XJ 12, 1996 Lotus Elise, 1985 Subaru, 1988 Isuzu Impulse, 1977 Volkswagen Scirocco, 1984 Mercedes-Benz, 1985 Subaru XT, 2011 Lotus vehicles.

In 1987, the Ford Product Engineering Office was still designing vehicles using drafting boards, triangles and compasses. The image shows a team designing the heater and air conditioner system for a forthcoming Ford F Series pickup truck.

The 1987 U.S. Postal Service Long Life Vehicles were fitted with fans to assist with defrosting and cooling the driver. The vehicles were not fitted with air conditioning. The vehicles are still in service in 2011 and the drivers still use the fans.

Federal Motor Vehicle Safety Standard (FMVSS) number 103 mandated windshield defrosters and defogging. Standard number 104 made windshield wipers and washers mandatory. Both standards became effective January 1, 1988.

The 1989 Ranger used a new two-slide lever instrument panel control. The control used plastic components except for the stamped steel levers. The Bowden cables snapped into place and the control included a cam mechanism. This control would be one of the last to use slide levers in Ford U.S. applications.

The 1990 General Motors Impact electric vehicle utilized a heat pump for climate control. This was early in the modern era of electric vehicles and new challenges were presented to the engineers to design high performance heating systems without the use of hot engine coolant or exhaust.

In 1991, Saab introduced R-134a refrigerant on some of the vehicles it built. This was posted on Saab's website saabhistory dot com.

The 1991 Nissan FEV electric vehicle also used a heat pump. Its brochure stated: "The FEV's electric air conditioner uses a rotary compressor directly driven by a dedicated motor. Use of a heat pump provides heating. An inverter drive increases air conditioning efficiency and thermal insulated glass reduces the load on the air conditioner. A vibration control mechanism maintains quietness. R&D is raising heating capacity."

An article in the March 4, 1992 issue of the Dallas Morning News stated that Ford Motor began installing R-134a refrigerant in small numbers of the Taurus equipped with 3.0-liter V6 engines on March 2, 1992. The article also stated that the Mercedes Benz and Jeep Grand Cherokee would begin using R-134a in April, 1992. The article quoted Ford Climate Control engineers Jay Amin and Norm Wood.

A running change was made on the 1992 Taurus to upgrade the temperature door control subsystem. An electric actuator and potentiometer replaced a Bowden cable and rack and pinion drive. This was done to improve feel and performance.

An article in the March 11, 1992 Detroit News described an infrared sensor system to turn on windshield wipers automatically when water was detected on the windshield. It was called Clear Advantage and Libby Owens Ford had developed it.

Robe cords were being fitted to luxury vehicles long after robes were no longer needed, as shown on the 1994 Lincoln. The cords were fitted as a decorative trim feature.

Dual zone climate control systems were used on the 1995 Lexus. This was an early application of the feature that was intended to let both front seat passengers select their own desired temperature.

In 1996, General Motors introduced its electric vehicle called the EV1. The climate control system featured a heat pump to both heat and cool the interior. A three-phase electric motor drove the compressor. Current flowing thru a metallic layer was used to defrost/defog the windshield. The back window used a resistor grid for defogging. The system utilized pre-heating/cooling to minimize power drain while driving.

Saab introduced ventilated seats in 1997. This was an early application of the feature in a production vehicle.

Chevrolet's 1997 ST Electric pickup had two heating systems. A heat pump was used in moderate weather and a diesel fuel-fired heater was used for temperatures below 40 degrees F.

The 1998 Ford Electric Ranger used PTC heaters for heating and defrosting. Cab temperatures were limited to 80 degrees F to 65 degrees F to conserve energy. A 3.5 kW AC motor drove the air conditioning compressor.

The 1998 Ford Escort employed an early application of the pull-pull control cable. This assembly contained two braided cables that ran side by side inside a plastic housing. As the instrument panel selector knob was rotated, it pulled on the applicable cable to move the door in the air-handling unit. Rotating the knob in the other direction pulled on the other cable to move the door in the opposite direction. This design was superior to the Bowden push-pull cable because it had less hysteresis and it was immune to kinking.

Ford's 1998 Mustang was fitted with an early application of the peanut refrigerant fitting. This fitting was judged to be more robust than the spring lock fitting even though it required power tools to tighten it. The fitting was so named because it had a cross section shaped like a peanut.

The 1999 Saab brochure stated that the company fitted a refrigerated glove box on some of its 2000 model year vehicles.

A patent was issued in 1999 for a windshield wiper device called "Wiper Shaker". It vibrates the windshield wiper blade against the glass to break up ice. It is hard-wired into the vehicle's electrical system. It is being offered as an accessory on several 2011 model year Ford and Lincoln vehicles.

The 1999 Lincoln Navigator was the first vehicle to use the Amerigon heated and cooled seat. The Amerigon thermoelectric technology is based upon the Peltier effect whereby passing current thru ceramic semiconductors results in heating or cooling, dependent on current flow direction. These seats are said to be in worldwide use in 2011, according to the Amerigon website.

The following is from a Toyota press release dated September 1, 2000, "Lexus has introduced a water-repellent coating for front door glass and mirrors on several 2001 models, including the GS 300, GS 430, and LS 430 sedans and the RX 300 SUV. The LS 430 also features standard rain-sensing windshield wiper operation. On the RX 300, a four-link wiper mechanism sweeps 89 percent of the windshield area".

The list of high technology climate control features on the 2006 Lexus IS 350/250 vehicle is impressive. The highlights include Neutral Net temperature calculation, variable capacity compressor, glow plugs in heater hoses to hasten warmup, humidity sensor, micro dust and pollen filter, smog sensor and heated and ventilated front seats. This list was compiled from the Lexus College Product Information for the 2006 Lexus IS 350/250.

Windshield wiper arms were styled items on the 2007 Subaru B9 Tribeca. The arms blended into the metal portion of the wiper blades that were also designed to be attractive as well as functional. Other vehicles have begun to use styled arms and blades such as the 2009 Nissan.

In 2008, Denso announced its new air conditioner air-handling assembly that featured package size reduced by 20%.

It is interesting to note that in 2011 small fans are still being manufactured and sold. There is a wide variety to choose from. This is despite the fact that essentially all cars and trucks are air-conditioned and have defrosters that clear windshields in time to meet or beat the worldwide standards.

JC Whitney is also still selling cool cushions. The latest version is called WAGAN BEAD/RATTAN COOL COVER SEAT CUSHION and it is listed on their website for $20.99.

The list of climate control features on the 2011 Lexus LS represent the state-of-the-art as this book is concluded. They include those listed above for the 2006 model, plus four seats heated and cooled, enhanced interior heater, four zone automatic temperature control with rear seat infrared sensors, power rear sunshades, rear seat air purifier, rear seat cool box and windshield wiper de icer.

The 2011 Ford Fusion parts manual pages are included to show how the parts were being fabricated in this era. They are interesting when compared to the many different designs that have preceded them over the past years. They are lighter in weight, smaller in size and their performance is significantly better than the systems that preceded them.

In 2011, Denso announced their new Ejector evaporator. The system was expected to reduce the power consumption of the compressor by about 25%. According to an article at the SAE Automotive Engineering web site sae.org/mags/aei/6741 the key to the device is two evaporator cores mounted face to face with an inlet plumbing system that is shaped to produce a venturi effect that pulls refrigerant from one side to the other.

The following is from a General Motors press release dated July 23, 2010:

**WARREN, Mich.** – General Motors Co. will introduce a new greenhouse gas-friendly air-conditioning refrigerant in 2013 Chevrolet, Buick, GMC and Cadillac models in the U.S. that keeps vehicle interiors as cool as today while reducing heat-trapping gases in the atmosphere by more than 99 percent.

The biggest benefit of the new refrigerant, (HFO-1234yf) supplied by Honeywell, is that it breaks down faster in the atmosphere than the refrigerant currently used (R-134a), On average, R-134a refrigerant has an atmospheric life of more than 13 years, giving it a global warming potential (GWP) of over 1,400.

## CONCLUSION
Future automotive climate control engineers will continue to design systems and components that are lower in cost, lighter in weight, smaller in package size with performance equal to or greater than the systems they replace. This book has shown how the climate control systems have improved tremendously over the past 116 years. We can only imagine, in our wildest dreams, how they will evolve in the next 116 years.
I hope that the engineers who read this book find it useful in some way as they overcome the enormous challenges that they will encounter while forging ahead and continue to make the occupants of automobiles comfortable as they travel in any type of weather.

The Automobile and Horseless Vehicle Journal   May 1897

Lap robes were used to keep warm in the beginning.

The Autocar   September 15, 1900

Raincoats, hats and gloves were worn prior to heaters.

### A Good Carriage Heater.

Lehman Brothers, at 10 Bond Street, New York, are manufacturing a carriage heater that has unusual merits, and some 150,000 are in actual use—an average yearly sale of

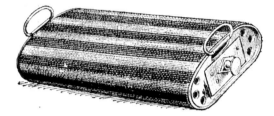

15,000—with a special coal of which they sell a million pieces a year. The Lehman heater gives a continuous heat for fifteen hours. It operates at an expense of ¼ cent per hour; it is absolutely unbreakable being constructed without solder or castings, and is perfectly safe. The Editor of THE AMERICAN AUTOMOBILE heartily recommends this heater.

American Automobile   December 1901

The popular Lehman charcoal heater used directly from horse drawn vehicles.

### "NEVER FREEZE"

SOLUTION OF
CALCIUM CHLORIDE

ALL READY FOR USE

This solution has been prepared especially for use in Gasolene Automobiles as a substitute for cooling water.

It makes freezing impossible at 0° F.

It also raises the boiling point to 225° F., thereby decreasing evaporation.

It can also be used in Acetylene Lamps to prevent freezing.

Price in 5 gallon cans, 35c. per gallon, and 10 gallon cans, 30c. per gallon.

No charge for cans.

Price in 100 gallon drums, 25c. per gal. Drums charged at $8 each and returnable at same price.

Manufactured by the

Merrimac Chemical Co.
75-77 BROAD ST.,        BOSTON, MASS.

American Automobile   December 1901

Calcium chloride antifreeze by Merrimac Chemical Co.

Automobile Magazine   March 1902

A bulky fur coat was needed to keep this driver warm.

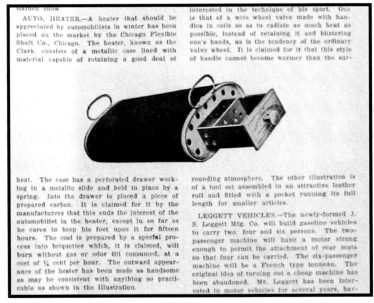

The Automobile & Motor Review   December 1902

The Chicago Flexible Shaft Company made this Clark model charcoal heater.

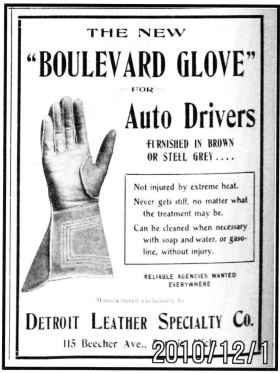

The Automobile March 1902 both images
Raincoats by The Post & Lester Co., gloves by Detroit Leather Specialty Co.

Automobile Magazine   August 1903

J. Stevens Arms & Tool Co. offered a top to protect the driver from rain and snow on their Stevens-Duryea vehicle.

Horseless Age   1903

"Old Pac" a 1903 Packard in Nevada on its crossing the United States after leaving San Francisco on 20 June 1903 and arriving in New York City on 21 August 1903. An umbrella served as protection from the sun.

Automobile Magazine   September 1903

A rain apron by The Automobile Equipment Co.

The Automobile   October 1903

Rain aprons protected these Packard engineers on their test trip.

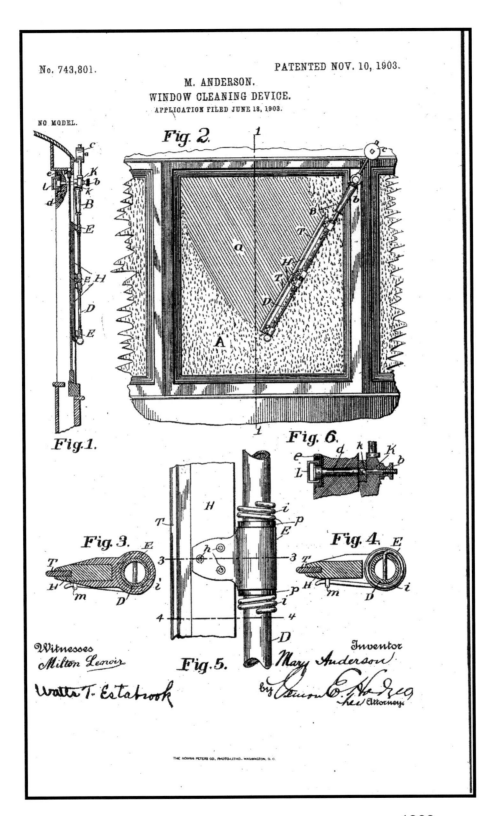

1903

Mary Anderson received a patent for a windshield wiper.

*Automobile Magazine   November 1903*

Brook Brothers offered robes, coats, capes, boots and trousers to motorists.

*The Automobile   December 1903*

Demmerle & Co. offered vests, long leather trousers, caps, coats and boots.

The Automobile   December 1903

The Davis Robe Co. offered an airtight bag lap robe.

Automobile Review   December 1904

Lehman had sold 250,000 charcoal heaters according to this J.W. Erringer ad.

Motor October 1905

The Scott Muffler Co. offered numerous types of neck mufflers.

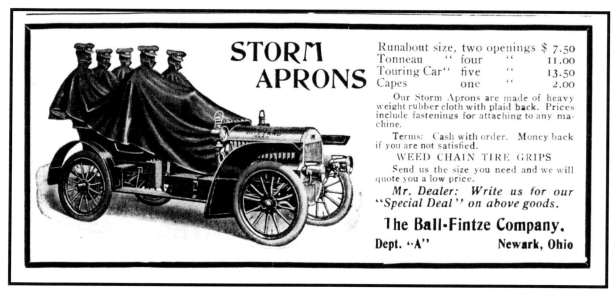

Motor December 1905

The Ball-Fintze Co. offered several styles of storm aprons.

Motor   October 1906

Morrison, McIntosh & Co. offered Rist-Fit gloves.

Motor   November 1906

This is a description of how the Lehman charcoal heater worked.

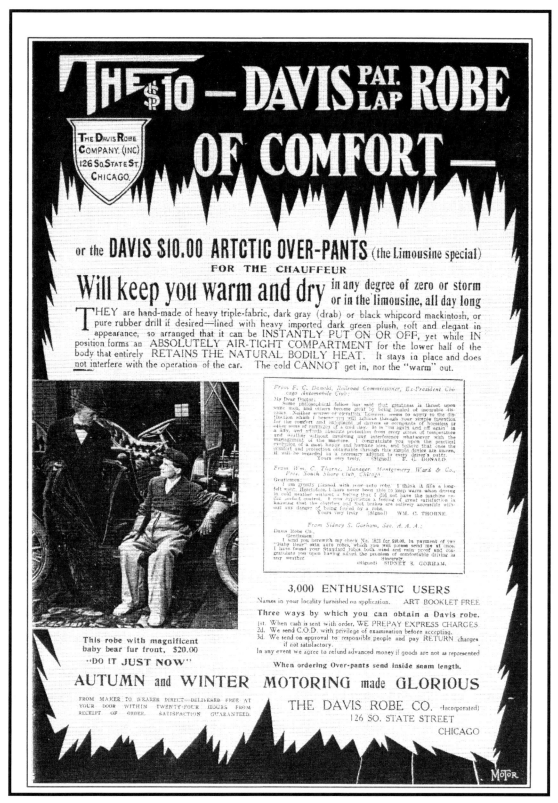

The Davis Robe Co. featured Artic Over-Pants for the chauffeur.

One of the earliest automobile specific exhaust heaters by The Motor Car Heater Co.

Sears Catalog 1908

Sears Roebuck Co. offered a broad selection of lap robes and charcoal heaters.

The Mandeville Steering Wheel Muff Co. offered several models of muffs.

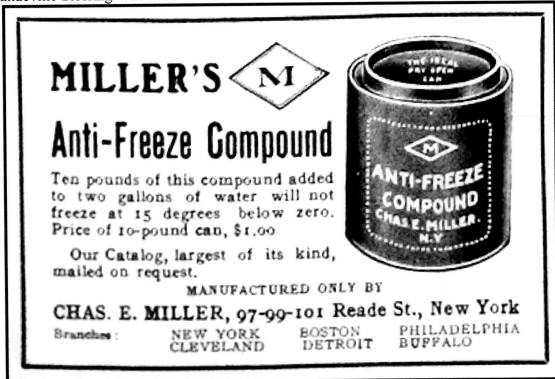

Chas. E. Miller featured an antifreeze compound at 10 pounds for $1.00.

Courtesy of Candy Waugaman and Fountainhead Antique Auto Museum

1908 Franklin fitted with storm curtains and windshield.

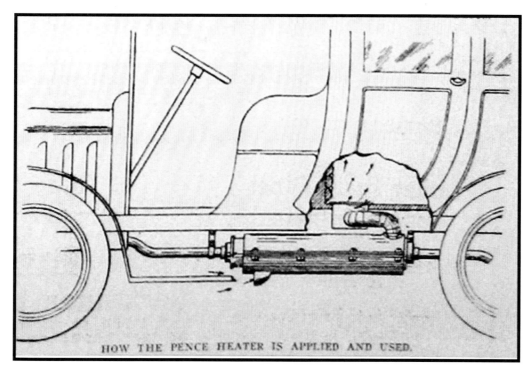
The Automobile   January 1908

The Pence exhaust heater was another early example of automotive specific heater technology.

## GOGGLES FOR DRIVING IN SNOW AND RAIN.

As all automobilists are aware, nearly every type of goggle yet invented has unfortunately to be abandoned when it is desired to drive fast through rain or snow. If the goggles are retained it soon becomes impossible to see through them, and if they are abandoned the driver suffers much inconvenience, the effect of rain or snow on the eyes at any speed above 25 miles an hour being painful.

To overcome this inconvenience a French inventor has produced a special type of goggle known as the L. I. M., an illustration of which is shown herewith, but which can be more completely understood by a reference to the line drawing, reproduced from *L'Automobile*.

The goggles consist of two opaque tubes $a$ joined at the forward end so as to be convergent. The two tubes are mounted on the usual soft leather guard designed to fit close to the face, so that no currents of air can penetrate to the eyes. The upper half of the face of each tube is closed by

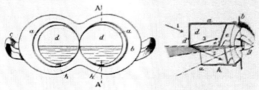

Sectional View of L. I. M. Rain-proof Goggle.

an opaque plate $d$, leaving the lower half open. To increase the range of vision the lower surface of each tube is cut away. Within each tube and behind the forward shield is a second inclined opaque shield $e$ and a glass shield $f$, there being sufficient open space between $e$ and $f$ to give the neces-

Goggles Designed to Protect Against Bad Weather.

sary range of vision under all ordinary circumstances. This range of vision is increased in wet weather by another range through the glass $f$, giving a view of the dashboard, bonnet and wheels of the car.

Protection against rain is obtained in the following manner: rain falling on the shield $d$ in the direction of the arrow, or in a more or less oblique direction, drips down to the lower edge of $d$, under the combined influence of the wind and their own weight, the drops of rain are detached from $d$ and carried inward until they strike the lower inclined glass shield $f$, then run down the surface of this to the lower part of the tube and finally escape through a hole $h$ made for this purpose. In the rare case where a drop of rain managed to pass through the opening between the two guards, its rate of travel would be so slow that it would not reach the eye. Three different types of goggles are made with ranges of vision for speeds of 90, 40 and 25 miles an hour.

The Automobile March 1908

A French inventor's storm goggles for safer driving in bad weather.

September 30, 1909 — THE AUTOMOBILE

## ANSWERED AND DISCUSSED

### MORE OIL COOLING

Editor THE AUTOMOBILE:

[2,029]—I would like to learn through your department of "Letters Interesting, Answered and Discussed," if a light oil, such as Havoline, would be safe to use instead of water in the radiator of a car with a four-cylinder vertical engine, of the thermo-syphon system of cooling? Last winter I used a mixture of one-third alcohol, glycerine and water, but found it troublesome, owing to the evaporation of the alcohol. I am inclined to try oil this winter, and want to know what you think of it.

J. A. RENÉ.

Superior, Wis.

This is a point that was covered in the letter (2019) of E. W. J., which was published in our Sept. 16 issue, except for the thermo-syphon point. This latter will be sufficient to bar the use of oil, as the latter is so heavy that it will not circulate except through the energetic action of a good pump. If you are in a position to add a pump to your circulating system, oil will do very nicely, as brought out in the letter referred to. We cannot give you a table of the properties of oil at low temperatures because no such table is available.

In case you did not like the mixture used last winter, why do you not try a calcium chloride solution, with which any desired temperature within reason may be obtained, and which is free from the objectionable evaporation you mention? That is, when using this, only the water evaporates, and this may be replaced by simply adding more water. Then, about once a month, the specific gravity of the mixture should be checked up. The only reason for this lies in the fact that if allowed to get too strong, the calcium chloride will precipitate out and this precipitate, in the form of a white powder or crystals, will clog up the system and prevent the remaining liquid from circulating. As the accompanying diagram of the freezing points of various strengths of calcium chloride solutions will show, the temperature desired may be attained by varying the amount of calcium.

The best way to make the solution is to first make a saturate solution of the chloride, and then use this, by adding water to it until the desired mixture is reached. This is done by taking half a gallon of water to 8 pounds of chloride for each gallon of saturate solution desired. It is a good idea to make a gallon or so extra to have on hand. You can tell if this makes a saturate solution by the fact that some of the crystals must remain in the bottom undissolved. If this is not the case, add more crystals until some of them will not dissolve. This solution is made applicable to the cooling system by adding to it more water, and finally when the right proportion is obtained, a handful of lime to render it slightly alkaline. The latter is done as this solution is said to have an acidic action on the metals of the whole cooling system. As the latter has never been proven, the simple precaution of the lime will counterbalance it. The diagram shows plainly what temperatures may be obtained by this solution.

### SAVANNAH WINNER'S SPEED

Editor THE AUTOMOBILE:

[2,030]—Will you kindly publish in the next issue of your valuable paper, "The Automobile," the average speed of the winner of the heavy car race at Savannah Thanksgiving Day, 1908?

JOHN A. HAMILTON.

Columbia, S. C.

At Savannah, the heavy car race was won by Wagner driving a Fiat. He covered the 16 laps of 25.13 miles each, a total of 402.08 miles, in 6 hours 10 minutes and 31 seconds, which is an average of 65.08 miles per hour. It might be interesting to you to know some of the details of the car. It was a four-cylinder motor of 155 mm bore by 160 mm stroke, practically 6.1 by 6.3, and was rated at 120 horsepower. The car had 50-inch tread, 107-inch wheelbase, and weighed 2,750 pounds. The tire and wheel equipment was 105 by 870 fronts and 120 by 880 demountable rear tires, all being Michelins. These sizes converted into inches and fractions were: 4 1-8 by 34 1-4 fronts and 4 3-4 by 34 11-16 rears. The car has double chain drive and Bosch magneto.

### USE OF UNIVERSAL RIMS

Editor THE AUTOMOBILE:

[2,031]—Will you please inform me on the following matters? I have regular clincher rims on my car. If I have universal rims put on can I use my same old casings with safety, there, of course, being no lugs on the universal rims?

W. SEVERANCE.

Stanford, Ky.

Yes, when universal rims are fitted to wheels for the express purpose of using old clincher tires on those rims, special flanges are supplied which hold the clincher tire safely on the rim. Then when the old clincher type of tire is worn out, and new ones are purchased, it is possible to obtain another set of different flanges which will hold the regular style of tire intended to be used with those rims. That is, with the exception of the flanges, which represent a very small expense, the universal rims are just what the name implies, universal, any kind of tire may be used with them, and, of course, used with safety, which is implied.

### WANTS LARGER TIRES

Editor THE AUTOMOBILE:

[2,032]—I have a car which is several years old and equipped consequently with very small tires. I have added a number of things to the extra equipment, as well as changing several parts of the mechanism. All of these changes have added to the weight of the car, so that now I feel that the weight per tire is too large for the tires now on the wheels. This contention is carried out in driving by the very low mileage which I obtain from the various tires. Now, I need another set of new tires and instead of getting the same old too-small size, I want to get the new larger sizes. To do this without buying new wheels, which I do not care to do now, I must use odd sizes. Can you give me a list of the odd sizes now obtainable, as well as a list of makers who make them, and local dealers who handle them?

G. J. O'HARA.

New York City.

Your decision to use larger tires is a very wise one, and is to be commended. The so-called "odd sizes" have not been made very long, and even now are not

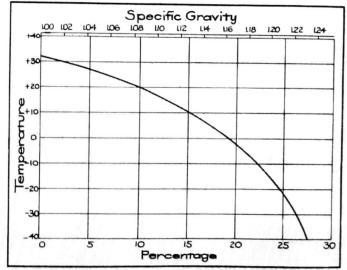

Diagram of Freezing Points of Various Solutions of Calcium Chloride in Water

The Automobile  September 1909

An article discussing the use of Havoline light oil and calcium chloride as antifreeze.

any cement at will, the quantity added being measured only by the required speed in drying. Then there are cases where certain degrees of tenacity are required. For these, other gums are added as rosin, mastic, gumlac, etc. These, however, should be used only when needed, and much discretion should be used in adding them to an already very satisfactory cement.

### MILE-A-MINUTE FOR HIM

Editor THE AUTOMOBILE:

[2,044]—Will you please answer the following questions through "Letters Interesting, Answered and Discussed"?
1. Had I ought to have a complete knowledge of an automobile to become a racing driver?
2. Do automobile companies give inexperienced men a trial?   L. M. S.

Denison, Kansas.

Considering the unusual number of race meets which have been held during the present year, and the equally unusual number of fatalities attending the same, at least an average of one man killed per meet, it is surprising that anyone would have the temerity to want to take up racing as an occupation. However, to answer the above questions:

1. Yes, for this purpose you cannot have too complete and thorough a knowledge of not only the machine you expect to drive, but all others as well, and a very clear insight into the principles governing the action of all of the various parts. Not only should you be able to simply drive the car, but you should be able to assemble, dissemble, repair, machine, or do any other kind of work necessary to make any one of the pieces entering into the complete car, and to put it into place, or take it out under any and all conditions. That is, you should be an expert on engines, transmissions, clutches, and other components, as well as on driving. As far as simple driving is concerned, a man of ordinary intelligence can learn to drive in the course of a single day, granting only a car to drive, and someone to show him how. Beyond that comes the superlative skill to permit a man to get more speed out of any given car than any one else can. More even than personal skill, a cool head and steady nerves are required. Without the requisite nerve and coolness, necessary to drive, say at the rate of a mile a minute carrying your gasoline tank in your lap, any amount of skill in handling a machine is of no possible use. More than this, accurate judgment is required since the tight places in which a racing driver is sometimes placed require not leisurely consideration, but instant and immediate decision as to what is best to do, and equally fast work in doing it. An ideal way to prepare yourself for this work would be to go into some automobile factory and work at least six months in each and every department there, closing with about a year of outside testing work. In that length of time (it would doubtless have proved that you possessed the requisite skill, judgment, nerve, and cool head. At that time you would find little trouble in obtaining a situation as a racing driver, since the supply is always less than the demand.

2. Nearly every firm finds it necessary to hire inexperienced men, who are not hired for the purpose of making racing drivers of them, but to learn the machinist's trade. The way for you to start in will be as an apprentice, then when you have had a little automobile shop experience, change to a position in which you would learn, for instance, to assemble engines, later one in which you would test them, etc. If money is no object, you may obtain a position in nearly any automobile factory in the country.

### PROPER ALCOHOL SOLUTION

Editor THE AUTOMOBILE:

[2,045]—Will you please answer the following through "Letters?" What proportion of denatured alcohol and water is proper for an anti-freezing solution? Is there anything in this same line which you consider better than alcohol, and if so, what is it?
TRUMAN B. PEIRCE.

Providence, R. I.

As a matter of fact there is no proper proportion, as any one of a number of proportions will give satisfactory results. There is this to be said, however, that the weakest solution which will stand the climate in which you are located will give the most satisfactory results. The reason for this is that the alcohol evaporates out from the solution, and the stronger the solution, the more there is to evaporate, the easier it evaporates, and the greater the influence of this evaporation upon the solution left. Accompanying this is given a diagram of the freezing points of various solutions of denatured alcohol in water. From this diagram select the lowest temperature which you are sure to meet and that will give you the strength of mixture to use. In making this selection, remember the advice given above. Since zero is seldom met with, you might try a 38 per cent. solution which will not freeze until that temperature, 0 deg., is exceeded.

Unless you are particularly desirous of using denatured alcohol, you will note from the diagram that wood alcohol gives a much lower temperature, the percentage mentioned above yielding about minus 22 degrees with wood alcohol. No particular one of these solutions is recommended and all of them have their drawbacks. So, the best way to do is to try one and if this does not suit you for any reason, try another. In this connection, see the letter of W. T. A. [2,040] elsewhere in this issue, as well as J. A. Rene [2,029] in the Sept. 30 issue, and that of E. W. J. in the Sept. 16 issue. Both of the two latter have tried various cooling solutions, and this winter are about to try the use of a light oil.

Do not think that because nothing has been said about other cooling fluids than the two kinds of alcohol and oil, that none of the others is as good. This is not the case; these were dwelt upon because you asked about the one, and the other was the subject of the most recent letter on this same subject. Calcium chloride, the curve of which was presented very recently, is very satisfactory to some, as is just plain salt to others. This latter, however, yields but zero degrees, so is not available where temperatures lower than that are common. Then the salts, which are not so well known as potassium carbonate, alone and in combination, are much used.

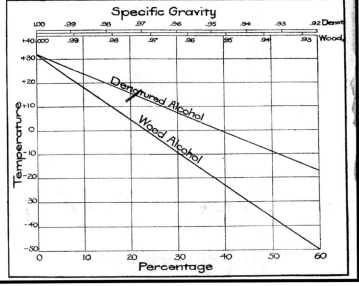

The Automobile October 1909

An article comparing denatured alcohol and wood alcohol for antifreeze

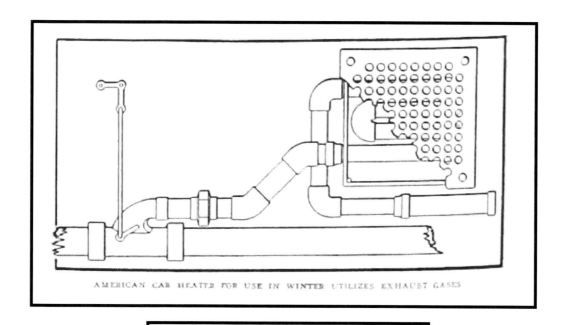

"American" Auto-Heater—This is a device for heating the car in Winter by the exhaust gases. It consists of a receptacle 10 inches square by 3 inches deep, fitted in the floor of the car, which contains a coil of radiating tubes. One end of the coil connects to the exhaust pipe, in front of the muffler; a valve is provided which is clamped over a small hole cut in the pipe, without disturbing the fixtures at either end. The valve flap which, when the device is out of use, covers the opening into the coil, swings down into the main exhaust pipe when opened, so as to deflect more or less of the gases into the coil. The other end of the coil has a small muffler to deaden any sound. The heater valve is operated by means of a small hickey which is usually placed on the riser of the front seat.

In any of the closed types of automobile bodies the temperature in zero weather can be brought up to 70 degrees in a few minutes. In an open car, of course, the heat is not so well retained, but by keeping the heater constantly turned on the car can be made quite comfortable. The heater is sold by Flavius C. W. Sudrow, 15 West Swan street, Buffalo, N. Y.

The Automobile   December 1909

The American exhaust heater offered by Flavius C. W. Sudrow.

The Automobile   December 1909

Victor Auto Supply Mfg. Co. and their Vasco windshield.

The Automobile   December 1909

The Gilliam Manufacturing Co. offered a hood cover/lap robe/ground robe/cape.

The Clark charcoal heater in several styles by the Chicago Flexible Shaft Co.

Image Circa 1988

A recent image of a Clark charcoal heater taken at a flea market. The asking price was marked $50.

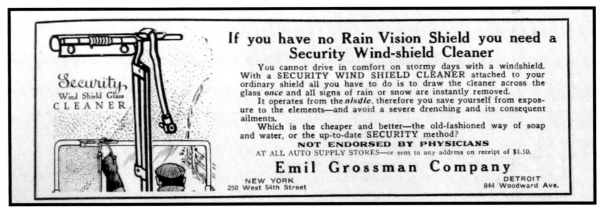

The Automobile   January 1912

One of the earliest windshield wipers by the Emil Grossman Co.

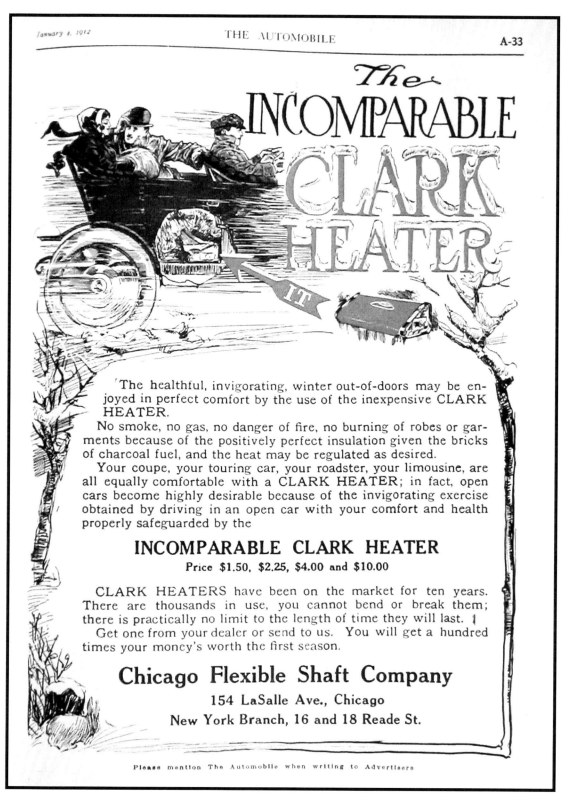

The Automobile  January 1912

The Clark charcoal heater was 10 years old and was offered for $1.50 to $10.00.

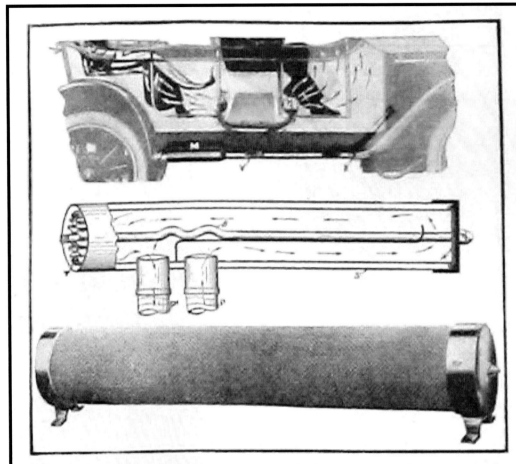

Figs. 3-5—Detroit automobile heater and parts

## The Detroit Automobile Heater

The Detroit Auto Heater Company, Detroit, Mich., manufactures the new heater, Figs. 3, 4 and 5. The same consists of a two-way valve which directs the exhaust either directly to the muffler or leads it through the heater and of the heater proper. The latter consists in turn of a corrugated, gastight pipe T through which the exhaust gas flows and which is surrounded by another pipe, an air space being formed between both. The exhaust enters through P and leaves through P1, and the whole device is inclosed in a screen which is equipped with feet, permitting of keeping the apparatus standing on the running board. The air which passes through the space between exhaust and outer pipe then flows into a manifold and to the front and rear compartment of the car, to increase the comfort of driver and passengers.

The Automobile December 1912

This Detroit Automobile Heater Co. model was the exhaust type.

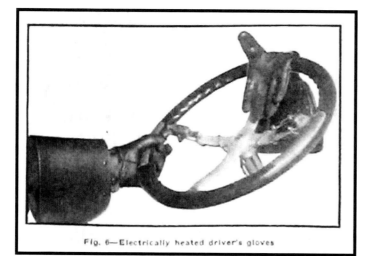

Fig. 6—Electrically heated driver's gloves

### Electrically Warmed Driver's Gloves

One of the many new accessories which the present winter has called forth is the outfit shown in Fig. 6, the electrically warmed glove made by Carron & Company, Inc., 1784 Broadway, New York City. This invention which after being tested for a long time is now ready for the market consists of a pair of brass plates which are mounted on the steering-wheel rim and are connected to the poles of a 6-volt battery. The gloves are equipped with two button contacts which, when pressed against the two brass plates, permit a current to flow through a wire resistance laid through the glove so as to be in contact with the back and palm of the hand and to surround the fingers; the effect is that if the wire is heated, the hand and its every part is warmed to any desired degree, the heat being regulated by means of a rheostat mounted on the spoke of the steering wheel. There is a rheostat for every glove, the two of which are independent of one another, so that one hand may be warmed and the other not, or both to varying degrees. The wire resistance consists of a very great number of hair wires wound with silk so as to form a cable; so that many wires may break and yet the working of the glove remains undisturbed. On the other hand, the wires are very strong and tough, so that breakage is not a frequent occurrence and need not be considered. Of course, the plates may be attached at any portion of the steering wheel rim and their position may be changed to suit the driver's requirements. As Fig. 6 shows, the appearance of the gloves is not different from an ordinary good type of glove and there is no difference between the use of this glove and any other. Furthermore, if it is desired to stop the warming of a hand or both, it is merely necessary to interrupt the contact between one of the glove buttons and the brass plate which it touches. This shows the simplicity and usefulness of this new accessory.

The current consumption is small, being approximately .70 ampere, so that about 10 watt are needed to keep the driver's hands warmed. It might not be needless to add that cold hands in winter are not only the source of a more general discomfort, but frequently the cause of dangerous situations, as control is made very difficult if the hands are not in normal condition.

The Automobile December 1912

Carron & Co. offered these electrically heated drivers gloves.

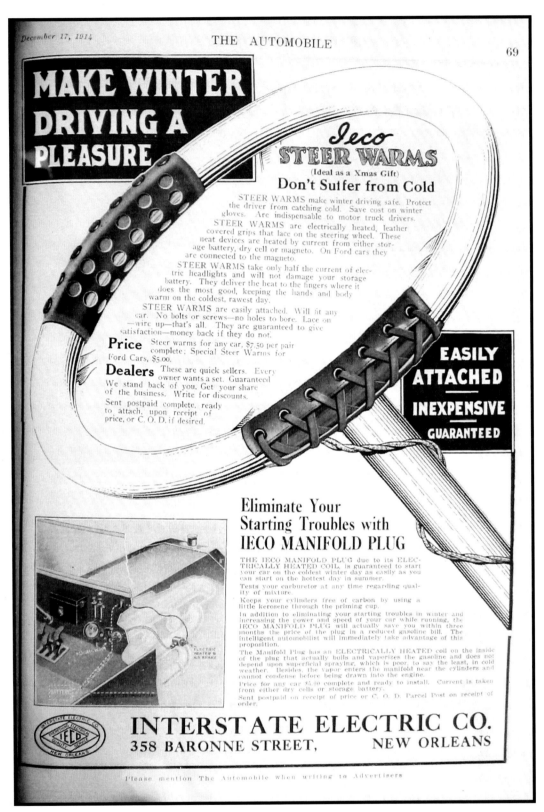

The Automobile December 1914

Interstate Electric Co. offered a device to warm the steering wheel electrically.

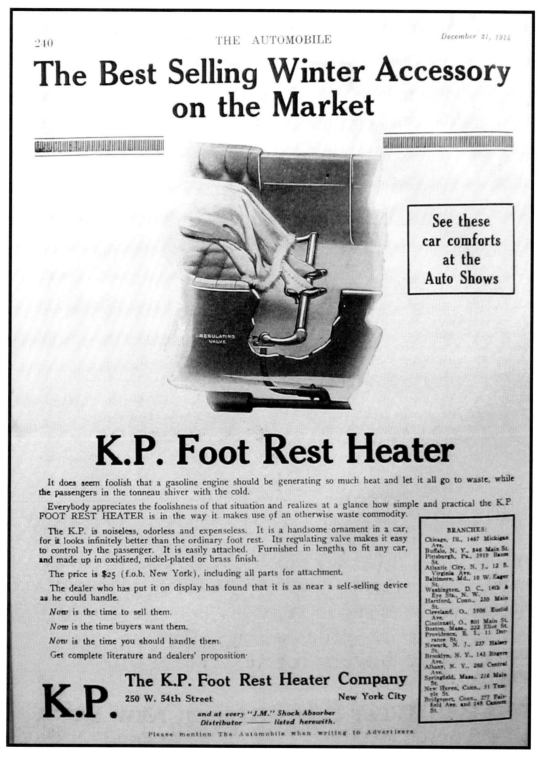

The Automobile  December 1914

The K.P. Foot Rest Heater Co. featured an exhaust type heater for $25.

Image taken at the Buick Gallery & Research Center

1915 Chevrolet Baby Grand fitted with two hand operated wipers.

Vehicle owner Jack Beatty

1916 Detroit Electric with ventilation register and electric windshield wiper motor.

## It's time now to prepare for next winter's business

Frosty days are here and every little nip serves to remind motorists that cold days are coming, when they will want their cars heated. Now is the psychological time to call their attention to the advantages of the

## Utility Protected Heater
### To Over 3,000,000 Waiting Prospects

Here is a device that has a tremendous sale wherever dealers talk it up to their customers. It is just what motorists need, and there is nothing on the market that serves the purpose so well. To show it and explain it is to make sales, and these sales mean a worth-while profit to the dealer.

### The Utility is an Ornamental Foot Rail That Radiates Heat

It is a practical, clean and durable heating device which keeps the car comfortable, regardless of the outside temperature, throughout the winter, and acts as an ornamental foot rail in the summer. It takes up no additional room in the car, costs nothing to operate, can't get out of order, and is absolutely weathertight. It utilizes the heat from the exhaust and is free from odor, noise, back pressure, escaping gas and loss of power. Heat is regulated by a convenient valve.

The Utility has been improved and protected so that it cannot burn the most delicate fabrics even at the highest temperature. Thousands are in use and giving unqualified satisfaction. Every customer who drives his car in the winter is a good prospect for the Utility, and you can make many sales if you get busy now and push it.

You can carry Utility Heaters in stock to meet immediate demands, as the only special parts are the exhaust pipe fittings which are instantly detachable and interchangeable. Send in your order now and be ready for the big winter business.

**PRICE COMPLETE $15 LIBERAL DISCOUNTS**
The Utility Jr. Heater for Fords Sells for $7.50

### The Hill Pump Valve Company
Archer Ave. and Canal St., Chicago, Ill.
Manufacturers of Famous UTILITY Auto Specialties

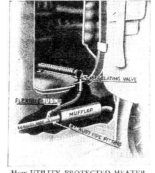

How UTILITY PROTECTED HEATER Is Attached to the Exhaust

Automobile Trade Journal   November 1916

The Hill Pump Valve Co. offered an exhaust heater for $15.

Thurston auto Supply Mfg. Co. offered its Brickey indirect exhaust heater for $3.90 for Ford cars and $4.90 for other cars.

# Hudson Fits Dash Radiator Control

Illustrating operation of the Hudson radiator shutter control by means of pull rod from the dash

At the left is a view of the Hudson radiator with the shutter closed and at the right with the shutter partly open

Left—Closed position of shutter link. Right—Open position

To remove the necessity for using padded covers, etc., on the radiators of Hudson cars, the company has brought out a device which acts as a radiator shutter and which is being listed as special equipment. The new device is operated from the dash by the driver and permits him, by means of a simple plunger, to govern the amount of air that comes in contact with the radiating surface, and hence puts the temperature of the cooling fluid under his control.

The device acts very much as a shutter, with a series of vanes which open and close in accordance with the movement of the controlling device on the dash. In summer the mechanism need not be detached, as it does not cut down materially the area of the radiator when it is full open, so that in warm weather it is simply left untouched.

In order that the device will be complete, it is necessary for the driver to know the temperature of the cooling water accurately enough to determine whether the shutter is open or closed to the correct degree. This part of the equipment is provided by a Boyce Moto-Meter mounted in its customary position on the radiator cap.

The device is valuable in warming the car in very cold weather, as the air can be entirely shut off from the radiator until the water temperature reaches its efficient temperature, which in the case of the Hudson is stated to be 130 deg. Fahr. The shutter vanes can then be opened to the desired amount by simply operating the pull-rod on the dash.

From a maintenance standpoint the device is simple, as there is nothing to it except the shutter and the rod and bell-crank operating mechanism. There is no need for any adjustments, and none is provided. The installation is simple, and is made possible by using a longer radiator shell which not only houses the radiator, but also the shutter. It can be fitted to all Super-six Hudsons, the cost of installation being $25, with Moto-Meter complete. If the owner has already a Moto-Meter the shutter assembly costs only $15.

The accompanying illustrations give a very clear idea of the construction and operation of the device, the neat and simple pull rod mounting on the dash being both conveniently situated and by no means detrimental to the appearance of the instrument board. The front views of the radiator showing the shutter both closed and in a partly open position make it evident that appearance has not suffered in this direction; in fact, if any comment is to be made on the matter, it may be said that the new design is more attractive than the old and far less likely to become clogged with dust and debris from the road.

## Goodyear Employees Paid for Ideas

ONE of the most popular operative features at the plant of The Goodyear Tire & Rubber Co., Akron, Ohio, is the suggestion system, which enables employees to capitalize their ideas for improving machinery, conditions, methods and product.

The workmen who make Goodyear products are given every opportunity to advance their ideas, and all suggestions adopted are well paid for. Interest in the system is well sustained, and thousands of dollars have been paid to employees for their ideas.

The suggestion box method is used, with boxes placed throughout the plant in convenient places. Employees are asked to put their ideas in writing and deposit them in the suggestion box. The suggestions are gathered each day and referred to the proper persons for consideration. All are carefully gone over, however inconsequential they may seem.

*The Automobile November 1916*

Hudson vehicles could be fitted with genuine radiator shutters operated with a pull rod on the dash.

Dayton Welding Co. offered a welded exhaust heater.

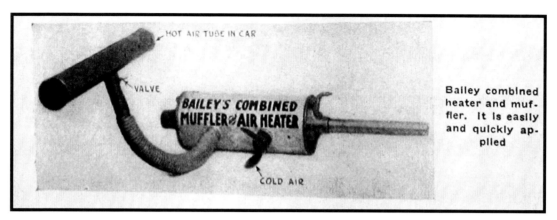

The Automobile   December 1916

The Bailey exhaust heater was combined with the muffler.

The Automobile   December 1916

The Dayton Top Improvement Co. Blackmore patented door curtain openers.

The Craftsman Motor Corporation featured the Autocraft-Bovey exhaust heater with an air-to-air heat exchanger.

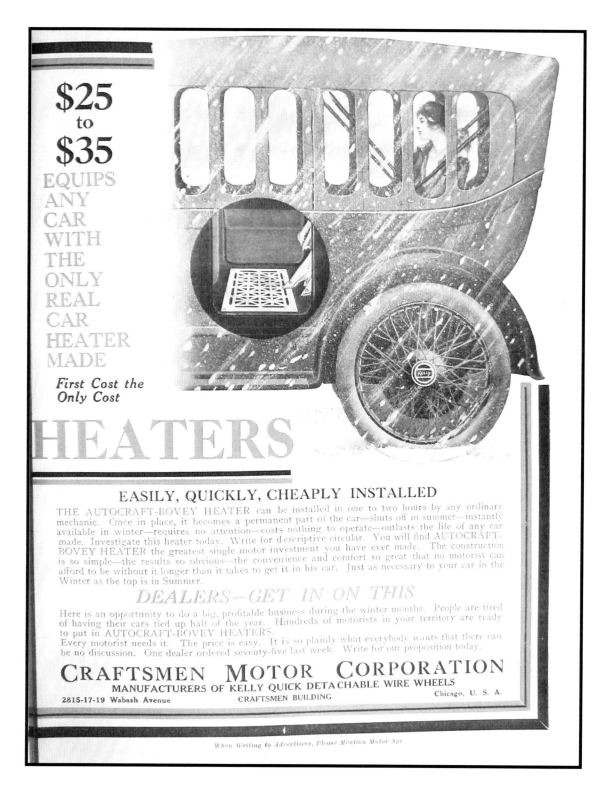

**$25 to $35**

EQUIPS ANY CAR WITH THE ONLY REAL CAR HEATER MADE

*First Cost the Only Cost*

# HEATERS

## EASILY, QUICKLY, CHEAPLY INSTALLED

THE AUTOCRAFT-BOVEY HEATER can be installed in one to two hours by any ordinary mechanic. Once in place, it becomes a permanent part of the car—shuts off in summer—instantly available in winter—requires no attention—costs nothing to operate—outlasts the life of any car made. Investigate this heater today. Write for descriptive circular. You will find AUTOCRAFT-BOVEY HEATER the greatest single motor investment you have ever made. The construction is so simple—the results so obvious—the convenience and comfort so great that no motorist can afford to be without it longer than it takes to get it in his car. Just as necessary to your car in the Winter as the top is in Summer.

### DEALERS—GET IN ON THIS

Here is an opportunity to do a big, profitable business during the winter months. People are tired of having their cars tied up half of the year. Hundreds of motorists in your territory are ready to put in AUTOCRAFT-BOVEY HEATERS.

Every motorist needs it. The price is easy. It is so plainly what everybody wants that there can be no discussion. One dealer ordered seventy-five last week. Write for our proposition today.

## CRAFTSMEN MOTOR CORPORATION
MANUFACTURERS OF KELLY QUICK DETACHABLE WIRE WHEELS

2815-17-19 Wabash Avenue    CRAFTSMEN BUILDING    Chicago, U. S. A.

*When Writing to Advertisers, Please Mention Motor Age*

Motor Age October 1916

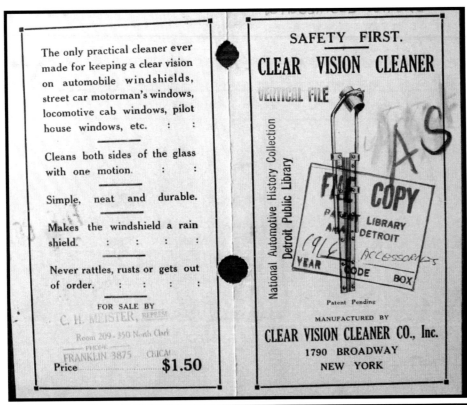
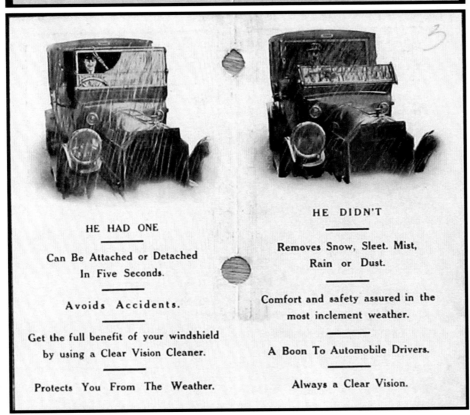

1916 Brochure

The Clear Vision Cleaner Co. offered a hand-operated windshield wiper to wipe both sides.

# Three Inclosed Cars Introduced by Jordan—Heater Part of Equipment

JORDAN cars are now available in limousines, coupes and sedans. In the design and production of these three models the Jordan policy of luxurious appointments has been adhered to to the smallest detail. In the limousine there is a surprising amount of expensive fitting for a car selling for $3,000. The color is in unusual shades of blue or green.

The front seats of the limousine are trimmed in long-grained hand-buffed leather in full French plaits without buttons. The upholstery in the rear compartment is of expensive French tapestry, the pattern selected to harmonize. The woman occupant is appealed to with a clearance between the rear seat and the back which is wide enough so that big hats will not interfere. The seat is tilted for comfortable repose. The glass throughout is sashless with mechanical lifts except in the rear quarters which are provided with lace lifts. There are dome lights, rear corner reading lamps and running-board lights. To be found in the equipment are: Flexible cloth robe rail, toilet set, watch in the front-seat back, motor phone and an inclosed heating system.

*The Jordan convertible sedan, which sells for $2,350. A heating apparatus is part of the equipment*

The custom-style sedan sells for $3,350. In this body too there is space provided so that a woman's hat will not hit the back of the top. For interior finishing there is a choice of rich domestic upholstery or French tapestry. The lines are, of course, low as befits Jordan cars. The top is of the convertible type with windows which disappear for warm-weather driving. This model, too, includes a heating apparatus as part of the regular equipment.

The coupe, listing at $2,300, is a four-passenger job. The top is of the permanent type. The four passengers are seated with a staggered two-seat arrangement and there is a carrying compartment large enough for two suit cases and a traveling bag in the rear deck. Another smaller compartment behind the seats will accommodate small parcels, doctor's cases, etc.

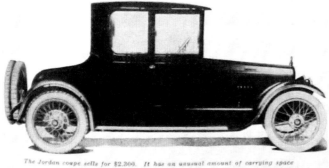

*The Jordan coupe sells for $2,300. It has an unusual amount of carrying space*

*The Jordan limousine has the long, low lines which characterize all bodies offered by this maker*

### SWEDEN'S MOTORING CONDITIONS

New York, Jan. 5—Sweden is a country naturally attractive to the motorist on account of its scenery and the macadam roads which wind and twist through the mountain and agricultural districts, but Sweden has been having trouble in getting supplies since the war began and of her 9000 cars more than half the taxicabs and a large percentage of the private cars are in storage.

Many of the cheaper makes of United States cars have been represented in Stockholm for years. Sweden herself has only two makes, one of which is produced in very small quantities. The Scaniavabis is made in two models, selling at about $2,500 and $3,500. About 300 are made each year. Imports, however, are retarded by excessive taxation, 15 per cent of the value at the port of entry, the tax including the freight, which now is from $500 to $600 a car. A tire is taxed at the rate of about 35 cents for 2½ lb. Accessories have several custom schedules.

Good roads lead 600 miles north from Stockholm, and satisfactory hotels are plentiful. The speed law is 25 m.p.h., and it costs about $9 to register. This is a continuous registration, the registration being for the car and not for the owner as it is here.

Motor Age   January 1917

The Jordan was one of the earliest vehicles to offer factory installed heaters.

Motor World October 1917

The Detroit Weatherproof Body Co. offered a detachable weatherproof top.

Motor World October 1917

The U-Auto-C Corp made the Klear-Sight manual windshield wiper.

## WORLD — WINTER BUSINESS

Oakland Robe Lock

### Oakland Robe Lock

The Oakland robe lock is a large clamp which fits around both robe and robe rail and is locked so that the robe cannot be removed. It is made of phosphor bronze, nickel plated and has a tumbler lock. Any thickness of material can be locked. Price, $2.50.—Robe Lock Co., Oakland, Cal.

Motor World   October 1917

The rob lock by the Robe Lock Co. sold for $2.50.

The Crew Levick Co. manufactured the Victor Fresh Air Heater that featured an air-to-air heat exchanger.

## Ottofy

The Ottofy heater consists of a tube which fits around the exhaust manifold of the car and is open at the front end. Cool air passing in at the front is forced back by the action of a fan to radiators which may be placed in both front and rear compartments. The amount of heat may be regulated. Price, for Ford roadster, $7.50; touring cars, $10; other roadsters, $12; touring cars, $18.—Ottofy Auto Heater Co., St. Louis.

Motor World   October 1917

This exhaust type heater by the Ottofy Auto Heater Co. collected radiant and convected hot air from the exhaust manifold and pushed it into the passenger compartment using the engine fan.

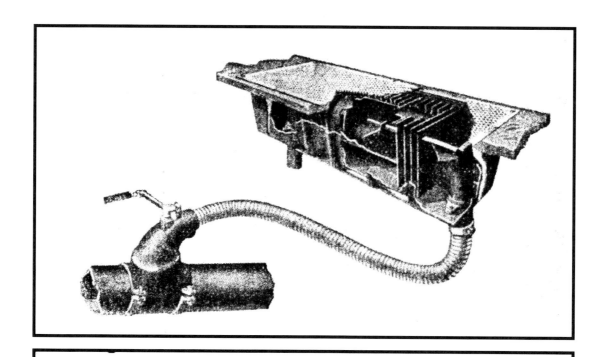

## Vesuvian

The Vesuvian heater is all steel with a seamless tubing core in which are welded a number of semi-circular baffle plates which, in addition to acting as a muffler, also assist in radiating the heat. Outside the tube are a series of fins which also act as radiators. The heater is connected to the exhaust pipe with a flexible tube, and there is a valve to regulate the degree of heat. Two styles are made, one setting flush and the other attached above the floor boards. Both types are $20. A special Ford type is $10.—The Reliable Auto Heater Co., Cleveland, O.

Motor World   October 1917

The Vesuvian exhaust heater by the Reliable Auto Heater Co.

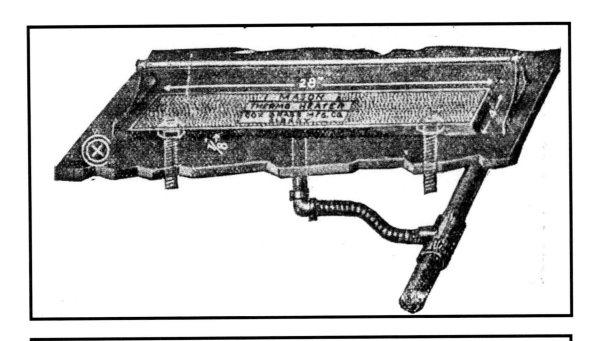

## Cox Thermo

The Cox Thermo Autoheater has been considerably improved and an aluminum model added. The heaters are made in two sizes, 28 x 4 oblong, and 8 x 14 square. The heater fits on the floor of the car without sawing or cutting and gets its heat through a flexible tube from the exhaust pipe. The heat is controlled by a thermostat, avoiding the possibility of overheating. Two types are made, polished cast aluminum and cast iron black enamelled.—Cox Brass Mfg. Co., Albany.

Motor World October 1917

The Cox Brass Mfg. Co. manufactured the Cox Thermo exhaust heater.

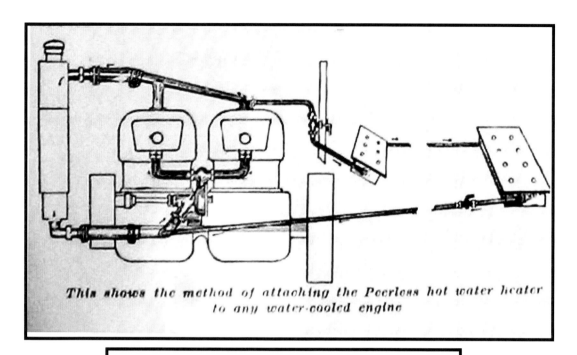

*This shows the method of attaching the Peerless hot water heater to any water-cooled engine*

**Peerless**—This is a hot water heater which derives its heat from the water circulation or cooling agent of the engine. The inlet and outlet pipes leading from the radiator to the engine are tapped and piping leads back to the two heaters, one in the front and one in the rear compartment as shown in the illustration. The heat radiating surfaces are constructed of copper and the air chamber is of heavy zinc, lined with asbestos and over all is a cover of a very thin sheet metal. Hot water is taken from the water jacket of the cylinders nearest the dash and is circulated through the heater from whence it returns to the bottom of the radiator. The flow of water is controlled by a stop cock on the intake line. These heaters are made in two sizes, the No. 1 having six tubes and selling for $35, and the No. 2 eight tubes, selling for $40. It is made by the Peerless Radiator Co., Inc., Gibbs, Idaho.

Motor Age   January 1917

This Peerless Radiator Co. system is one of the earliest hot water heaters.

**Konserv**—The Konserv is an electric heater made in three styles for which current is supplied by the regular electric system of the car. It is designed to give a steady heat, but does not get hot enough to burn or scorch. It is finished in nickel, and the three sizes are 14 in., 22 in. and 34 in., the prices respectively being $5, $6 and $6.50. There also is an electric heater designed to keep the engine and radiator warm when the car is left in a cold garage. It is placed under the hood. In one design, current may be taken from a 110-volt lighting socket and in another a combination is found that may be used either on 110-volt or 6-volt battery circuit. Current consumption is 100 watts per hour and the price of the first one mentioned is $5 and the second $7.50. They are made by the Konserv Electric Co., Cleveland, Ohio.

Motor Age   January 1917

An image of the Konserv electric heater was not included in the article. Electric was a rare type of heater from the early era.

**J. M.**—This is a footrest heater with a water jacket around the exhaust passage and the heater itself is a long cylinder through the center of which is found the water jacket. The exhaust from the engine is diverted and the amount of heat may be controlled by opening or closing a valve in the exhaust line. It is made by the J. M. Shock Absorber Co., 210 South Seventeenth street, Philadelphia, Pa.

Motor Age   January 1917

An image of the J.M. Shock Absorber Co. water heater was not included with the article.

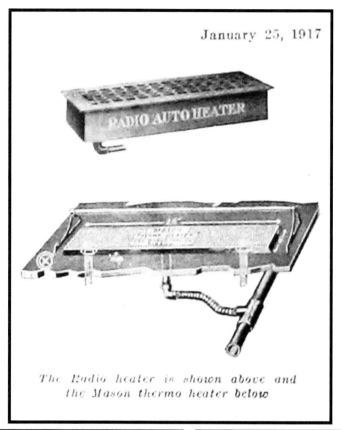

*The Radio heater is shown above and the Mason thermo heater below*

**Mason Thermo**—The feature of this heater is that it has a thermostat to regulate the temperature. It is 28 in. long and 4 in. wide and rests on top of the floor so that no cutting is necessary to install it. A V-shaped hole is cut in the exhaust pipe between the engine and the muffler and attachment is made by clamping a brass coupling over this hole. Three holes are bored in the floor opposite the openings at the bottom of the heater for pipe attachments. The heater is made of cast iron with black enamel finish and sells for $10. There is also one made of bronze, nickel-plated, at $15. It is made by the Cox Brass Mfg. Co., Albany, N. Y.

**Radio**—This exhaust heater is somewhat improved over last year, the main change being in material used for construction, sheet steel having taken the place of cast iron, which reduces the weight and materially adds to the heating efficiency. The heater measures 14½ by 5 by 3 in. Heat is controlled by a lever at the side. Installation is a job for a mechanic and requires from 1 to 2 hours. The price is $8 and it is made by the Milwaukee Specialty Co., 705 Chestnut street, Milwaukee, Wis.

Motor Age   January 1917

Cox Brass Mfg. Co offered the Mason Thermo exhaust heater with a thermostat.

The Milwaukee Specialty Co. manufactured the Radio exhaust heater.

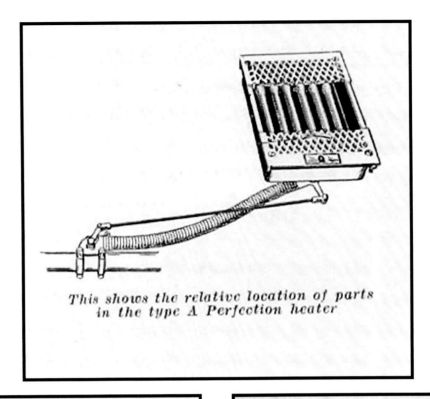

*This shows the relative location of parts in the type A Perfection heater*

**Perfection**—Perfection heaters are made in several sizes and at prices ranging from $15 to $25. The type A heater is installed flush with the floor and sells for $25, while the type B, which is a vertical type, is especially adaptable for commercial cars and sells at the same price. The type B is a footrest heater and in addition to the three mentioned, there are seven other types now being marketed. The floor space required for the type A is 8½ by 12⅝ in., and 4½ in. clearance below the floor is necessary. The top of the heater is aluminum and the heat control lever is built in the floor plate. The type AB is identical with type A except that heat control is a separate unit and may be located so that the chauffeur can turn the heat on or off as desired. In the type B the heating tubes are protected by a perforated metal housing and the heat control is separate and may be located at a point convenient to the driver. It is designed especially for commercial cars. Type C is similar to type B, only smaller. The type D is a footrest heater, designed for use in either open or closed cars, with heat control separate. With oxidized brass housing it sells for $25 and with enameled steel $20. The type F, which is for Ford cars, sells at $15. In installing the Perfection heater, a small hole is cut in the exhaust pipe ahead of the muffler. A butterfly valve having the same radius as the exhaust pipe is placed over the hole and the flexible tube connected between the valve leads back to the position of the heater in the floor of the car making a similar connection to the heater inlet. It is made by the Perfection Spring Service Co., Cleveland, Ohio.

Motor Age   January 1917

The Perfection Spring Service Co. exhaust heater featured a control lever "that the chauffeur can turn the heat on or off as desired".

This convertible coat and goggles was "the latest motoring costume or disguise."

The Automobile January 1917

A manual windshield wiper by John W. Jepson.

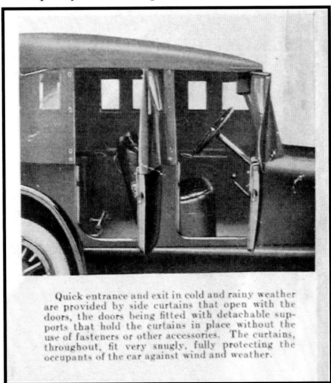

1917 Oldsmobile brochure

Side curtains were attached to the doors with integral supports.

Automotive Industries   October 1918

A cowl-top vent for the Ford Model T by the N. A. Petry Co.

1918 Oldsmobile brochure

An exhaust heater/footrest combination was offered by Oldsmobile.

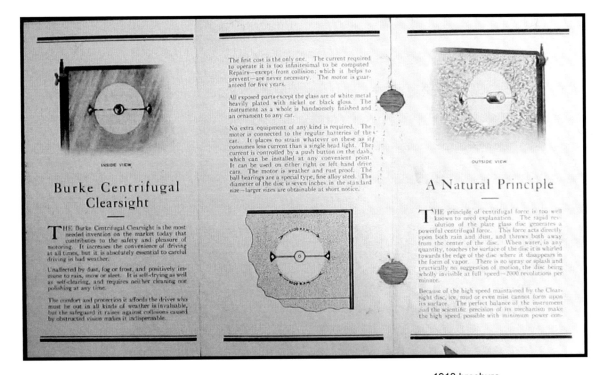

1918 brochure

This unique circular electric windshield wiper was manufactured by the Automobile Devices Co.

Ford Motor Company image

This 1918 Ford Model T featured canvas side curtains that provided some protection from the weather.

1919 brochure

A manual windshield wiper by Stadeker Metal Specialty Co.

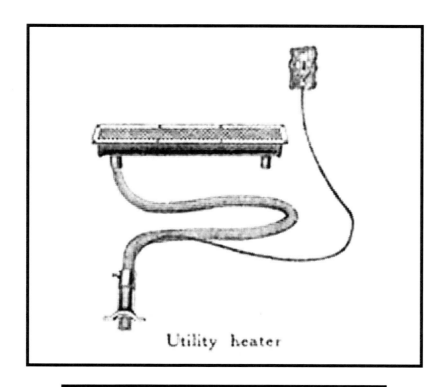

## Utility Heater

With cold weather approaching, attention will be given to winter accessories, and one of the first announcements of these winter comforts is made by the Hill Pump Valve Co., Chicago, which has just placed on the market a new Utility register-type heater. It is designed for cars too small to use the Utility foot-rail type, or for those who wish a heater sunk beneath the level of the floor yet combining all the qualities of the other type.

All that shows is the register, nickel plated, similar to registers in the home. The heater is easily controlled with one finger from the driver's seat. The 27 in. length, nickel finish, sells for $30; the Utility Jr., 17 in. long for coupes, Fords, tonneaus or other small cars and front seats of large cars, $22.50; and the Utility Jr., for Buicks, $25.

The company also has put on the market an improved design of the old model Utility foot-rail heater. This in the 27 in. length sells for $20; the 17 in. for front seats, coupes and small cars, $12.50; the Utility for Fords, $9, and the Utility DeLuxe for eight-cylinder cars, $25.

Motor Age  October 1920

The Hill Pump Valve Co. offered the exhaust Utility Heater with one touch control.

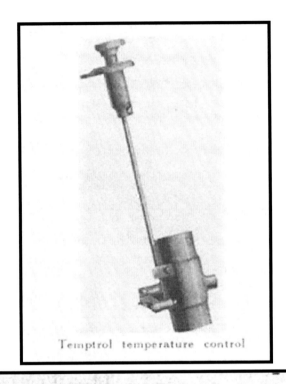

Temptrol temperature control

## Temptrol

The Johnson Co., Detroit, manufactures the Temptrol, a simple device for regulating temperature and which is operated by hand from the instrument board. It is claimed to quickly warm up the motor after starting, maintain an efficient operating temperature regardless of weather conditions, and to conserve the heat in the waterjacket.

The device consists of a butterfly valve in the circulation pipe and an operating valve running to the instrument board. To operate Temptrol, press the control on the dash which temporarily holds the circulating water in the waterjacket where it is quickly heated to an efficient temperature. Slowly opening the valve will gradually bring the circulating water to the proper temperature and hold it at a good temperature while running. The price of this device is $7.50.

Motor Age   October 1920

An engine coolant thermostat by The Johnson Co. was hand controlled.

Circa 1920 ad

The lower left block in this ad lists the vehicle manufacturers fitting these exhaust heaters by The Perfection Heater & Mfg. Co.

Vehicle Circa 1920   Image 2010

The plaid lap robe is stowed on the robe rail.

Dykes Encyclopedia 1922

Celluloid was the transparent material used for side curtains.

96

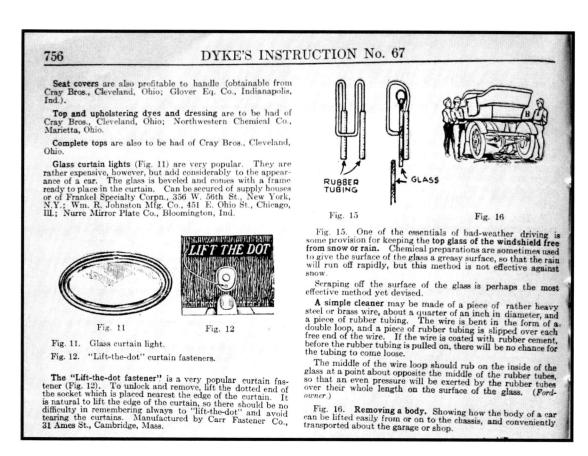

Dykes Encyclopedia 1922

Figure 11 shows a glass curtain light. "They are rather expensive, however, but add considerably to the appearance a car."

Note the homemade hand operated windshield wiper in Figure 15. It was to be made of quarter inch steel or brass wire bent into a double loop with two pieces of rubber tubing attached to the open ends with rubber cement.

Ford Motor Company image

A 1923 Ford Model T fitted with side curtains

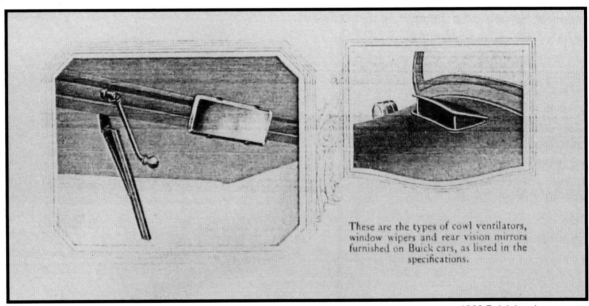

1923 Buick brochure

Buick featured a hand-operated windshield wiper and a cowl-top ventilator.

Motor Age January 1924

The Hill Pump Valve Co. offered two types of exhaust heaters- direct and indirect.

Motor Age   January 1924

The Fulton Co. Helzen pedal slot closers were intended to keep cold air out.

Motor Age   January 1924

The Presto-Felt hand-operated windshield wiper with double sided blades.

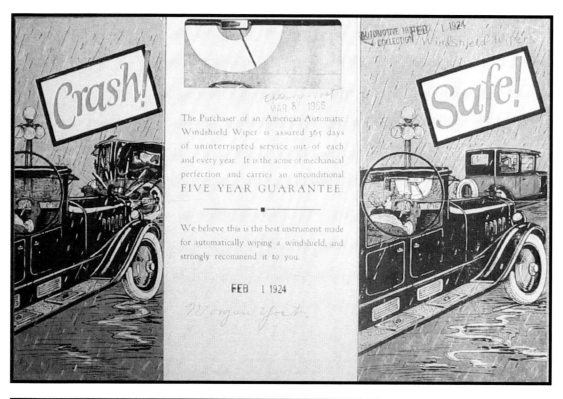

1924 brochure

A vacuum motor windshield wiper manufactured by the American Automatic Wiper Co.

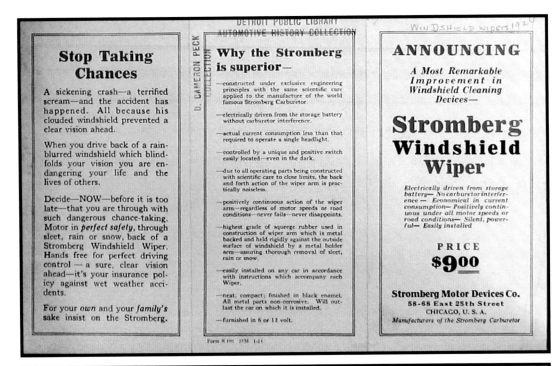

1924 brochure

An electric windshield wiper system manufactured by Stromberg Motor Devices Co.

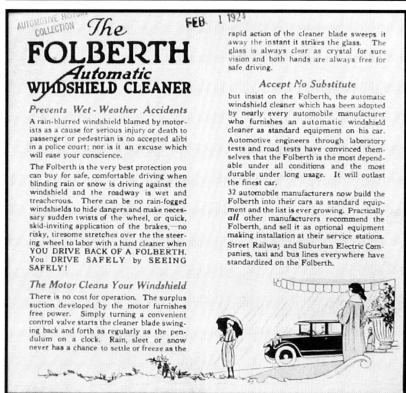

1924 brochure

The Folberth Auto Specialty Co. manufactured this vacuum windshield wiper motor and 32 vehicle manufacturers fitted them.

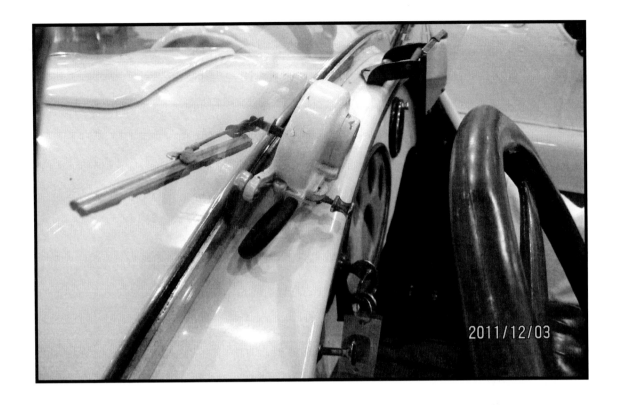

Stahls Collection

1925 Kissel fitted with a vacuum windshield wiper motor. Note the shaft passing thru the glass.

Stewart-Warner Speedometer Corp. built exhaust heaters that sold for $15.

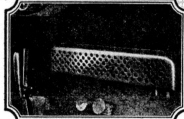

An exhaust heater by the Kokomo Electric Co. with the Kingston name.

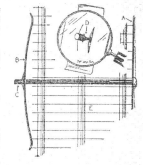

Ford Dealer and Owner December 1925

The U-Auto-Shutter by the Haines Mfg. Corp.

Ford Dealer and Service Field   September 1926
The Nathan Novelty Mfg. Co. side curtains with non-breakable glass (top).
The Kunkle Manufacturing Co. exhaust heater (above).

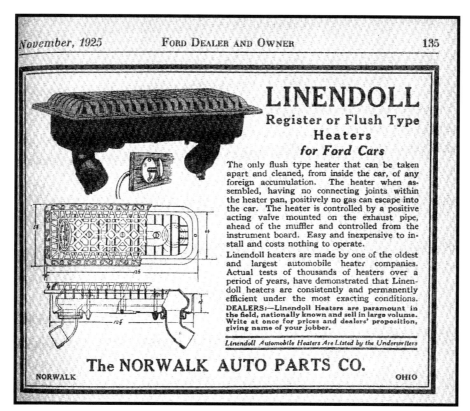

Ford Dealer and Owner   November 1925

The Norwalk Auto Parts Co. exhaust heater with air-to-air heat exchanger.

Vehicle Owner Ron Balint

1925 Chevrolet with roll up windshield.

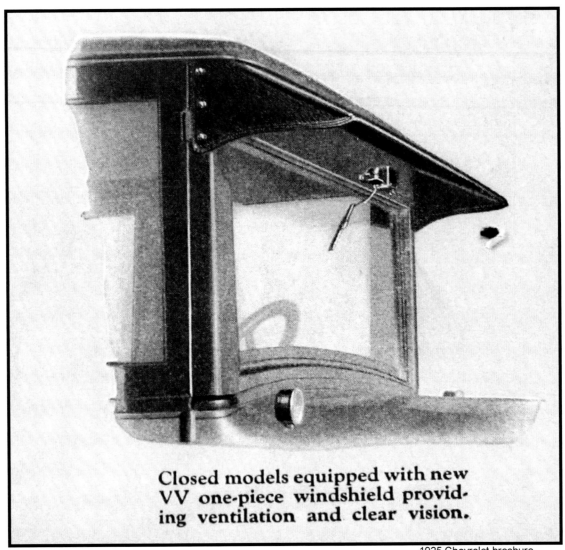

1925 Chevrolet brochure

Note the windshield wiper mounted forward of the glass.

1926 Chevrolet brochure

The lower left caption reads:
Cowl ventilation is obtained by opening the windshield glass to the level of the dash. When raised above this point perfect car ventilation is obtained.

The lower right caption reads:
The Fisher VV windshield is operated by a conveniently located Ternstedt regulator.

1926 Chevrolet brochure

Chevrolet offered as standard equipment a robe rail and "Snug fitting pedal closures make the driving compartment comfortable in cold weather."

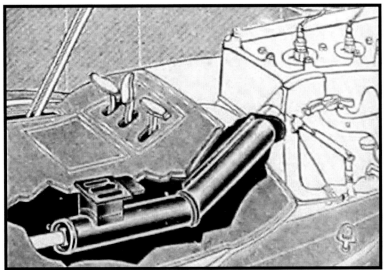

Ford Dealer and Service Field   November 1926

The Cooper brand exhaust heater.

Richard Kughn Collection

1927 Pierce Arrow with parallel vertical vacuum-powered windshield wipers.

1927 Ford Model T brochure

This Ford Model T featured close fitting side curtains fitted to the doors.

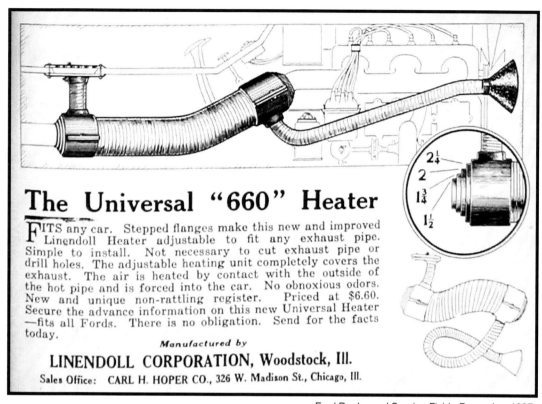

The Linendoll Corporation manufactured this Universal 660 exhaust heater.

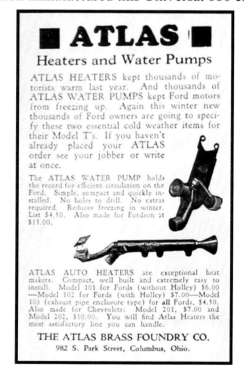

The Atlas Brass Foundry Co. manufactured this exhaust heater.

Ford Dealer and Service Field   October 1927

The Kokomo Electric Co. offered a variety of exhaust heaters.

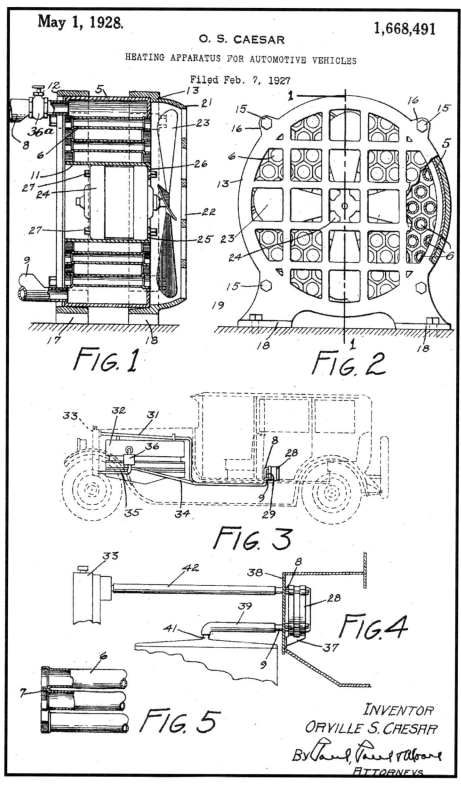

1928

The patent for the hot water heater. This system has been used on virtually every automobile since the 1930's. Specifically hot water with a heat exchanger and an electric fan. The patent was assigned to Tropic-Aire Inc.

Vehicle Owner Jim & Barb Morningstar

1928 Chevrolet with heater floor register in the shape of the Bow Tie logo.

Vehicle Owner Douglas Ives

1928 Whippet with wiper motor mounted forward of windshield. Note the wire arm and attachment to the blade with a cotter pin.

Ford Dealer and Service Field   May 1928

Ventilating eaves for the Ford Model A by the Hubbard Products Co.

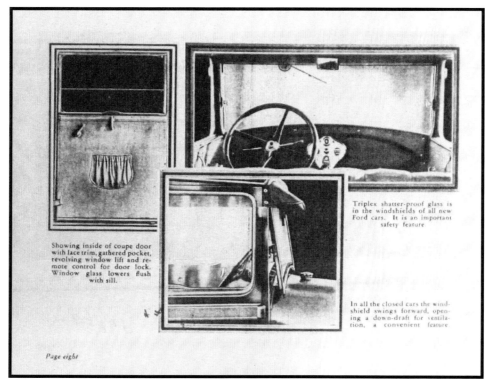

1928 Ford Model A brochure

In all closed cars the windshield swings forward, opening a down-draft for ventilation, a convenient feature. Triplex shatterproof glass windshields were also featured.

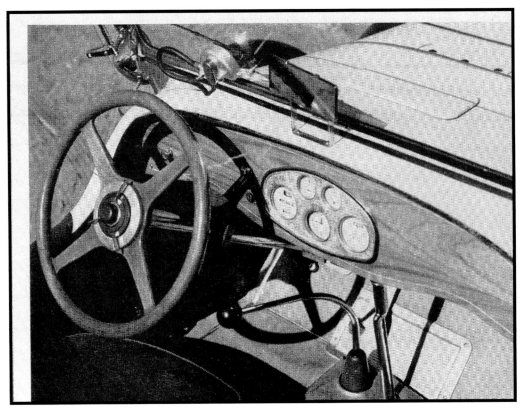

Car Life August 1963

1929 Kissel with its wiper motor mounted to the inside of the glass.

Woodward Dream Cruise August 2011

A 1929 Chevrolet with restoration of the windshield lifter mechanism in progress.

Ford Dealer and Service Field   January 1929

Door window ventilators for the Ford Model A by Indian Sales Corporation.

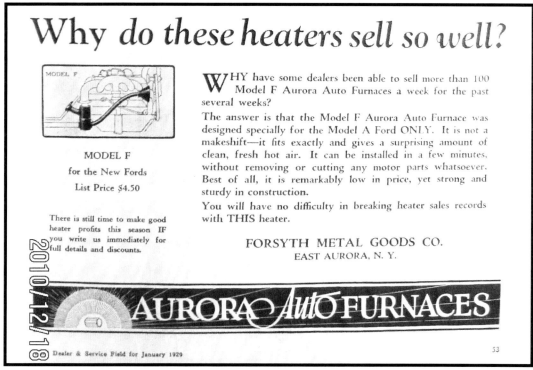

Ford Dealer and Service Field   January 1929

Forsyth Metal Goods Co. built Aurora Auto Furnaces exhaust heaters.

Ford Dealer and Service Field   January 1929

The Greb Co. manufactured the Sleetoff, an electric windshield warmer.

Ford Dealer and Service Field   June 1929

A hot water heater manufactured by Tropic-Aire, Inc.

Ford Dealer and Service Field   June 1929

The Liberty Foundries Co. built the HaDees brand hot water heater.

Ford Dealer and Service Field   August 1929

The Red Cat brand exhaust heater was built by the G.A. Roth Mfg. Co.

Ford Dealer and Service Field   September 1929

This Hottentot brand hot air heater by the Donaldson Co. used heat from the vehicle's radiator to warm the interior.

Ford Dealer and Service Field   September 1929

The Heat Blaster Co. manufactured the exhaust heater fitted with an electric fan.

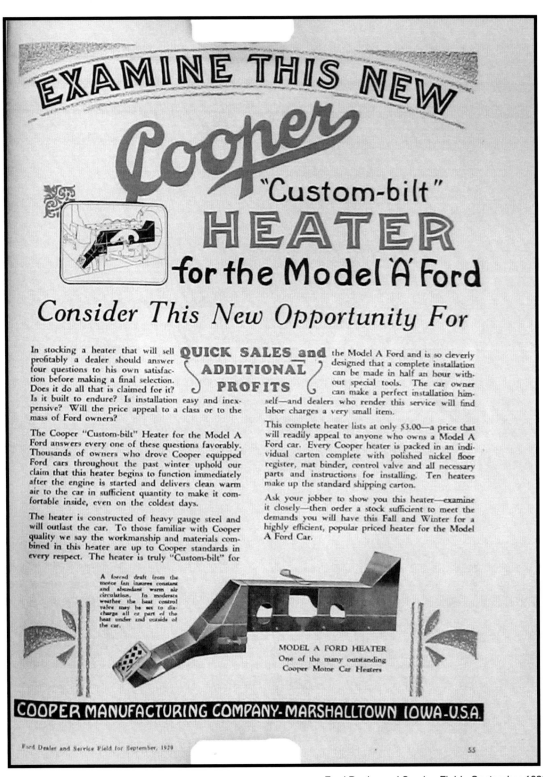

Ford Dealer and Service Field   September 1929

The Cooper Manufacturing Co. offered this exhaust heater for $1.00.

Ford Dealer and Service Field   October 1929

Dunn Manufacturing Co. sold their exhaust heater for $3.50 list.

Ford Dealer and Service Field   October 1929

This Peters and Russell, Inc. exhaust heater ad featured graph of BTU's / Time

Ford Dealer and Service Field October 1929

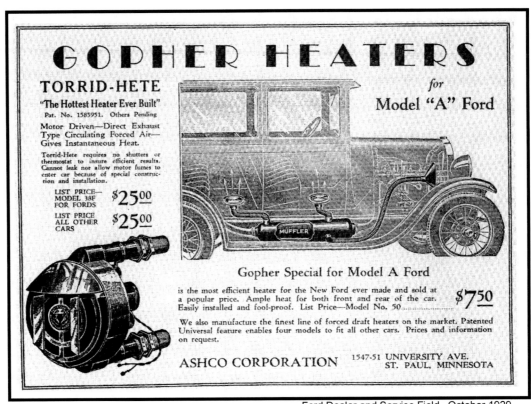

Ashco Corp. sold two exhaust heaters- TORRID-HETE and GOPHER.

A replacement gear for the electric windshield wiper by John C. Hoof Co.

# BOSCH WINDSCREEN WIPER

**ROBERT BOSCH A.-G.**
STUTTGART AND FEUERBACH

---

## The Bosch windscreen wiper

being driven by means of an electric motor is independent of the speed of the engine and can be operated even when the car is standing,

moves silently backwards and forwards with uniform regularity using very little current — scarcely more than the tail lamp — and needing no attention except switching on when required,

has adjustable squeegee with spring device to ensure even pressure on the glass,

has a squeegee arm which traverses a large angle, approximating a semi-circular area, thus clearing a large field of vision for the driver and, in the case of the two-arm pattern, for the front seat passenger also,

is easily fitted to any windscreen,

is of attractive appearance and small dimensions.

---

A windscreen wiper is **a sine qua non** in rain, snow or fog, but it must function reliably and efficiently or it is worse than useless. The Bosch windscreen wiper can be depended upon under any conditions to give that same measure of service and satisfaction which has been associated with the name Bosch in all parts of the World since the inception of motoring.

1929 brochure

Robert Bosch A.-G. manufactured an impressive electric windshield wiper motor.

## Construction.

The Bosch Windscreen Wiper, designed for either 6 or 12 volt circuit, consists of a small electric motor with switch and gears and the arm which holds the squeegee. The motor is fed from the lighting installation, and is set in motion by means of the switch situated at the end of the motor case. On the end of the motor shaft is a worm gear which drives a worm wheel. On the worm wheel shaft is a gear wheel which drives the gear wheel to which is fixed eccentrically a toothed brass strip which converts the rotary motion of the motor to an oscillating motion of the wiper arm.

Fig. 1 Bosch Wiper

## Wiring.

There are two terminals fitted, positive and negative, the positive being insulated, the negative being connected to the frame of the motor.

## Dimensions.

Length $6\frac{1}{4}''$, largest diameter $2\frac{1}{4}''$, Weight 2 lbs.

## Bosch Wipers for Omnibuses

In the case of omnibuses, a particularly large field of vision for the driver is often desired. For this purpose two special types of Bosch Wipers are supplied which meet this requirement.

Neither of these Wipers should be fitted with auxiliary attachment.

### When ordering please state:

1. Voltage (12 or 6 volts)
2. Whether with one or two arms
3. Method of fixing (see figs. 3—9)
4. Whether for right or left hand steering.

## Fitting.

The fitting of the Bosch Wiper depends upon the type of windscreen to which it is to be attached. The various methods of fixing are shown on pages 5 and 7.

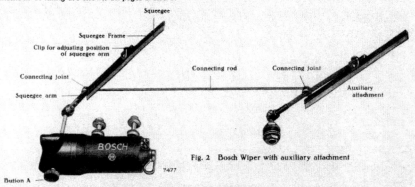
Fig. 2 Bosch Wiper with auxiliary attachment

### 1. Cases where the windscreen need *not* be drilled.

If there is a frame under the windscreen the Bosch Wiper is screwed on to it (Fig. 3).

If there is no lower frame, the Bosch Wiper can be fixed on to one of the uprights of the screen and the wiper arm can be fitted on a lever arm (Fig. 4).

Fig. 4 also shows how the Bosch spotlight can be easily fitted.

In the case of a windscreen having a framework all round, the fixing is best effected on the top of the frame, when the wiper is either clamped (Fig. 5) or screwed on to the frame if it is sufficiently wide (Fig. 6).

1929 brochure

## Methods of fitting Bosch Wiper with auxiliary attachment.

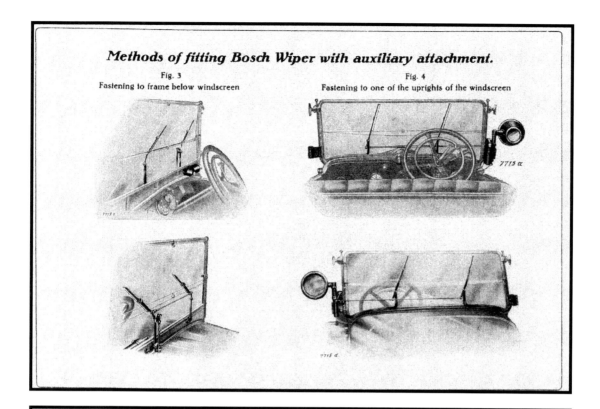

Fig. 3
Fastening to frame below windscreen

Fig. 4
Fastening to one of the uprights of the windscreen

### 2. Cases where the windscreen <u>must</u> be drilled.

In the case of two-piece windscreens, the Bosch Wiper can be suspended on two bands on the lower half of screen as shown in Fig. 7. In this case a hole must be drilled in the glass to take the wiper arm.

For one-piece windscreens, with no top frame, three holes must be drilled at distances of 50 mm (Fig. 8).

When fixing is effected in accordance with figs. 3, 4, 7 or 8, the wiper arm is below the squeegee. In this way the wiped-off water flows away downwards better than in the reverse method of fixing (Figs. 5, 6, 9), that is to say, on the return movement of the wiper the water does not trickle down again over the cleaned surface.

When the Bosch Wiper is fastened directly on to the glass, all three holes are drilled with a diameter of 10 mm, but when the wiper is attached to the frame, the hole for the wiper arm is made with a diameter of 9 mm, the two other holes being drilled to a diameter of 7 mm.

When fixing the Wiper, care should be taken that the wiper arm does not graze the edge of the hole, and that the distance between casing and glass is not less than 6 mm, as otherwise the wiper arm cannot be disengaged (Fig. 8).

The wiper arm must be so adjusted that the pressure on the glass amounts to 100—150 g.

The Bosch Wiper is best fitted in one of the workshops listed on page 8, which are equipped with the necessary tools and have experience in drilling the glass.

### Fixing sets.

If desired, fixing sets can be supplied for fixing the single Bosch Wiper and the Bosch Wiper with auxiliary attachment in accordance with the methods of fixing shown in Figs. 3, 4 and 9.

## Changing the squeegee arm.

If the squeegee becomes worn or unusable owing to the action of the weather, the squeegee frame with squeegee can be obtained as a spare part and changed.

1929 Brochure

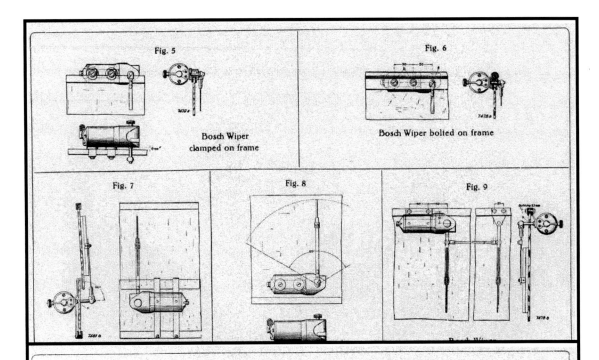

Fig. 5 — Bosch Wiper clamped on frame
Fig. 6 — Bosch Wiper bolted on frame
Fig. 7
Fig. 8
Fig. 9

# REPRESENTATIVE HOUSES

## Europe

**Amsterdam C**, N. V. Willem van Rijn, Keizersgracht 171
**Athens**, G. Paléologue & Co., 20, Rue Santarosa
**Barcelona**, Equipo Bosch S. A., Mallorca, 281
**Berlin**, Robert Bosch A.-G., Verkaufsbüro Berlin, Charlottenburg 4, Bismarckstraße 71
**Berlin SW 48**, Eisemann-Werke A.-G., Zweigstelle Berlin, Friedrichstraße 225
**Breslau II**, Eisemann-Werke A.-G., Zweigstelle Breslau, Tauentzienstraße 35
**Brussels-Midi**, Allumage-Lumière S.A., 23/25, Rue Lambert Crickx
**Bucharest**, Leonida & Cie. S.A., Calea Victoriei 53
**Budapest, VIII**, Bosch Robert korl. fel. társaság, Vas-utca 16
**Cologne**, Eisemann-Werke A.-G., Zweigstelle Köln, Mastrichter Straße 13
**Constantinople**, Constantin Dassira & Georges Dassira, Galata, Rue de Banques 66, 68, 70
**Copenhagen**, A./S. Magneto, Gammel Mont 12
**Danzig**, Alfred Bauch, Langer Markt 32
**Frankfort on the Main West**, Robert Bosch A.-G., Verkaufsbüro Frankfurt a. M., Moltke-Allee 47-53
**Geneva**, Robert Bosch S.A., 78, Rue de Lausanne
**Glasgow, C. 2**, J. A. Stevens Ltd., 218/222 Bothwell Street
**Hannover**, Eisemann-Werke A.-G., Zweigstelle Hannover, Marienstraße 49
**Helsingfors**, A.-B. Walfrid Alftan O.-Y.
**Leipzig**, Eisemann-Werke A.-G., Zweigstelle Leipzig, Gottschedstraße 18
**London, W. 1**, J. A. Stevens Ltd., 21/22 Upper Rathbone Place
**Luxemburg**, Romain Lecorsais, Ing., Grand'rue 51
**Madrid**, Equipo Bosch S.A., Calle Viriato 18
**Milan (126)**, S.A. per il Commercio dei Materiali Bosch, Via Londonio, 2
**Oporto**, Roberto Cudell, Rua Passos Manuel 41-1°
**Oslo**, A./S. Automagnet, Kongensgate 16
**Paris XVIe**, Fernand Péan, Ing. A.M., 97, Boulevard Exelmans
**Prague XII**, Robert Bosch, Marš. Foche 8
**Rome (27)**, S.A. per il Commercio dei Materiali Bosch, Via Novara, 8-14
**Sevilla**, Equipo Bosch S.A., Paseo de Colón 4 dupl°
**Sofia**, Léon Arié, Uliza Targovska 15
**Stockholm**, Aktiebolaget Robo, Birgerjarlsgatan 25
**Stuttgart-Berg**, Robert Bosch A.-G., Verkaufsbüro Stuttgart, Stuttgarter Straße 17
**Turin (10)**, S.A. per il Commercio dei Materiali Bosch, Via A. Vespucci, 52-54
**Vienna, IX**, Robert Bosch G. m. b. H., Spittelauerlände 5, bei der Friedensbrücke
**Warsaw**, J. Kestenbaum, Ul. Wilcza 29
**Zagreb**, Frank i Drug, Gundulićeva 40
**Zurich**, Robert Bosch A.-G., Utoquai 57

## America

**Barranquilla**, A. Held, Correo Apartado 127
**Buenos-Ayres**, Robert Bosch S.A., Calle Rivadavia 1857-61
**Chicago**, Robert Bosch Magneto Co., Inc., 1122, South Michigan Avenue
**Habana (Cuba)**, Albert Eppinger, Avenue Belgica, 10
**Mexico D. F.**, Sommer, Herrmann & Cia. Sucs. Departamento Maquinaria, Apto. 299
**Montevideo (Uruguay)**, Eugenio Barth y Cia., Uruguay, 757
**New York**, Robert Bosch Magneto Co., Inc., 3601 Queens Boulevard, Long Island City
**Rio de Janeiro**, Steinberg & Cia., Avenida Rio Branco 31/33
**Santiago**, Chili, Saavedra, Bénard y Cia., Lda. Sociedad Comercial
**São Paulo**, Steinberg & Cia., Rua Barão de Itapetininga, 16
**Valparaiso**, Saavedra, Bénard y Cia., Lda., Sociedad Comercial, Avenida Brazil 929

## Asia

**Bangkok**, Windsor & Co.
**Beyrouth**, Eastern Engineering Co., B. P. No. 257
**Calcutta**, Martin & Co., Department Bosch Service, 58, Free School Street
**Canton (China)**, Jebsen & Co., 10, Western Bund
**Colombo (Ceylon)**, Freudenberg & Co., De Mel Building
**Hongkong**, Jebsen & Co., 12, Pedder Street
**Jaffa**, Gebrüder Wagner, P. O. B. 249
**Kobe**, C. Illies & Co., 84b, Yedomachi
**Penang**, N.V. Straits Java Trading Co., Weld Quay
**Shanghai**, Jebsen & Co., 7, Hankow Road
**Singapore**, N. V. Straits Java Trading Co., 114, Cecil Street
**Soerabaia**, N. V. Willem von Rijn's Technisch Bureau, Kahasin 15
**Tokio**, C. Illies & Co., 1 Yurakucho Ichome Kojimachi-ku
**Tsingtau**, Henzler & Co., P. O. Box 230

## Africa

**Alexandria and Cairo**, Équipements Électriques d'Automobiles
**Cairo**, 11, Rue Ganieh Charkass
**Alexandria**, 42, Rue Fouad Ier
**Johannesburg**, F. Hoppert, 86, Marshall Street
**Nairobi**, Kenya Colony, Africana Ltd.

## Australia and New Zealand

**Melbourne and Sydney**, Robert Bosch Supply & Service Co., Pty. Ltd.
**Melbourne**, 256/258, Latrobe Street
**Sydney**, 249, Elizabeth Street
**Wellington**, Jas J. Niven & Co., Ltd., 152-72, Wakefield Street

„The above branches and agencies possess well-equipped workshops fitted with all appliances and tools for repairing and mounting Bosch products. They employ specially trained mechanics, who have come from the Bosch workshops or were taught there. They also hold permanent stocks of Bosch parts and accessories. To ensure satisfactory results, we recommend that repairs should only be put into the hands of these branches and agencies."

1929 Brochure

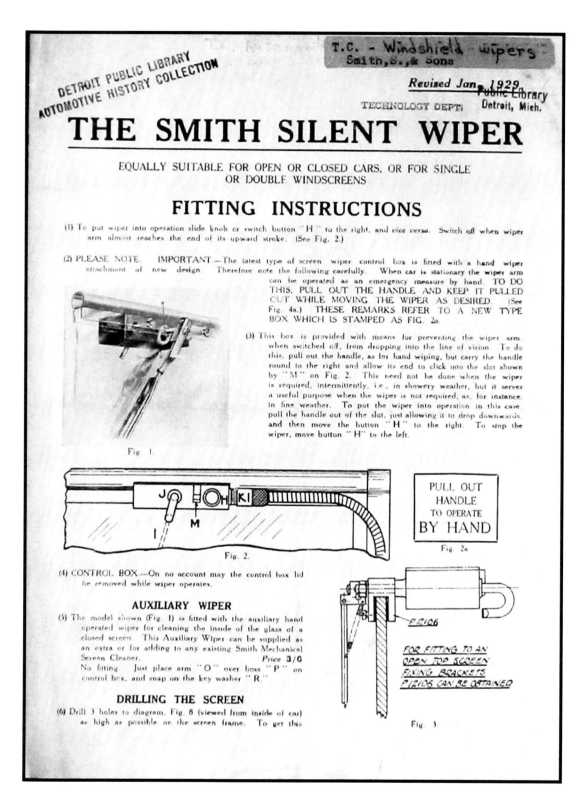

1929 brochure

This speedometer cable driven windshield wiper system was built by S. Smith & Sons (M.A.).

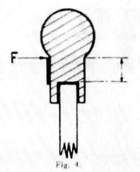

Fig. 4.

result, the top edge of the diagram should be laid as close as possible to the edge of the channel section—see "F" (Fig. 4). IMPORTANT: MAKE SURE THAT THE SPINDLE OF THE CLEANER ARM DOES NOT CHAFE OR BIND IN THE HOLE THROUGH THE SCREEN FRAME, and is free to move to and fro not less than 1/8 in. (See Fig. 4a.)

(7) If the distance between the edge of the glass and "F" (see Fig. 4) is thought to be limited, drill the holes with an 1/8 in. drill and open out to the correct sizes. If the larger size drills touch the glass edge, operate them from both sides of the screen. The remaining metal and the edge of the glass can then be cleared out with a round file aided by turpentine; it is an advantage to use a little camphor with the turpentine.

(8) The standard fixing studs are 3/4 in. long; if longer studs are required, ask for two studs, P. 12031.

(9) On some screens, mostly on Saloons, the top half is designed in such a way that the control box cannot be fitted to the metal frame. The top edge of the frame may fold into an edging, or when the top half is opened the spindle holding the wiper arm might foul the overhanging roof of the car. In such a case it is necessary to fit the control box to the glass screen itself. In fixing, use 1/8 in. thick fibre washers on each side of the glass.

(10) Various methods of drilling glass have been employed, and we can supply special drills. A three-cornered file with the end pointed is a rough substitute. Use turpentine mixed with a little camphor when drilling, and if possible remove the screen and lay it flat on some form of pad during the operation. DRILL THE GLASS FROM BOTH SIDES.

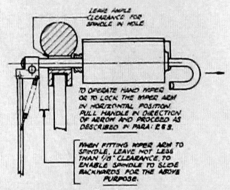

Fig. 4a.

(11) When fitting the cleaner arm "I" to the spindle, it should be provisionally set in line with and in the same direction as the handle "J" on the opposite end of the spindle, and after test it should be finally set to clear an even arc of the glass. (See Fig. 2.) The square-headed screw (or cotter pin of the latest type) must then be well tightened.

(12) The above applies when the box is fitted with the drive entering from the right as illustrated. If it is necessary to fit a control box so that the drive enters from the left, the box should be fitted "upside down." This latter method of fitting will mostly apply if two wipers are fitted, one to each side of the screen. In this case, of course, one flexible shaft should be led up each outside screen pillar.

(13) For fitting to bottom of Screen to operate above the box, the arm "I" should be assembled opposite or away from the handle "J," otherwise as above.

## TENSION OF CLEANER ARM ON SCREEN

(14) The spring should be tensioned so that the rubber just leans over and makes edge contact as it wipes. It should not turn right over. Adjustment can be made by sliding the spring mount along the spindle. IMPORTANT.—MAKE SURE THAT THE SPINDLE OF THE WIPER ARM CANNOT CHAFE OR BIND IN THE HOLE THROUGH THE SCREEN FRAME OR THE GLASS, AND ALSO MAKE SURE THAT THE SPINDLE CAN MOVE TO AND FRO AT LEAST 1/8 in. (See Fig. 4a.)

## SPECIAL FITTING FOR SINGLE SCREEN

(15) The illustration (Fig. 3) clearly shows the special fixing brackets which can be obtained for fitting the Cleaner to an open top or single screen. As will be seen from the illustration, these brackets fit over the top of the glass and are clamped firmly in position by means of clamping screws.

1929 brochure

## HOW THE DRIVES ARE FITTED

### SPEEDOMETER DRIVE (Fig. 5.)

(16) Connect Speedometer Flexible shaft end "A" (originally connected to Drive "D") to "B" on the Screen Cleaner gear box. Connect "C" on the latter to "D." NO FITTING! Just interpose the Cleaner gear box. Screw all collars up tight.

### "G" TYPE GEAR BOX FOR CONTINENTAL SPEEDOMETER DRIVE (Fig. 6.)

(17) A nick or slot must be made in the shank "C" of the gear box to take the cotter screw shown in "D" when the gear box is placed in position.

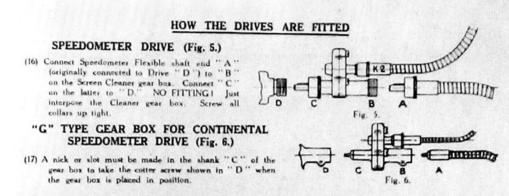

### INDEPENDENT PULLEY DRIVE

(18) On certain Cars, i.e., A.C., 14 h.p. Crossley, 10 h.p. Swift, 11.4 h.p. Standard, etc., the gear box two-way drive as described above cannot be used, chiefly owing to obstruction. In such cases an ordinary pulley drive such as is used for driving a speedometer can be utilised to drive the wiper. Fitting instructions therefore would be similar to those applicable to a speedometer drive. The split pulley should be firmly fixed to the shaft, and the driven pulley bracket firmly fixed to the chassis frame by two set screws.

In fitting the Cardan arm to the chassis, care should be taken to see that the brake rods, etc., do not foul the bracket; they should, in fact, have some clearance, so that under vibration of driving no noise or rattling takes place. The pulleys should be carefully set in line, and the bracket arm should be given a "set" so that the two pulleys are in the same plane; this will, of course, be governed by the slope of the Cardan Shaft.

The bracket should be applied so that the belt lays as near as possible parallel with the ground; this prevents the arm pulley tugging the belt in the vibration of the Shaft.

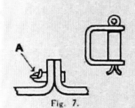

Fig. 7.

(19) JOINING A FLAT BELT.—The belt should be joined as follows :—Pierce each end of belt for the staple with an awl (or penknife). Bring the ends together and drive staple home (a block should be interposed between hammer and staple to avoid damaging the latter). Insert collar and split pin and turn the split pin ends well back in the direction shown at "A" (see Fig. 7). The tension on Belt is effected by suspending a coil spring from hole in pulley-arm to hole in bracket arm. For necessary tension, stretch the spring to about 1½ times its idle length.

(20) JOINING A SPRING BELT.—A spring belt has to be very carefully joined, and to do this, one end must be held while the other end is screwed backwards by one's fingers for about six complete twists or turns. This end is then placed in the other end and screwed forward six turns to complete the join. Unless this method is adopted the belt join will unscrew itself in turning and fall off.

### FLEXIBLE SHAFT

(21) The lead of the Flexible Shaft to the Screen Wiper should be as direct as possible, bearing in mind that there should be the very least possible number of bends in the shafting. The radius of bends should be not less than 6 in. where sharp bends are unavoidable, and not less than 9 in. otherwise.

---

CUT THIS OUT AND USE AS A TEMPLATE.

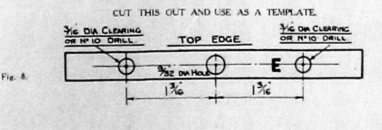

Fig. 8.

1929 brochure

THE LEAD OF THE SHAFTING NEED NOT FOLLOW THE SPEEDOMETER DRIVE SHAFTING. IT SHOULD GO BY THE SHORTEST WAY FROM THE DRIVE TO THE CONTROL BOX ON THE SCREEN. AND IN MANY CASES IT CAN BE LED UP THROUGH THE FLOOR BOARDS AT A POINT JUST BELOW THE RIGHT HAND SIDE OR END OF THE DASHBOARD.

(22) The flexible shaft connections are not alike. The widest tang piece is to be connected to the control box, and the longest tang piece must be connected to the gear box or drive.

The flexible shaft should be held to the frame by one clip, P.11839, and, if necessary, one or two small clips P.12078.

If the flexible shafting passes down the side of the screen and then through part of the coach work, it is advisable to partially cover this hole with a neat ebonite bush, and we can supply this fitting, i.e., P.11840, on request.

## LUBRICATION

(23) GEAR BOX.—Move shutter piece, and insert thick oil or thin grease about every three thousand miles.

(24) FLEXIBLE SHAFT.—Disconnect two collars "K1" and "K2," remove slotted washer at "K2"; the flexible shaft can then be pulled out a short distance at "K1." Force some thick oil down flexible shaft casing; this should be done about every three thousand miles (see Fig. 2 and Fig. 5).

(25) SCREEN CONTROL BOX.—This is filled with grease when manufactured, and should not need lubricating under ten thousand miles. To lubricate, take out six small screws, remove lid, and use thin grease.

(26) INDEPENDENT PULLEY DRIVE BRACKET.—This should be re-greased every three or four thousand miles.

## WINDSCREEN WIPER—GEAR BOXES

(27) "I" Type Gear Boxes to be used generally with Smith Speedometers with the exception of the following cars:—
12 H.P. and 20 H.P. Austin, and late 14 H.P. Rover Cars.
"LL" Type. Generally for Watford (North & Sons) Speedometers.
"L" ,, 12 and 20 H.P. Austin Cars.
"G" ,, For Cars equipped with Standard Continental Speedometer Drive, and generally for Jaeger and O.S. Speedometers and late 14 H.P. Rover Cars.
"A.C." or "S" Type. Generally for Cars equipped with Stewart Speedometers.

(28) The above list will be found to be generally correct, but owing to changes in equipment by the manufacturers, it may be necessary to forward to us the following particulars:—

Make of Car ...........................................................................

Date and H.P. of Car ...........................................................................

Make of Speedometer ...........................................................................

Method of Driving Speedometer ...........................................................................

Length of Speedometer Flexible Shaft ...........................................................................

# S. SMITH & SONS (M.A.), LTD.
# CRICKLEWOOD : : : : LONDON, N.W.2

London Showrooms: 179-185, GT. PORTLAND STREET, W.1

*Also at*

| BIRMINGHAM | MANCHESTER | GLASGOW | BELFAST | DUBLIN |
|---|---|---|---|---|
| 122, Alma St. | 14a, Jackson's Row. | 19, West Regent St. | 18, Sussex Place. | 34, Lower Abbey St. |

Printed in England. No. 115 S. & C. 239 29—5,000

JUN 2 1929

1929 brochure

Ford Dealer and Service Field   October 1929

Eveready Prestone brand antifreeze was manufactured by the National Carbon Co.

## APCO THERMOSTAT AND CRANKCASE VENTILATOR

Nearly every car built today is factory equipped with a device to raise the engine temperature quickly either by closing the radiator shutter or reducing the water circulation by means of a thermostatic valve.

The Apco Thermostat for Model A Fords will insure the proper temperature in three minutes and maintain it regardless of the weather with the result that you have a smooth, economical operation within three minutes after starting the motor.

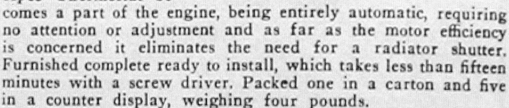

Positive temperature control means less dilution of crank case oil, therefore oil saving, better lubrication and longer life of parts. Running continually with a rich mixture, which is required when engine is cold, results in carbon and consequent knocks and frequent valve grinding. Once started, even on the coldest days, the carburetor control can be set at "lean" and remain there—a very definite gasoline economy which means that the Apco Thermostat pays an actual cash dividend.

When installed the Apco Thermostat becomes a part of the engine, being entirely automatic, requiring no attention or adjustment and as far as the motor efficiency is concerned it eliminates the need for a radiator shutter. Furnished complete ready to install, which takes less than fifteen minutes with a screw driver. Packed one in a carton and five in a counter display, weighing four pounds.

A very definite demand among Ford dealers for a crank case ventilator has caused the development of the Apco Crank Case Ventilator that not only removes the odors but prevents oil splashing over the engine and also provides a handy self-closing oil filler cap.

Ford Dealer and Service Field  December 1929

Apco engine cooling thermostat for Model A Ford vehicles.

## The Breeze Filter Windshield Screen

HERE is a windshield ventilating screen that will add worlds of comfort to your business or pleasure driving. The opening of the windshield is the best way to get the benefit of the fresh, exhilarating air. With a Breeze Filter protecting the opening, you may drive through swarms of insects without discomfort, and through clouds of dust and dirt without danger to your eyes from the small particles always in the air.

The Breeze Filter is a screen similar to that used in Pullman car windows. It is coarse enough to give ample ventilation and yet fine enough to break the force of the wind. It is below the line of vision, finished to harmonize with the instrument board and kept free from vibration noise by rubber cushions.

Furnished for closed car Type VV Windshields only. For real comfort the Breeze Filter is worth many times its cost.

**No. W-241**—Breeze Filter for Cadillac.

**No. W-242**—Breeze Filter for La Salle.

*Be sure to specify model and body type when ordering by mail.*

*The Breeze Filter Windshield Screen*

## The Sedan Fan

YOU need a Sedan Fan for city driving, because city driving means poking along through crawling, congested traffic over hot paved streets in a broiling, stuffy atmosphere, filled with poisonous exhaust fumes.

You need a Sedan Fan for touring over dust filled country roads. With a Sedan Fan turned on, you are as cool and as airy as if you were making forty on smooth concrete.

You need a Sedan Fan while parking—you know how much of your driving time is spent standing still. You know, too, how unbearably hot a motionless car can be. Learn the comfort and relief a Sedan Fan will give you.

You need a Sedan Fan even in winter, for in cold weather it will keep your windows clear of condensation and frost. It will circulate heat. It will make smoking permissible, even with windows closed.

The Sedan Fan runs on the battery of the motor car. It uses less current than a single headlight.

Finished in indestructible Zapon Lacquer Enamel (Olive Green), or in Satin Silver. Double swivel joint allows fan to be set at any angle. Easy to install.

**No. F-100**—Sedan Fan, Green Zapon, 6 Volts.
**No. F-101**—Sedan Fan, Satin Silver, 6 Volts.

*The Sedan Fan*

1929 Cadillac accessory brochure

Cadillac offered a Breeze Filter for the VV windshield to prevent insects from entering the passenger compartment. A small fan was also offered in Green Zapon or Satin Silver.

## De Luxe "Visionall" Windshield Cleaner

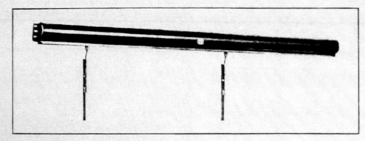

*The De Luxe "Visionall" Windshield Cleaner*

PROPELLED entirely by the surplus suction of your motor, the Visionall twin blades sweep back and forth at a steady speed across the entire windshield, providing driver and passenger with a complete clear vision.

The blades may be regulated to operate jointly or separately. By their horizontal movement in the vertical position, rain and snow is carried off by gravity. Their uniform and thorough action cleans your windshield to perfection. The body of this cleaner containing the motor is hidden away under the visor.

You will find your investment in a Visionall, money well spent.

## De Luxe Sport Side Wings

*De Luxe Sport Side Wings on Front and Rear Windshields*

DE LUXE Sport Side Wings are custom made for Cadillac and La Salle open sport body styles having that type of front windshield which folds forward. They are made of the best grade of heavy plate glass. Brass frames and trimmings are heavily nickel plated. A patented locking device insures firm, dependable adjustment in any desired position.

These DeLuxe Sport Side Wings give protection against disagreeable cross currents. They are available for the front windshield and also for the rear windshield on those styles of sport cars having the double cowl. In placing orders by mail, be sure to specify model, body style, and also whether for front or rear windshield.

**No. W-237—DeLuxe Sport Side Wings.**

*De Luxe Sport Side Wings on Front Windshield*

26

1929 Cadillac accessory brochure both images

Cadillac offered a Visionall vacuum motor powered windshield wiper with parallel vertical blades. Rain and snow was carried off by gravity.

## Closed Car Side Wings, Regular

CLOSED Car Wings permit the opening of the front windows in practically all kinds of weather. By deflecting air currents from the sides of the car they produce a suction that carries out gas fumes, smoke, and foul air, insuring ample ventilation at all times without draughts.

These wings also give protection against dust, rain, and snow entering windows. In hot summer weather, they may be instantly adjusted to divert cooling air currents into the car.

Constructed of the best grade of clear plate glass, cut and beveled in beautiful designs. Nickel plated brackets.

No. W-228—La Salle Closed Car Wings.

No. W-229—Cadillac V63 and Series 314 Closed Car Wings.

No. W-232—Cadillac 341 Closed Car Wings.

(Be sure to specify model and body type on all mail orders.)

*Closed Car Wings (Regular)*

*Closed Car Wings (De Luxe)*

### Closed Car Side Wings, De Luxe

THIS type of closed car side wings attaches to the door hinge. The installation is a very easy matter. These wings combine all of the features of the regular type wings described above, except that they cannot be reversed to divert air currents into the car.

Glass is best grade of French plate. Brackets are of solid brass, highly nickel plated and polished.

Side wings of this type make a beautiful, unobtrusive installation on both Cadillac and La Salle. In placing orders by mail, be sure to specify the model of the car for which desired.

No. W-234—Closed Car Wings, DeLuxe Type, for Cadillac Series 314 and La Salle 303.
No. W-235—Closed Car Wings, DeLuxe Type, for Cadillac 341.

### Plate Glass Ventilating Eaves

WITH Ventilating Eaves installed on a closed car, the windows may be left open at the top in all kinds of weather without unpleasant draughts and cross air currents. The suction caused by a car in motion removes all foul air, gases, and smoke from the interior of the car and at the same time does not permit rain and snow to enter. The efficiency of the heater is greatly increased by the free circulation of air made possible, without discomfort, by the use of these ventilators.

Ventilating Eaves are made of clear, polished plate glass with nickeled brackets, a pleasing addition to any car.

Many motorists install them on all doors and windows.

No. V-162—Ventilating Eaves.

(Be sure to specify model and body type and whether installation is to be made on front or rear doors or windows.)

*Plate Glass Ventilating Eaves*

### Perfection Special Open Car Windshield Wings

FOR the open Cadillac Series 314 and 341, and also the La Salle body types. These windshield wings are designed especially for the sport windshield which folds forward over the cowl. They are made of the best grade of heavy bevel plate glass. Brackets are of bronze, nickel plated. The trim appearance of these windshield wings, together with their small unobtrusive brackets, make them an ideal installation, much admired by Cadillac and La Salle owners. Attachment to windshield frame is made by means of tap screws. The installation is a simple matter.

No. W-226—Special Perfection Windshield Wings for Sport Windshield.

*Perfection Special Open Car Wing*

1929 Cadillac accessory brochure

Plate glass ventilating eaves were offered to improve ventilation and prevent rain and snow from entering the passenger compartment. The glass side wings were forerunners to the Fisher Body ventipanes.

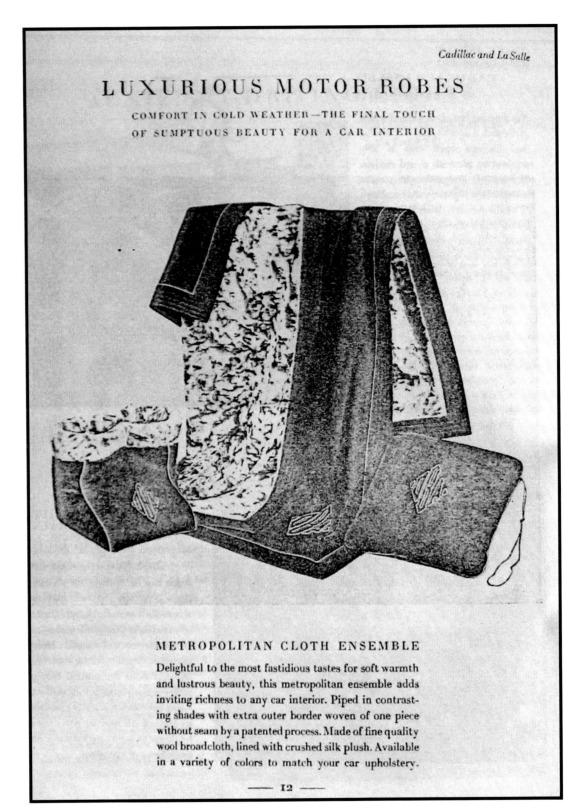

1930 Cadillac accessory brochure

The Cadillac lap robes sold for $37.50 to $85.00, monograms 7.50 each, pillows $12.50 to $15.00, foot muffs 22.50 to 27.50.

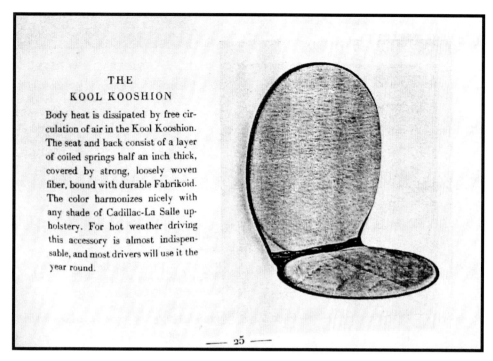

This Cadillac Kool Kooshion sold for $3.50.

1930 Cadillac accessory brochure

These Cadillac tire chains sold for $8.50 to $9.50 in steel and $9.00 to $10.25 in rubber

> *Cadillac and La Salle*
>
> # EFFICIENT CAR HEATERS
> ### ASSURANCE OF COMFORT IN WINTER DRIVING
>
>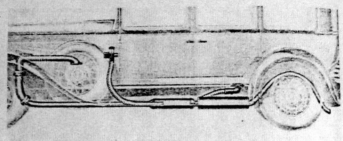
>
> ### CADILLAC-KELCH RECIRCULATING HEATER
> ### FOR CADILLAC AND LA SALLE
>
> The Cadillac-Kelch recirculating heater takes the sting out of winter driving. No more disagreeable trips because of cold weather. Designed by Cadillac engineers so that it will efficiently and quickly heat the automobile, the Cadillac-Kelch heater positively discounts the coldest winter days. Furnished with a single register in the rear compartment, or with two registers, one in front.
>
>
>
> ### CADILLAC-LA SALLE HOT WATER HEATER
> ### CIRCULATES CLEAN, UNIFORM, COMFORTABLE WARMTH
>
> In coldest weather the Cadillac-La Salle Hot Water Heater will maintain a steady, comfortable temperature in your car. It is connected to and forms a part of the water-circulating system of the engine. The hot water is brought to a small radiator just beneath the instrument panel. A fan behind the radiator then circulates clear, steady heat to all parts of the car. The speed of the fan is controlled by a simple, two-speed switch on the dash and the fan requires only as much power as a single tail light. A chromium-plated deflector is easily adjusted to send the heat in any direction. Takes up little room—neat and attractive in appearance. No odors—no fumes—no carbon monoxide gas.
>
> —— 16 ——

1930 Cadillac accessory brochure

Cadillac offered a hot water heater and a Kelch recirculating exhaust heater. The hot water version was priced at $42.50; the exhaust type was $42.50 plus $12.50 for the front register.

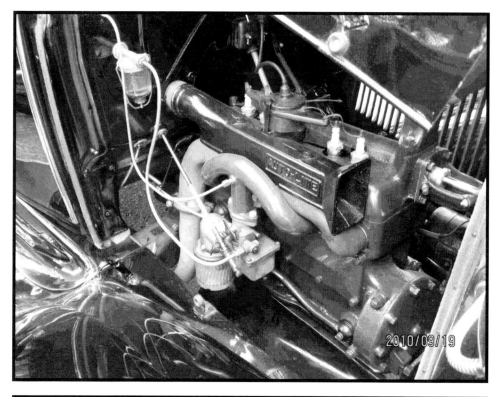

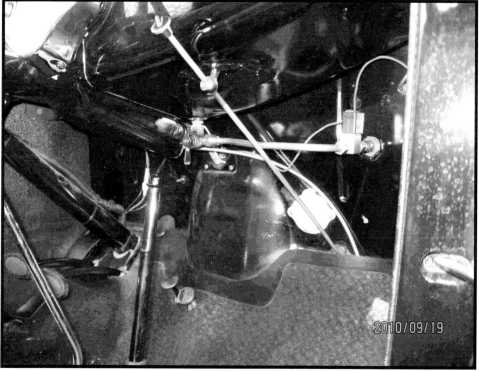

Vehicle Owner D. Gray

A 1930 Ford Model A fitted with an Auto-Lite brand exhaust heater and hand-operated rotating shut off door on the passenger side of the dash panel.

Vehicle Owner Greg Gardner

A 1930 Ford Model A fitted with an Autolite exhaust heater and distribution device located in the passenger compartment. The distribution device provided round-flanged elbows for attaching defroster ducts.

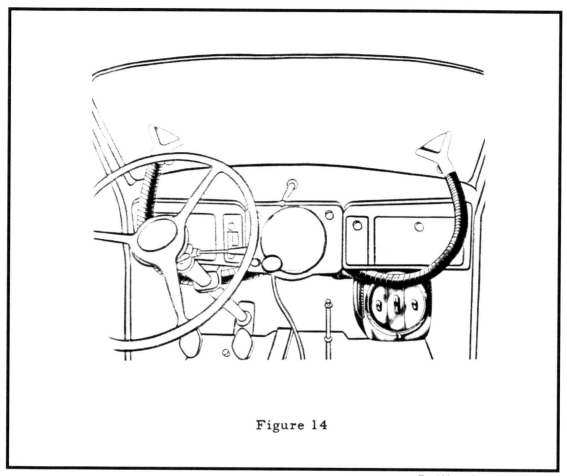

Ford Motor Company image

This is how the defroster ducts and nozzles were probably used with the distribution device shown above on the 1930 Ford Model A owned by Greg Garner.

The image is from Harold V. Joyce's paper *Progress of Automotive Heating Through the Years*.

Ford Dealer and Service Field   January 1930

The Kingsley-Miller Co. offered a hot air heater system with two ducts into the passenger compartment.

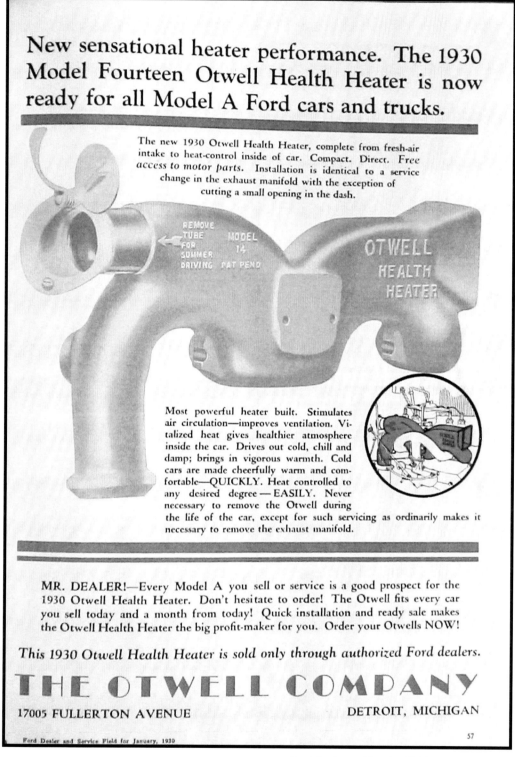

Ford Dealer and Service Field   January 1930

The Otwell Company manufactured this exhaust heater that was a unique exhaust manifold with a fresh air duct built in. The outlet tube was to be removed for summer driving. It was to be "sold only through authorized Ford dealers".

## EVER-TITE THERMOSTAT FOR BETTER WINTER OPERATION

Some automobile engines have pump circulation of the water in the cooling system, while others have thermo-syphon cooling. The Model A Ford has both. And consequently runs too cool for efficient winter use, unless some simple means, such as radiator shutters or a thermostat is used to maintain the cooling solution at more normal operating temperatures.

The Ever-Tite Thermostat is entirely automatic in its operation, including a thermostat which entirely closes a valve and shuts off the flow of the cooling solution until the engine warms up and reaches a temperature of 140 degrees or so. Then the valve gradually opens, as the cooling solution gets warmer, until the valve is fully open when a normal operating temper-

ature is reached. In this way, a more uniform engine temperature is maintained, resulting in less oil dilution, and less wear and tear on engine parts.

The installation of an Ever-Tite Thermostat not only saves gasoline when warming up the engine, but also makes it possible to set the carburetor adjustment for a leaner mixture for, when the engine has once been warmed up, the thermostat "shuts off" the flow of the cooling solution as soon as the temperature drops to about 140 degrees, thus preventing the flow through the radiator and keeping the engine warm for subsequent easy starts.

The Ever-Tite Thermostat sells at $2.25 list and is made by Ever-Tite Bolt Co., 4750 Sheridan Road, Chicago, who are also makers of the Ever-Tite universal hose clamps, quickly adjusted to any size.

Ford Dealer and Service Field   January 1930

The Ever-Tite Bolt Co. offered an engine-cooling thermostat for $2.25.

Ford Dealer and Service Field   January 1930

Portland Woolen Mills, Inc. offered a variety of lap robes.

Vehicle Owner Ryan Johnson

A 1930 Ford Model A fitted with an electric windshield wiper motor forward of the glass.

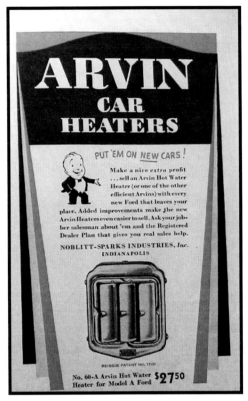

Ford Dealer and Service Field
July 1930 both images

Liberty Foundries Co. offered this version of a Liberty brand hot air heater. Noblitt-Sparks Industries, Inc. offered this Arvin brand hot water heater.

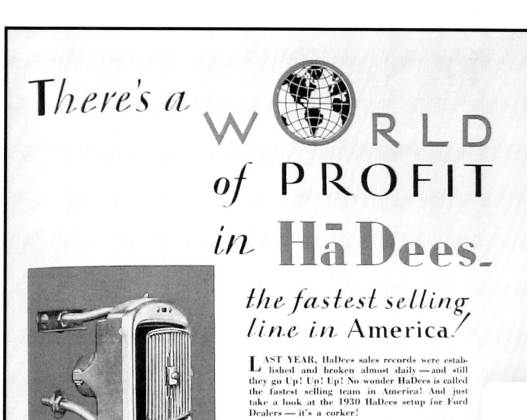

Ford Dealer and Service Field   August 1930

The Liberty Foundries Co. offered HaDees brand antifreeze as well as heaters.

Ford Dealer and Service Field September 1930
The Fulton Company manufactured the Frost Shield electric windshield defroster.

Ford Dealer and Service Field   September 1930

Noblitt-Sparks Industries offered exhaust heaters, hot air heaters and hot water heaters under the Arvin brand.

Ford Dealer and Service Field September 1930

The Francisco Auto Heater Co. offered this exhaust heater with two electric blowers.

Ford Dealer and Service Field   September 1930

The G.A. Roth Mfg. Co. offered an electric blower with this version of their exhaust heater sold under the Red Cat brand.

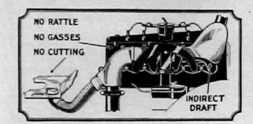

Ford Dealer and Service Field  September 1930

Progressive Brass Manufacturing Co. offered this exhaust heater.

Ford Dealer and Service Field   October 1930

The Metal Stamping Co. offered this radiator shutter under the Weather King brand

Ford Dealer and Service Field   October 1930

Cooper Manufacturing offered an electric blower with this exhaust heater.

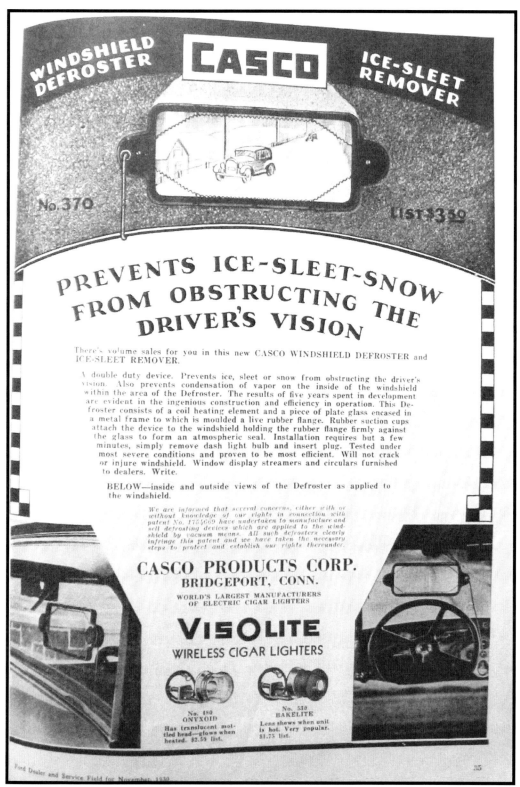

Ford Dealer and Service Field   November 1930

Casco Products Corp. built this electric windshield defroster with a list price of $3.50

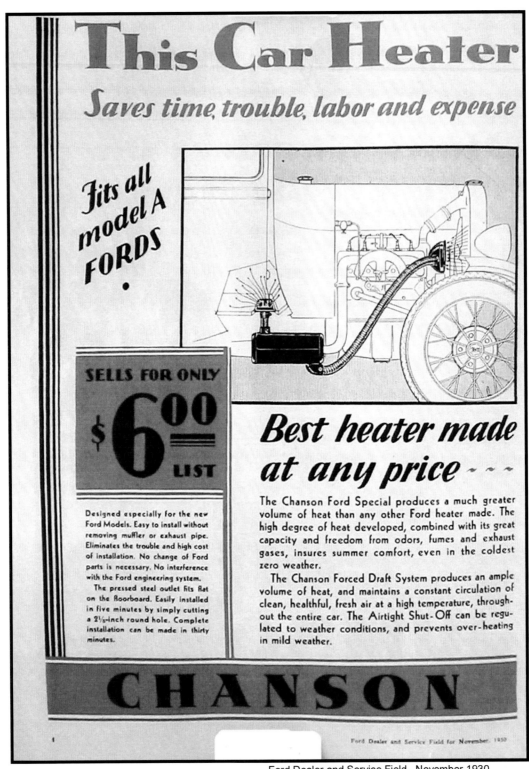

Ford Dealer and Service Field   November 1930

Chanson Division, Illinois Iron & Bolt Co. offered this exhaust heater with a fresh air duct from the engine fan. It sold for $6.00 plus optional rear outlet for $2.25.

Ford Dealer and Service Field   November 1930

Ford Dealer and Service Field   November 1930

Liberty Foundries Co. offered this hybrid combination hot water heater and exhaust heater. It was called the Ha Dees flash hot water heater and it had a list price of $27.50.

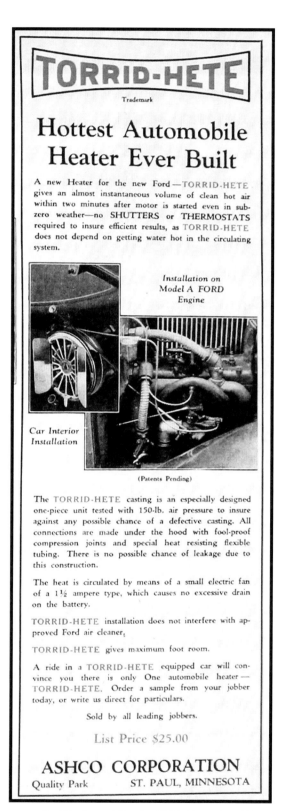

Ford Dealer and Service Field   November 1930

This Ashco Corp. Torrid-Hete brand exhaust heater had a heat exchanger in the passenger compartment with an electric blower.

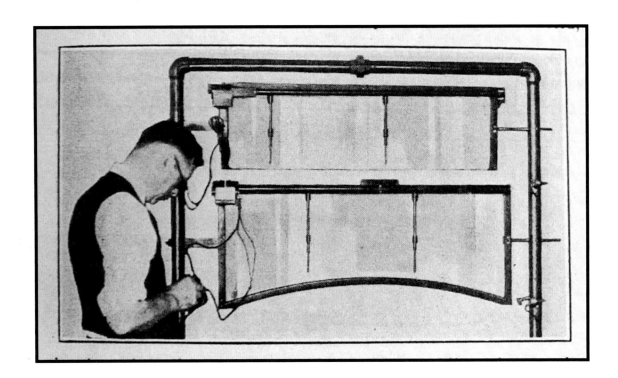

### New Windshield by Handy

Distributors of Handy Governors and allied Handy products are receiving first deliveries of the Handy Safety View Windshield Wiper a new and highly perfected device embodying several exclusive features.

The Handy Safety View is electrically operated, using current from the battery already installed on the car, bus or truck. It acts therefore at a steady speed, independent of throttle position. Its two blades travel horizontally across the windshield, along virtually its entire surface and providing clear, uniform vision in all directions.

Details of the Handy "Safety View" are extremely rugged. The Driving mechanism is simple and sturdy. The entire device has been tested with characteristic Handy thoroughness, by means of laboratory fixtures and equipment designed and built especially for the purpose. A test is shown in the illustration. An accepted production model has repeatedly proved its ability to survive an equivalent of use far more severe than that which it will receive in the whole lifetime on any vehicle on which it may be mounted. Complete information can be obtained from Handy Governor Corp., 3928 West Fort St., Detroit, Mich.

Ford Dealer and Service Field   November 1930

The Handy vertical parallel electric windshield wiper system.

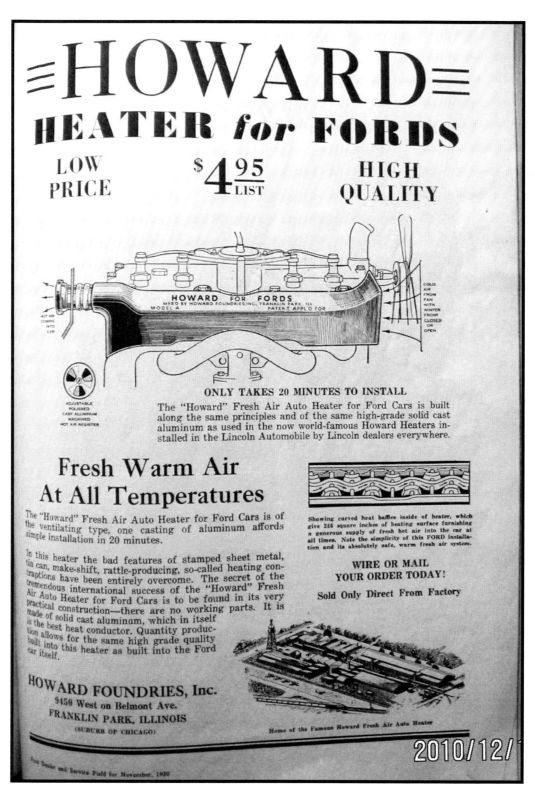

Ford Dealer and Service Field   November 1930

Howard Foundries, Inc. offered this exhaust heater that included and integral fresh air duct and a rotary hand operated register in the passenger compartment.

Ford Dealer and Service Field   December 1930

Coleman Manufacturing Co. offered this exhaust heater with a fresh air duct. It was sold under the Kunkel

# HALF MILLION
## CARS ENJOY THE PLEASURES OF
## WITH A KUNKEL

*Ten Reasons Why You Should Sell KUNKEL Heaters:*

1. Kunkel Heaters give more heat in two minutes than any other heater on the market.
2. They give more heat value per dollar than any other heater offered Ford owners.
3. Will last life of Model A car.
4. Can be regulated to any desired temperature.
5. Kunkel Heaters furnish 100% fresh air at all times.
6. Every Kunkel Heater sold sells more Kunkels—once a Kunkel user, always a Kunkel Booster.
7. It is the best honest-to-goodness heater offered Ford dealers today.
8. They are free from noise due to rough roads or engine vibration.
9. Made in types to fit all Model A or AA Fords.
10. Light in weight — simple in design — easy to install.

This view shows a Kunkel Heater from the right side of the car to illustrate the connection to the exhaust pipe.

Eastern Factory:
**COLEMAN MANUFACTURING CO.**
10709-11 QUINCY AVE.     CLEVELAND, OHIO

*More than* **40%** *of all cars sold are* **FORDS**

*More than* **40%** *of Hot Air Heaters sold to* **FORD** *Owners are* **'KUNKELS'**

## through Authorized Ford Dealers

Ford Dealer and Service Field   December 1930

Ford Dealer and Service Field   December 1930

Linendoll Corp. introduced an electric blower on this model of their exhaust heater. It recirculated the air in the passenger compartment. It sold under the Linco brand.

Ford Dealer and Service Field   December 1930

F. C. Purcell & Co. manufactured this Radi-Air Heater that was a hot air heater with a plenum behind the engine radiator to gather the heated air. A duct carried the heated air into the passenger compartment.

Ford Dealer and Service Field   December 1930

The Zenith Manufacturing Corp. offered this exhaust heater with a unique integral belt driven fan. It had a list price of $12.50.

### Genuine Chevrolet Thermostat $2.00 Installed

The Genuine Chevrolet Thermostat is especially useful in colder climates. It is very valuable to the driver in that it warms up the motor quickly —prevents crankcase dilution, and thereby makes the engine operate more economically and efficiently. This is especially true where a driver makes very short runs such as is usually the case in the city. At the price which this Thermostat is offered, there is ample allowance for installation charges and in addition a very satisfactory profit for the Chevrolet dealer.

### Genuine Chevrolet Heater $9.85

The Genuine Chevrolet Heater has proven itself to be a high profit item for Chevrolet Dealers. Since its introduction last fall, over 100,000 of these heaters have been sold to Chevrolet owners, and there is still a large market yet untouched. Chevrolet Dealers have complete selling plans for moving these Heaters quickly. Additional selling effort will result in more heater sales during January and February.

This heater will fit on all Chevrolet Sixes and consequently every owner of a Chevrolet Six is a logical prospect.

1930 Chevrolet brochure

Chevrolet offered a Genuine engine cooling thermostat for $2.00 and an exhaust heater for $9.85. "Since its introduction last fall 100,000 of these heaters have been sold to Chevrolet owners…"

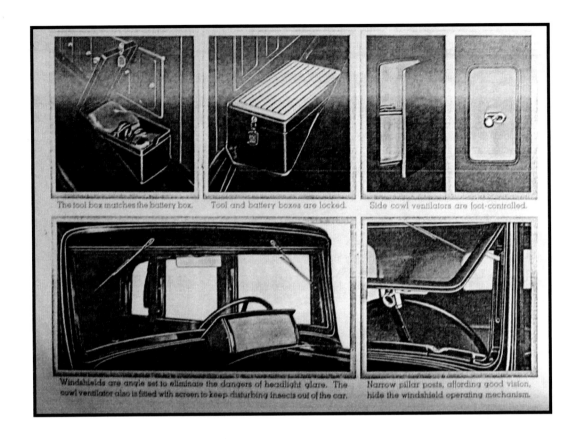

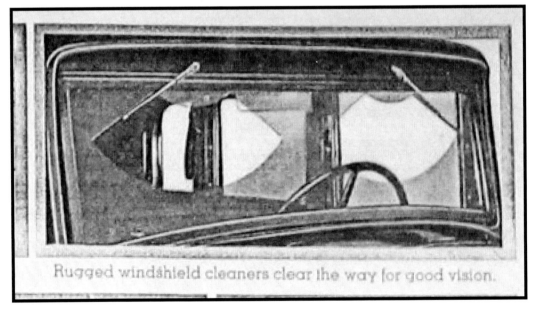

1931 Packard brochure

This Packard featured dual windshield wipers (cleaners), an opening windshield and a large cowl-top ventilation air intake scoop and side cowl ventilation air intake scoops. The side cowl doors were intended to be foot operated.

Stahls Collection

1931 Packard vacuum windshield wiper motor.

Patented Nov. 24, 1931

1,833,847

# UNITED STATES PATENT OFFICE

THOMAS MIDGLEY, JR., OF WORTHINGTON, ALBERT L. HENNE, OF COLUMBUS, AND ROBERT R. McNARY, OF DAYTON, OHIO, ASSIGNORS TO FRIGIDAIRE CORPORATION, OF DAYTON, OHIO, A CORPORATION OF DELAWARE

## HEAT TRANSFER    REISSUED

Application filed February 8, 1930. Serial No. 426,974.

This application relates to the art of transferring heat from one point to another and specifically to the art of refrigeration.

Heretofore, as far as we are aware, refrigerants and heat transfer agents have been chosen chiefly for their boiling points and stability in the refrigerating or heat transfer cycle irrespective of other desirable properties, such as non-inflammability and non-toxicity.

It is the object of our invention, on the other hand, to provide a process of refrigeration and, generically, a process of heat transfer in which these desirable properties, such as non-inflammability and non-toxicity, are obtained in combination with the desired boiling points.

Broadly stated, the part of our process which deals with the controlling of the properties of the refrigerating or heat transfer agents consists in replacing hydrogen by fluorine or other halogen, or both, in aliphatic hydrocarbons in which at least one hydrogen has already been replaced by fluorine.

Broadly stated, the part of our process which relates to the transfer of heat or the production of refrigeration comprises changing the physical state of, for example, by condensing or evaporating, a halo-fluoro derivative of an aliphatic hydrocarbon, and dissipating to, or withdrawing from, an object to be heated or cooled, the latent heat necessary for changing the physical state of the said derivative. By a halo-fluoro derivative of an aliphatic hydrocarbon we mean a derivative containing more than one fluorine atom with or without other halogen atoms, or one fluorine atom with one or more other halogen atoms.

Referring now specifically to our mode of controlling the properties of the refrigerating or heat transfer agent, aliphatic monofluorides form the structural nucleus on which the agents are built. Broadly speaking, if in the structural formula $CH_3F$ we increase the fluorine content (number of atoms) by the substitution of fluorine for hydrogen, the boiling point decreases, stability increases, inflammability decreases, and toxicity decreases. If we keep the fluorine content constant and substitute another halogen for hydrogen in the nucleus, the boiling point increases, the stability decreases, the toxicity increases, and the inflammability decreases. The degree to which these variations take place depends on what the other halogen (chlorine, bromine, or iodine) is. As the ratio of the halogen content to the hydrogen content increases the inflammability decreases.

Because there are several variables, and because of the value of relative proportions, we have placed the compounds of the group discussed on plots wherein

Fig. 1 is a plot applying the rules of substitution to typical groups having one carbon atom,

Fig. 2 is a plot applying the rules to groups having two carbon atoms, and

Fig. 3 is a key to Fig. 2, showing the radicals corresponding to the numbers used in Fig. 2.

Referring to the plots generally, the dashed lines indicate fluorine substitutions and the solid lines indicate chlorine substitutions. Similar plots are obtained with bromine and iodine in place of chlorine except that the plot is elongated in the direction of higher temperatures with bromine, while with iodine the temperatures are still more elevated. The amount of elongation is readily determined by applying the boiling points of some of these compounds.

Referring specifically to Fig. 1, this plot contains all the compounds which can be derived from $CH_3F$ by chlorine and/or fluorine substitutions, together with data which assist in the formation of the plot. On the base line appear the numerals zero to four which show halogen content, and the vertical line gives the approximate boiling points in degrees centigrade. At each point of intersection is given the chlorine and fluorine content and the complete formula of the corresponding compound is found by making this halogen substitution for hydrogen in the formula $CH_4$. We have drawn a horizontal dashed line at about $-25°$ centigrade to indicate approximately the optimum vapor pressure conditions which we desire for operating an air cooled refrigerator. It is obvious that one

This is the patent for R-12 refrigerant. It was awarded to Messer's. Thomas Midgley, Jr., Albert L. Henne and Robert R. McNary. It was assigned to Frigidaire Corp. It would later be called Freon.

Vehicle Owner Howard E. Reinke

1931 DeVaux fitted with a ventipane.

Image by Doug W posted on wikipedia

1931 American Austin with its wiper motor mounted in front of the windshield.

1931 Cadillac accessory brochure

The 1931 Cadillac heater was an exhaust type. It sold for $41.00 to $55.00.

Stahls Collection

1932 Duesenberg windshield wiper system used a vacuum motor mounted on the passenger side and a link to drive the driver side arm.

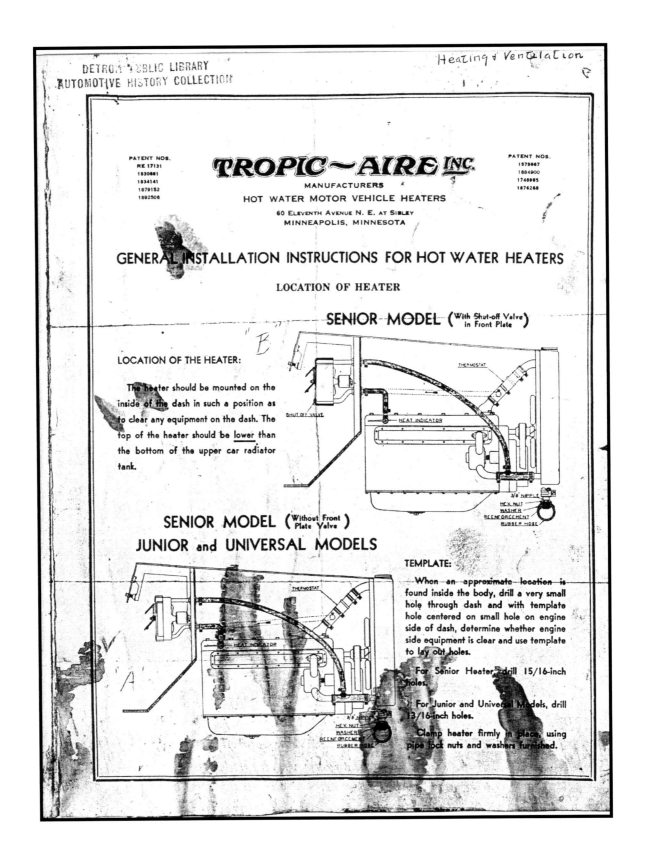

Tropic-Aire Inc. instructions for installing their hot water heaters.

# THERMOSTATS

There is a big difference in the thermostat required for use with the hot water heater and the one used to equip cars in production by car manufacturers.

The car manufacturer is interested in a thermostat that will build up the temperature of a motor to about 140 degrees and this temperature will give good performance from the motor for both winter and summer and can be installed at the factory and forgotten.

This kind of a thermostat will NOT DO AT ALL for a hot water heater.

For use with a hot water heater it is necessary to have a thermostat that will give as high a temperature as possible without undue loss of anti-freeze solution.

A thermostat that will give a maximum temperature of 140 degrees would not be suitable for hot water heater work because the heat from the face of the heater would not be sufficient.

As alcohol is the most widely used anti-freeze solution, it is necessary that the thermostat used should not cause an undue loss of alcohol. Alcohol boils at 172 degrees and at 165 degrees there is a very great loss by evaporation and a very considerable odor from the overflow of the radiator.

It has been found by experience from a number of winters that alcohol anti-freeze solution can be run at a maximum temperature of 160 degrees without undue loss of alcohol.

Therefore it is necessary to use thermostats, with alcohol, that crack open at 160 degrees or slightly lower.

Therefore, the setting of thermostats, for use with this solution or any other solution, if so arranged as to open at between 157 degrees and 160 degrees, will be satisfactory.

Thermostats that give this temperature to the hot water will enable the heater to work at high efficiency and give splendid results.

The thermostat that is set to open at 157 degrees and 160 degrees must also have the characteristic of opening very slowly and to maintain this slight opening against the pump pressure and it must be active enough to maintain the temperature of the main radiator of not over 160 degrees.

All of this calls for a very special instrument, engineered and built to the most exacting requirements.

This is the style of thermostat that is now being offered by Tropic-Aire, Inc., and is the result of a great deal of experimentation and embodies the experience of extensive use over the last five years.

This thermostat is built in two styles, the non-adjustable and the adjustable style.

The non-adjustable is the lower priced. It is set to open at 157° to 160° at the factory. It should never be tampered with. If in doubt as to the performance of this thermostat or if it is thought that it is not working properly, remove it from the car and immerse it in water at 160° temperature.

If the valve is slightly cracked open and the spring feels soft, the thermostat is all right. It should open to its fullest extent at 180 degrees.

The adjustable type thermostat is more expensive and also more convenient and satisfactory. It is made to begin where the equipment thermostat used by the manufacturers leaves off. In other words, it will give temperatures from 140 degrees to 180 degrees. The adjusting dial is marked for Summer running, for Alcohol and for Prestone. By means of the adjustable thermostat the operator can adjust his heat to the point where he gets the best results whether in summer or in winter. On very cold days he can adjust to a slightly higher temperature than he would have on warm days. In the summer he can set it back to the point where his engine will run as cool as with the equipment thermostats.

**TROPIC-AIRE INC.**

MANUFACTURERS

HOT WATER MOTOR VEHICLE HEATERS

60 Eleventh Avenue N. E. at Sibley

MINNEAPOLIS, MINNESOTA

Circa 1932

# Special Instructions For TROPIC-AIRE Heater Installation on Cars Not Equipped with Water Pumps

## IMPORTANT
### Instructions When Using Thermostat

The efficiency of hot water heaters is naturally improved if the water circulating system of the car is equipped with a thermostat for proper temperature control. We recommend the installation of a thermostat especially designed for hot water heaters in all cases where maximum efficiency is required.

On all cars in which a water-line thermostat is installed, the upper hose connection to the heater should be made below the thermostat.

When a special thermostat is installed, remove the equipment thermostat if there is one.

## INSTRUCTIONS FOR ATTACHING HOSE LINES FORD MODEL "V-8"

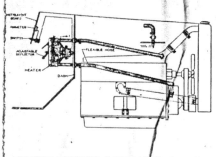

LOCATION OF HEATER—IMPORTANT—The heater is to be installed as LOW as foot room will allow.

WATER CONNECTIONS—The principal difference in this installation is that the water ENTERS the TOP of the heater and LEAVES the BOTTOM.

Do not have any bends in upper hose that will interfere with the free flow of the water and PARTICULARLY do not run THIS hose higher than the heater.

In some cases where the water circulation is sluggish a spoon can be installed in the cylinder head water jacket to deflect the water (see cut). Where it is necessary to install this spoon proceed as follows: Drill $\frac{23}{32}$-inch hole in the cylinder head water jacket for a $\frac{1}{2}$-inch standard pipe tap and screw in the spoon so that the water will flow into the spoon as shown. These spoons can be obtained from us at a small cost.

The above cut shows the installation of Hose lines under the hood. NOTICE that the hose from the bottom heater pipe attaches to the hose from the top of the motor to the radiator. This connection must be ABOVE the water pump and as CLOSE to pump as possible.

As shown, the valve, hose forming washer and hose reinforcement are used.

A different method would be to tap the water pump casting for $\frac{3}{8}$ pipe thread and screw the valve into the casting.

The OUTLET hose from the top heater pipe attaches to the LOWER radiator hose as CLOSE to the motor as possible.

This installation uses exactly 5 feet of hose.

BE SURE THAT THE SYSTEM IS FREE OF AIR

Circa 1932

## HOSE CONNECTION

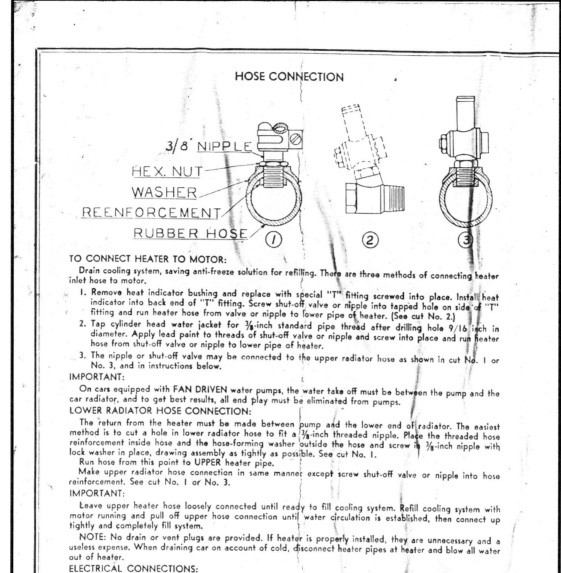

### TO CONNECT HEATER TO MOTOR:

Drain cooling system, saving anti-freeze solution for refilling. There are three methods of connecting heater inlet hose to motor.

1. Remove heat indicator bushing and replace with special "T" fitting screwed into place. Install heat indicator into back end of "T" fitting. Screw shut-off valve or nipple into tapped hole on side of "T" fitting and run heater hose from valve or nipple to lower pipe of heater. (See cut No. 2.)
2. Tap cylinder head water jacket for 3/8-inch standard pipe thread after drilling hole 9/16 inch in diameter. Apply lead paint to threads of shut-off valve or nipple and screw into place and run heater hose from shut-off valve or nipple to lower pipe of heater.
3. The nipple or shut-off valve may be connected to the upper radiator hose as shown in cut No. 1 or No. 3, and in instructions below.

### IMPORTANT:

On cars equipped with FAN DRIVEN water pumps, the water take off must be between the pump and the car radiator, and to get best results, all end play must be eliminated from pumps.

### LOWER RADIATOR HOSE CONNECTION:

The return from the heater must be made between pump and the lower end of radiator. The easiest method is to cut a hole in lower radiator hose to fit a 3/8-inch threaded nipple. Place the threaded hose reinforcement inside hose and the hose-forming washer outside the hose and screw in 3/8-inch nipple with lock washer in place, drawing assembly as tightly as possible. See cut No. 1.

Run hose from this point to UPPER heater pipe.

Make upper radiator hose connection in same manner except screw shut-off valve or nipple into hose reinforcement. See cut No. 1 or No. 3.

### IMPORTANT:

Leave upper heater hose loosely connected until ready to fill cooling system. Refill cooling system with motor running and pull off upper hose connection until water circulation is established, then connect up tightly and completely fill system.

NOTE: No drain or vent plugs are provided. If heater is properly installed, they are unnecessary and a useless expense. When draining car on account of cold, disconnect heater pipes at heater and blow all water out of heater.

### ELECTRICAL CONNECTIONS:

Drill 5/8-inch hole in instrument panel at location desired for the switch. Remove outside nut and insert switch from behind the instrument panel. Replace switch plate and nut, and turn switch to the proper position as indicated by the lettering on the switch plate. Adjust lock nut behind instrument panel and clamp switch firmly in place by tightening outside nut.

Important—When installing the pilot light switch be sure that it is thoroughly grounded (scrape paint around hole) otherwise it will not light.

Connect the longer heater motor wire to "Motor" terminal of the switch. Ground the shorter wire by fastening it to a grounded screw or nut.

On all cars except the Ford, use the extra length of wire furnished to connect "BAT" switch terminal to the ammeter post opposite storage battery terminal. On Ford cars bring the extra length of wire through the hole in dash back of the terminal box and connect it to the left hand terminal (facing engine side of dash).

It is suggested that the wires be connected to the switch terminals before switch is mounted on instrument panel.

Circa 1932

Vehicle Owner Doug Width

1932 Oldsmobile fitted with a chrome-plated vacuum wiper motor.

Note the wiper shaft passes thru the lower windshield frame.

Auction America Auburn, IN 2011 May

A 1932 Chevrolet with a vacuum-powered fan used to defog the windshield.

Ed Meurer Collection

A 1933 Ford fitted with both a Ford hot air heater and an aftermarket hot water heater

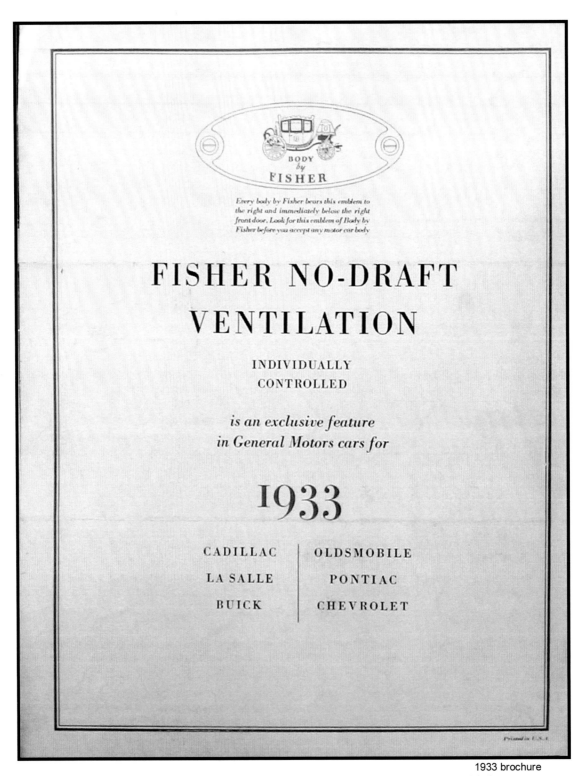

1933 brochure

Fisher Body introduced a new No-Draft Ventilation system on its 1933 vehicles. It featured fully integrated ventipanes and rain proof cowl-top air intake scoops.

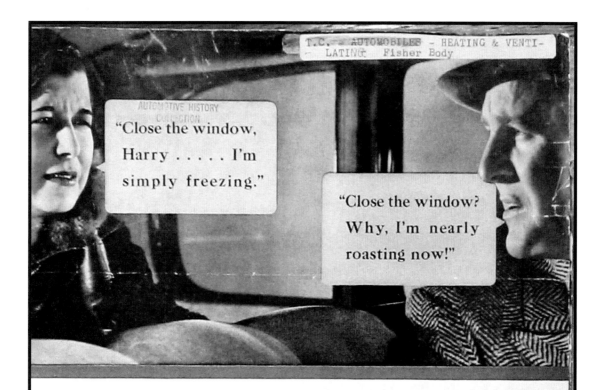

1933 brochure

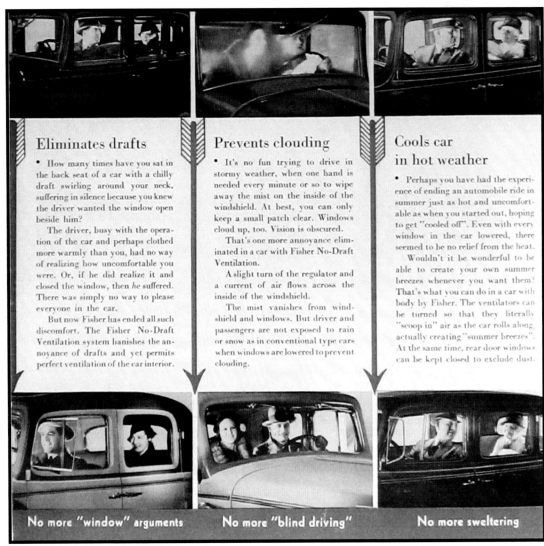

1933 brochure

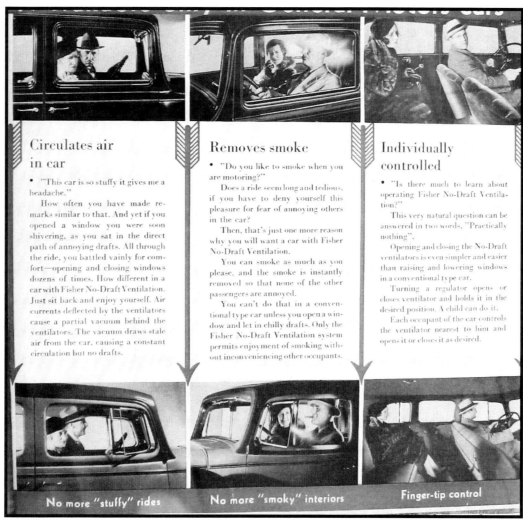

1933 brochure

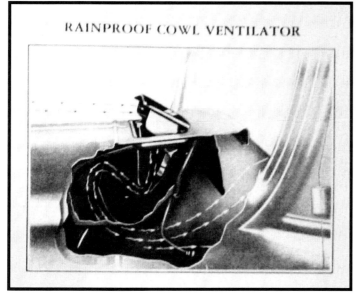

1933 Cadillac brochure

The 1933 Fisher Body ventilation air intake was rainproof per this brochure.

1933 brochure

Vehicle Owner Verl Wetherway

A 1933 Chevrolet with Fisher Body No-Draft Ventilation.

Vehicle Owner Glynette & Barry Wolk

A 1933 Continental Flyer fitted with a single cowl-mounted wiper.

1933 Packard brochure

The Packard ventilation system for the 1933 model year featured a swing out windshield, ventipanes and swing out side glass in the doors trailing the ventipanes.

## Packard Ventilation Control—New Health, Comfort, and Safety

*Packard controlled ventilation is simple in operation and practical in application.*

*Now, for the first time, healthful ventilation may be confined to the rear seat too.*

*OLD: Air rushing in the driver's window eddies around the necks and shoulders of the passengers*

*NEW: Proper ventilation for the front seat alone is controlled by the forward section of the front window*

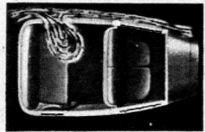

*NEW: Swinging out the rear quarter window can confine the air flow to the rear seat only*

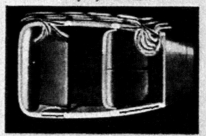

*NEW: Windows in this position give perfect ventilation, even in a rain or with occupants smoking*

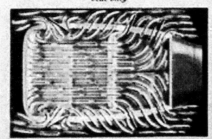

*NEW: Setting the windows wide catches all possible air and deflects it into the car for summer ventilation*

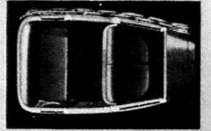

*NEW: Windows may be pulled shut and locked as on any former design to keep the car equally as tight*

1933 Packard brochure

Specialty Interest Autos   May-June 1971

1933 American Austin fitted with wiper motor forward of the glass.

1934 Ford brochure

This 1934 Ford featured a swing out windshield.

1934 Ford brochure

The Ford ventilation system featured side windows that slid back rather than ventipanes.

1934 Ford brochure

Note the head line "Air Condition your Ford for winter driving". This was an early offer of a genuine Ford accessory exhaust heater.

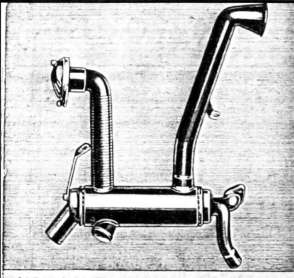

1934 Ford brochure

# QUICKER HEAT - HOTTER HEAT - MORE HEAT

## and this new FORD HEATER air conditions your V-8

HERE'S an entirely new principle in car heater design. It's built just like a boiler—with flues. 24 of them, each 13 in. in length, provide almost 500 sq. in. of heat radiating surface. That's where this new Ford Heater gets its extra high efficiency.

There's nothing cheap or flimsy about the way it's built. This rugged unit is really a rigid, integral part of the exhaust line. To eliminate vibration and exhaust rumble the tubes are forced through the six separators in a power press; the holes in the separators being slightly smaller than the diameter of the tubes. These flues have a greater total opening than the exhaust pipe so there's no back pressure to interfere with peak engine performance.

But the most exclusive feature is the DOUBLE HEADER construction. These headers are also pressed over the ends of the tubes which are then welded to the outer header. That makes the fresh air passage completely independent and outside of all welded joints. 100% SAFE!

The highest temperatures available for car heating purposes are obtained from the exhaust gases. As these hot gases pass through the flues a large percentage of their heat is absorbed by the walls. The fresh air blown into the heater by the engine fan passes outside and around these hot flues. This is instantly heated forced on through the heater and into the car. Thus, the heated air is as clean wholesome and odorless as that used for air conditioning modern homes.

1934 Ford brochure

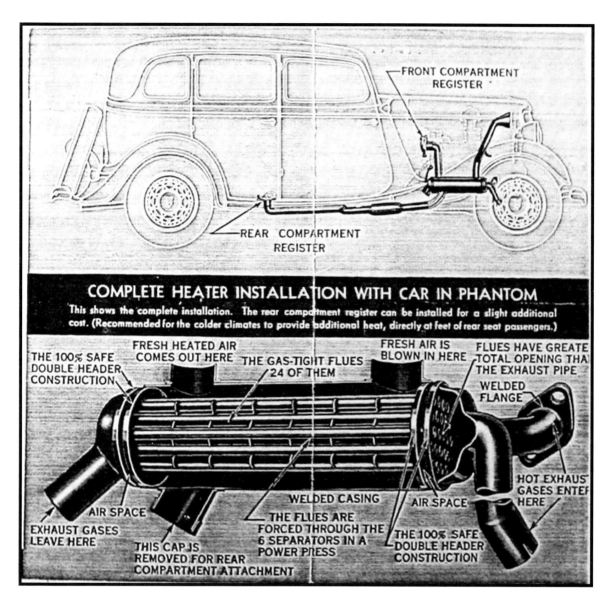
1934 Ford brochure

> All the advantages which Fisher No Draft Ventilation brings to Oldsmobile ownership might be summed up as follows:
>
> *Elimination of drafts*
>
> *Prevention of clouding or frosting of the inside of windshield and windows.*
>
> *Circulation of air in the car, keeping the air fresh at all times.*
>
> *Removal of tobacco smoke.*
>
> *Better cooling of car in hot weather.*
>
> *Individual control, which permits air circulation to be localized to as few passengers as desired.*
>
> The screened cowl ventilator is of a new type, opening toward the windshield. This construction was 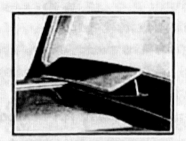 adopted because exhaustive tests show that air currents do not move in a straight line over the hood but lift gradually from radiator to windshield and then take a downward direction.
>
> As a result, Oldsmobile's new cowl ventilator provides even more adequate ventilation, while eliminating the discomfort from direct air currents common to the conventional type.

1934 Oldsmobile brochure

The 1934 Oldsmobile list of advantages of their ventilation system. Note the unique rear facing cowl-top ventilation air intake scoop.

The brochure also stated the body was well insulated against heat, cold and noise. The dash had ¼ inch of Celotex and ½ inch of jute. Door and body panels, the passenger and rear compartment floors were all insulated. The brochure stated the dash panel was designed and manufactured to receive a hot water heater "without the necessity of unsightly holes or mountings".

Two interior sun visors that are fully adjustable ... and two windshield wipers that won't slow down on a hill or while accelerating ... assure clear vision under all driving conditions.

Here's something new in cowl ventilation —

Air pressure close to the windshield is greater ... so more air comes in when the ventilator opening is reversed. A deflector beneath the cowl prevents blast.

1934 Buick brochure

The 1934 Buick noted their cars had "windshield wipers that won't slow down on a hill or while accelerating…".

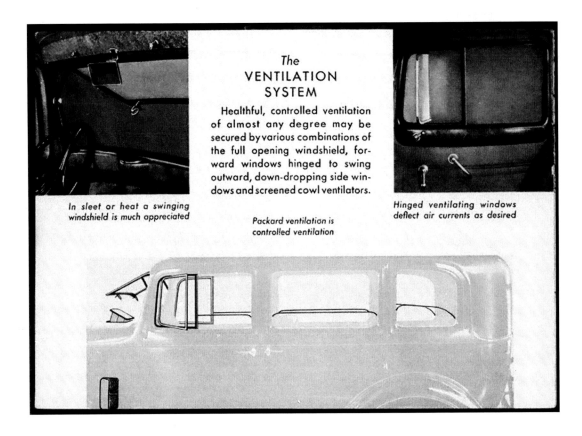

1934 Packard brochure

The ventilation system on the 1934 Packard had roll down side windows in the doors not the swing out type used previously.

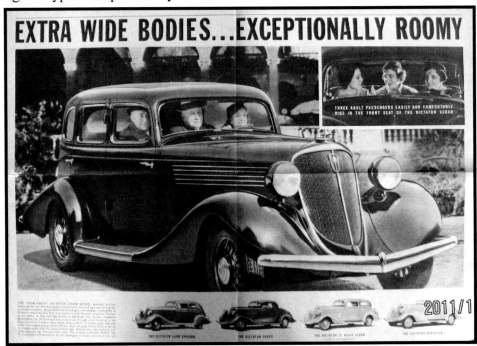

1934 Studebaker brochure

In 1934 Studebaker fitted its vehicles with cowl-mounted wipers.

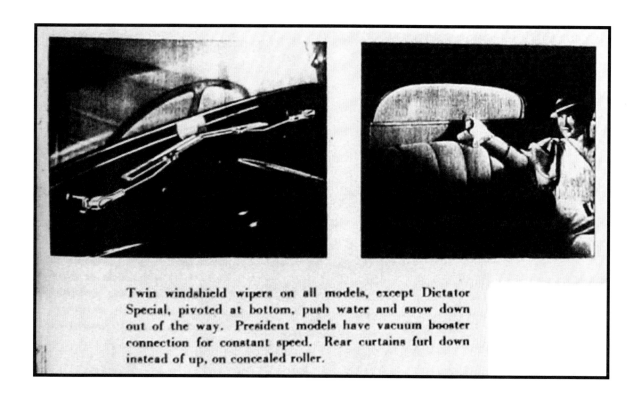

1934 Studebaker brochure

Note the permanent horizontal ventilation device that was intended to shed rain and snow. The windshield also opened. The twin cowl-mounted wipers were designed to "push water and snow down out of the way".

Vehicle Owner Bob Mantel

1934 Chrysler Air Flow with dual windshields and cowl-top ventilation scoops.

Stahls Collection

1935 Chrysler Air Flow with an electric fan fitted to the steering column pointed at the driver rather than the windshield.

1935 Oldsmobile brochure

One of the first Fisher Body vehicles with cowl-mounted windshield wipers.

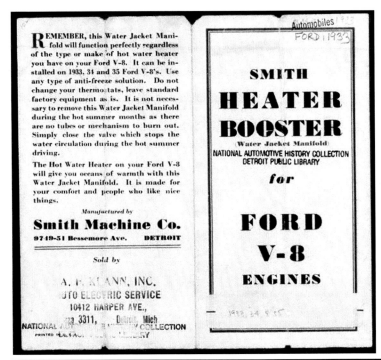

Circa 1935 brochure

Smith Machine Co. offered this combination exhaust heater and hot water heater manifold for Ford V8 engines.

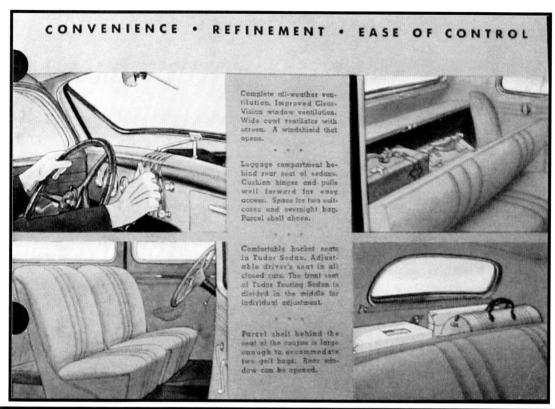

1935 Ford brochure both images

Vehicle Owner Jack Hazen

1935 Ford Coupe with roll down rear window.

Automobile Trade Journal   November 1935

Lion Products Co. offered a Rubberset Sleet chaser electric windshield defroster. Noblitt-Sparks Industries, Inc offered an Arvin steam heater as well as hot water models.

# HUDSON MOTOR CAR COMPANY
# ACCESSORY AND PARTS MERCHANDISING
## 1935 SERIES BULLETINS

No. 8
Date 12/10/34
Subject: ROBES AND PILLOWS

TO ALL DEALERS:

As a feature of the preview of 1935 models in Detroit on November 13th, Robes and Pillows to match the upholstery of the various models were displayed. So much favorable comment was received from distributors and dealers that we decided to include them in our Accessory Group.

The robes and pillows are made in our own Trim Department of the same cloth and as is used in car production, assuring you of an accurate match of color and material with the car in which they are to be used.

A set of one robe, two square pillows and one wedge pillow, placed in cars in your showroom will make a splendid display and will immediately interest your customers.

We have not set any list price on these robes and pillows, believing that You will prefer to set your own. There will be occasions when these may contribute to new car deals, and without any established lists you can sell them at any figure you desire without code violation providing the selling price is always above cost.

Part numbers and net prices for various models are listed.

| Part No. | Name | Dealer Net |
|---|---|---|
| | For all Terraplane Models: | |
| 113900 | Wedge Type Pillow | $2.25 |
| 113906 | Square Type Pillow | 2.25 |
| 113912 | Robe | 9.76 |

1935 Hudson Accessory Bulletin

Hudson offered genuine accessory lap robes and matching pillows.

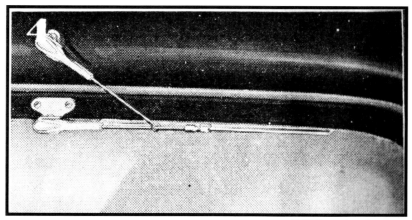

1935 Terraplane brochure

Terraplane vehicles were fitted with auxiliary windshield wiper arms.

Vehicle Owner Jim Amman

1935 Hudson fitted with auxiliary windshield wiper arms.

Vehicle Owners Scott & Linda White

1936 Hudson Terraplane fitted with auxiliary windshield wiper arms.

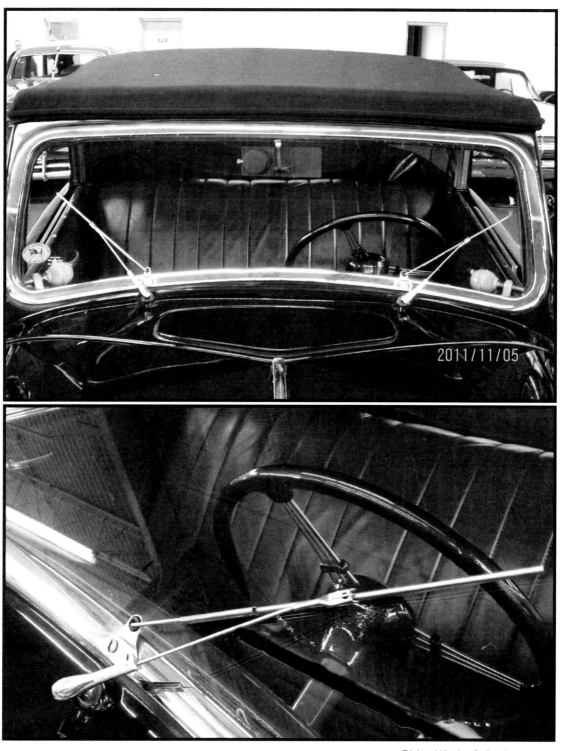

Richard Kughn Collection

1936 Plymouth DeLuxe convertible coupe fitted with auxiliary windshield wiper arms.

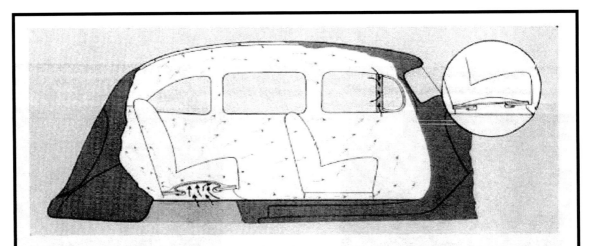

1936 Hudson brochure

This Hudson ventilation air intake was under the rear seat and featured a filter.

Ford PDC car show 2011 July

A 1936 Ford with a link to connect the dual windshield wipers.

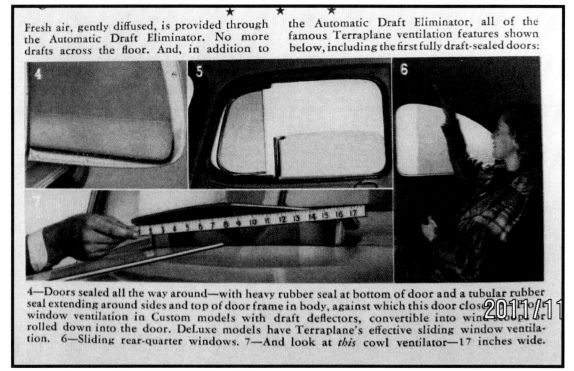

1936 Terraplane brochure

The Terraplane ventilation system featured a cowl-top air scoop 17 inches wide.

1936 Terraplane brochure

Another view of the Terraplane auxiliary windshield wiper arm.

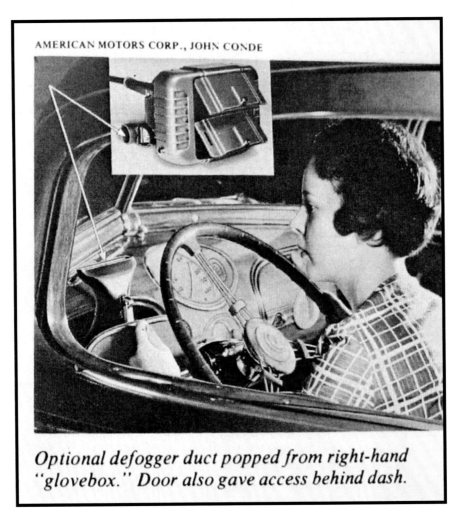

1936 Terraplane defroster nozzle was stowed behind the instrument panel access door.

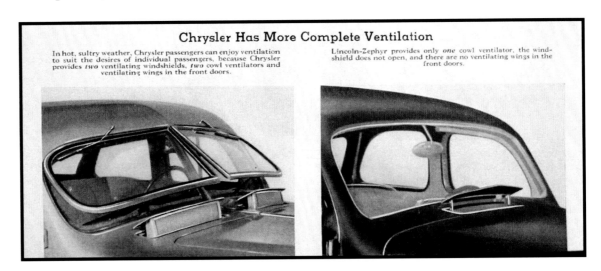

A comparison of the ventilation systems on the Chrysler Air Flow and Lincoln Zephyr.

Popular Mechanics November 1936

Windshield washers were introduced in 1936.

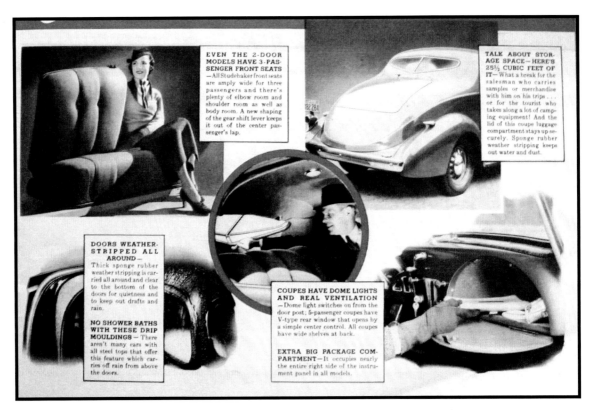

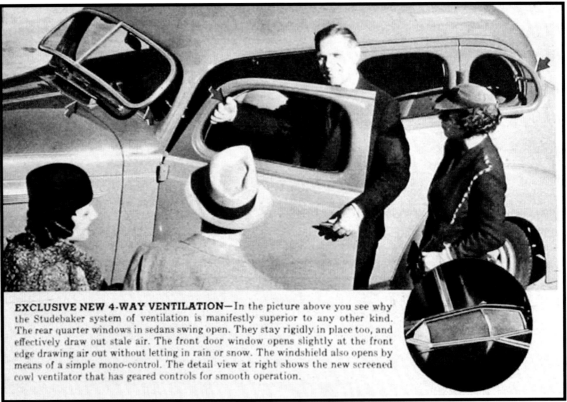

1936 Studebaker brochure

Studebaker's ventilation system had rear sliding side glass and opening windshields and an opening rear window on the coupe.

Vehicle owned and images by Richard & Regina Jandrey

A 1936 Studebaker coupe with swing open rear window.

1936 Studebaker brochure

Twin robe cords were fitted to these Studebakers

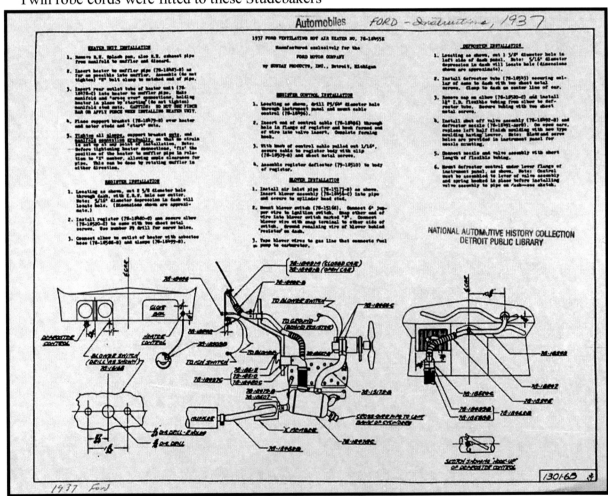

1937 Ford heater installation instructions

1937 Pontiac with trim logos on the hot water heater and radio

1937 Terraplane cowl-top ventilation scoop.

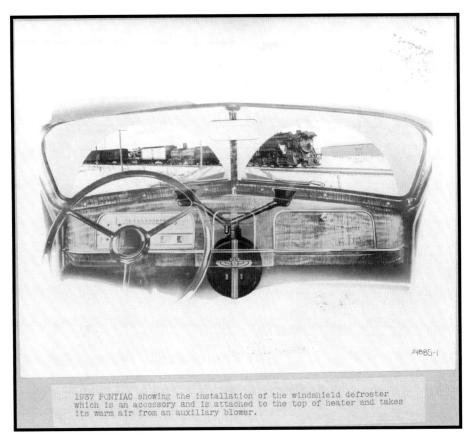

General Motors image

1937 Pontiac fitted with an early defroster system.

Vehicle Owner Chris Trimmer

1937 Pontiac with defroster nozzles projecting through the instrument panel.

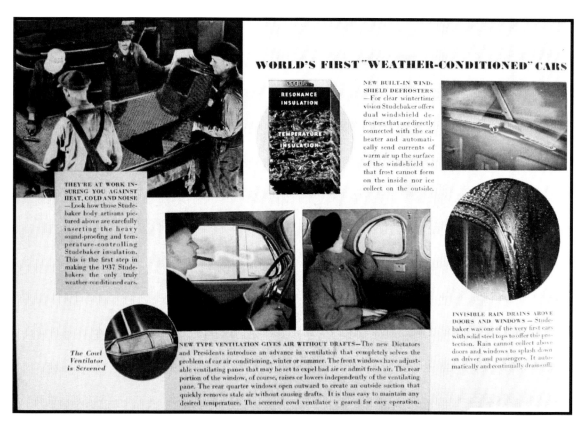

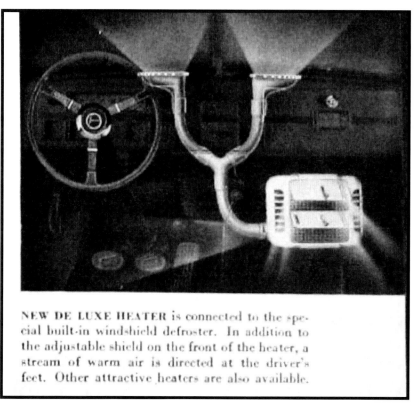

1937 Studebaker brochure

Integrated defroster nozzles and body insulation were introduced on this Studebaker.

Auction America Auburn, IN 2011 May

1937 Ford fitted with "dead air" devices to minimize fogging.

1938 Nash brochure

The 1938 Nash Weather Eye heater and defroster system was a major leap in technology for the era. It was the first automatic temperature control heating system using a water valve. Note that it also was integrated into the vehicle dash panel. The vehicle was designed from the start to accommodate the system. It also contained a serviceable air filter and a rain separator.

## Nash Conditioned Air System Brings You A NEW MIRACLE IN WINTER DRIVING COMFORT!

### 70° inside when it's zero outside

Never Again — Icy Drafts . . . Cold Rear Seats . . . Dust . . . Stale, Stuffy Air . . . or Dangerous Window-steaming. Now — a Healthful, Everfresh Stream of 70° Warm Air *Even at Zero* — in your Nash!

Everybody knew it would come some day. But Nash engineers give it to you *right now!*

A CONDITIONED AIR SYSTEM in your new Nash car!

Think what this means! No more numbing, miserable rides . . . no more icy drafts that cause colds . . . no more dust . . . no more stale air or monoxide fumes that make you drowsy and cause accidents.

All winter long, right through sub-zero blizzards, you and the back-seat passengers can enjoy 70° (80° if you like) warm, *conditioned* air. And the price of a regular heater is all you pay.

Nash Conditioned Air works on an entirely new principle . . . fundamentally different fom the old-time heater.

Other cars, in motion, have lower air-pressure inside than outside. That creates a vacuum which sucks in dust, cold drafts . . . no matter how well-built the car may be.

Nash reverses this . . . draws in fresh air at one inlet . . . cleans it of dirt, dust, excess moisture . . . then filters and heats it to desired temperature . . . and distributes it *under pressure* to all parts of the car. Air pressure *inside* is greater than outside, at all speeds. Stuffy air and smoke are forced out; cold air can't creep in.

You can easily prove this in a demonstration by holding a cigarette near a Nash door. The smoke is pulled *out*. Even with six heavy smokers, the air is constantly fresh and clean. From 300 to 400 cubic feet of conditioned air per minute are delivered at driving speeds; 100 cubic feet when car is standing still. This more than meets requirement of State Ventilating Codes for theatres, etc.

Not only in winter, but *all year around*, you'll enjoy Nash Conditioned Air. For turning a dial turns off heat, yet leaves the rest of the system effective.

You can drive all day on dusty roads, yet step out with spotless clothes and face.
In a rainstorm, you can enjoy fresh, circulated air without getting wet. You can shut out all bugs, bees, insects, in countryside driving.

With windows closed, road and city noises are hushed . . . you can drive in restful silence even in city traffic.

Already, health authorities and police commissioners are hailing Nash Conditioned Air as a great step forward in safer driving. Doctors agree that it should minimize the danger of catching cold in winter driving.

For the welfare of your family . . . for the sheer thrill of the experience . . . take a ride in a new Nash. You'll discover that Conditioned Air revolutionizes driving comfort! Drop in and see your Nash dealer.

---

**ONLY NASH CONDITIONED AIR GIVES YOU:**

1—Seventy Degrees comfort for both front and rear passengers in zero weather.
2—No chilling drafts.
3—Clean, filtered, fresh air at all times—even on a dusty road, or while it is raining or snowing.
4—No fogged or steamed windows or windshield to hinder your vision.
5—No drowsiness from stale air, or the possible presence of monoxide fumes.

### You Can't Beat a NASH!

1938 Nash brochure

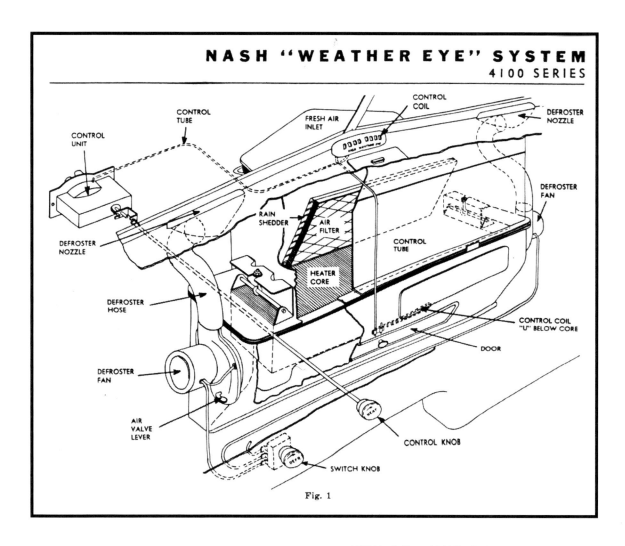

1938 Nash (from 1940 Nash shop manual)

Note the integral defroster nozzles.

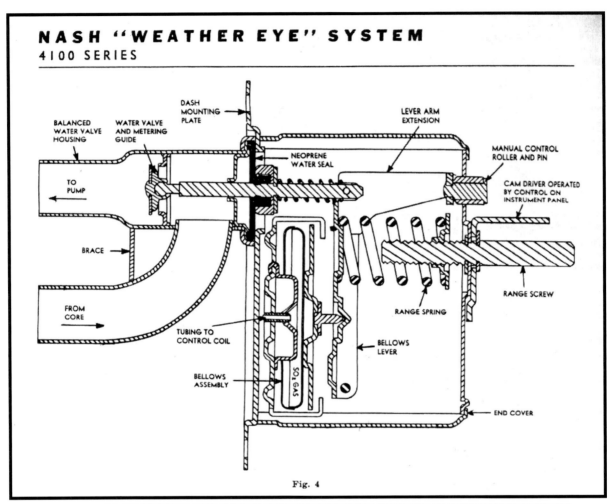

Fig. 4

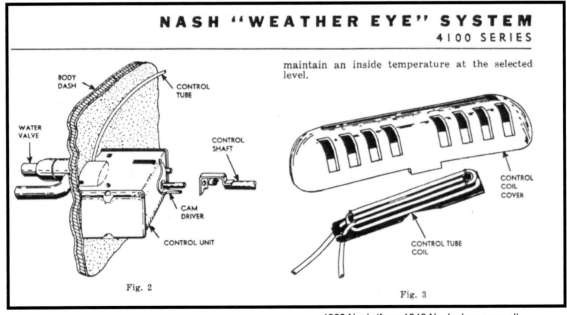

Fig. 2

Fig. 3

1938 Nash (from 1940 Nash shop manual)

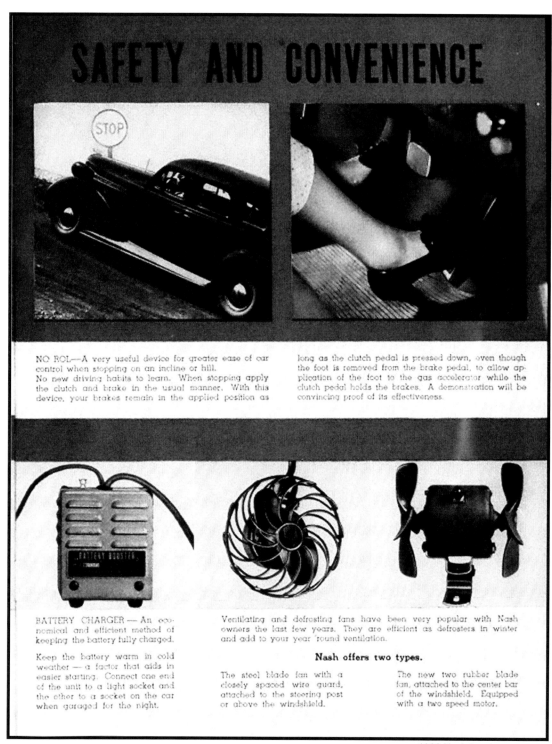

1938 Nash brochure

Note two fan types were offered as genuine accessories.

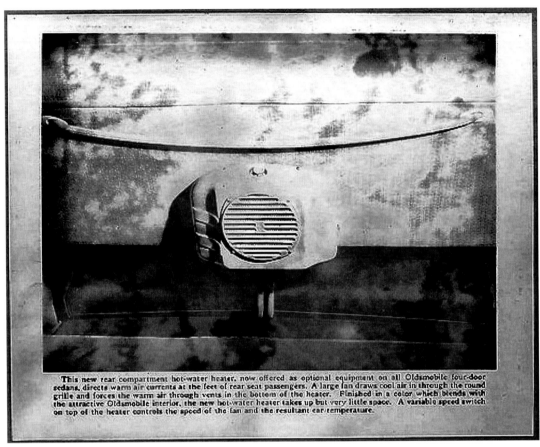

1938 Oldsmobile offered an early application of a rear seat hot water heater.

1938 Oldsmobile price list

## WINDSHIELD AND COWL

Divided windshields on all models are set at an angle that improves vision and reduces wind resistance, thus decreasing wind noise at the front of the car. This construction, by increasing vision from the front compartment, adds much to driving safety.

Windshield wipers are pivoted at the lower edge of the windshield and are operated by a concealed mechanism. More efficient wiper action, plus neater appearance is gained by this construction. Provision is also made for the installation of a windshield defroster which directs hot air from the car heater against the windshield, through two vents incorporated in the lower windshield moulding. The defroster is operated by a blower fan and the unit may be purchased to fit either the Oldsmobile standard or de luxe heater.

A large screened ventilator which opens from the front is built into the cowl. This ventilator, when open, provides more than ample circulation of air around the car floor in hot weather.

**ADVANTAGES—**
- Improved vision and reduced wind noise resulting from the use of the divided V-type windshield.
- Greater efficiency and neater appearance of windshield wipers mounted at the bottom.
- Built-in defroster vents provide clear vision for both driver and front seat passengers and materially promote safety.
- Adequate ventilation of front compartment floor in hot weather through the use of the large cowl ventilator.

1938 Oldsmobile Engineering Features

Note the windshield wipers are pivoted at the lower edge of the windshield and are operated with a concealed mechanism. Also note the integrated windshield defroster nozzles are fitted to the lower windshield molding.

1938 Oldsmobile brochure

Dual windshield defroster vents were a new feature on the 1938 Oldsmobile.

Vehicle Owner Jim Campbell

1938 Buick with defroster outlets in the lower edge of windshield trim moulding.

## "Packaged" Accessories Shipped with Cars from Factory, Assembly Plants or Zone Warehouse
## —Not Installed—

The following Accessory Groups may be installed on all 1939 models ("60"-"70"-"80"). Prices on these Accessory Groups are the same for all models and include E.O.H. These Accessories are not included in the amount subject to Retroactive Discount.

### "G" Accessory Group

|  | Dealer Net | List Price | Suggested Installed Price |
|---|---|---|---|
| Standard Heater | $ 6.65 | $11.95 | $13.95 |

(Above prices include E.O.H.)

### "H" Accessory Group

| | | | |
|---|---|---|---|
| DeLuxe Heater | $ 9.20 | $16.45 | $18.45 |

(Above prices include E.O.H.)

### "J" Accessory Group
#### Models "70"—"80"

| | | | |
|---|---|---|---|
| Dual Windshield Defroster | $ 4.10 | $ 7.25 | $ 8.00 |

(Above prices include E.O.H.)

### "JJ" Accessory Group
#### Model "60"

| | | | |
|---|---|---|---|
| Dual Windshield Defroster | $ 4.10 | $ 7.25 | $ 8.00 |

(Above prices include E.O.H.)

### "K" Accessory Group

| | | | |
|---|---|---|---|
| Rear Seat Heater | $ 7.90 | $14.95 | $19.95 |

(Above prices include E.O.H.)

### "L" Accessory Group

| | | | |
|---|---|---|---|
| Fender Lamps | $ 3.00 | $ 5.00 | $ 6.00 |

(Above prices include E.O.H.)

### "M" Accessory Group

| | | | |
|---|---|---|---|
| Hinged Rear Body Guard / Front Grille Guard | $ 2.70 | $ 4.50 | $ 4.50 |

(Above prices include E.O.H.)

Page 11

1939 Oldsmobile price list

# STUDEBAKER ACCESSORIES

 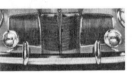    

**Touch control automatic tuning radio.** Station selection is entirely automatic. Only one button to operate. Tuning is instantaneous. It's a custom set that's priced with the lowest. Concealed antenna is high capacity type—adjustable for long or short range reception.

**Electric clock.** Precision jeweled, built to high standards, regulated and run-in like a fine watch. Installs in the package compartment door. Indirectly illuminated.

**Power defroster,** a component part of the Climatizer, draws dry, warm, fresh air from the front compartment and spreads it in a thin fast moving layer over the entire area of the windshield, providing ample defrosting capacity to keep windshield clear in all weather.

**Studebaker fog lights** insure effective results when driving in fog, rain or snow—they are available in pairs or singly with either amber or white lens.

**Chromium wheel discs** or stainless steel wheel mouldings are smart.

**Studebaker Climatizer** is available on all Champion models. Central location under front seat keeps foot space free. Fresh filtered clean air, thoroughly warmed when required, is evenly distributed to front and rear.

**Grille and trunk guards** of several designs are available. Guard illustrated above provides protection for fenders and lamps, front and rear. Guard shown at right may be installed front or rear.

**Luggage ensemble** styled in beautiful striped linen. Designed to fit Champion luggage compartments. Available singly or in sets of three.

**Controllable spotlight** may be focused in any desired direction from the car interior. Locates signs and numbers easily.

1939 Studebaker Climatizer (1940 brochure shown)

Note the 1939 Studebaker Climatizer system was fitted with an air intake filter.

A wind tunnel heater test being performed circa 1938.

1939 Ford service bulletin

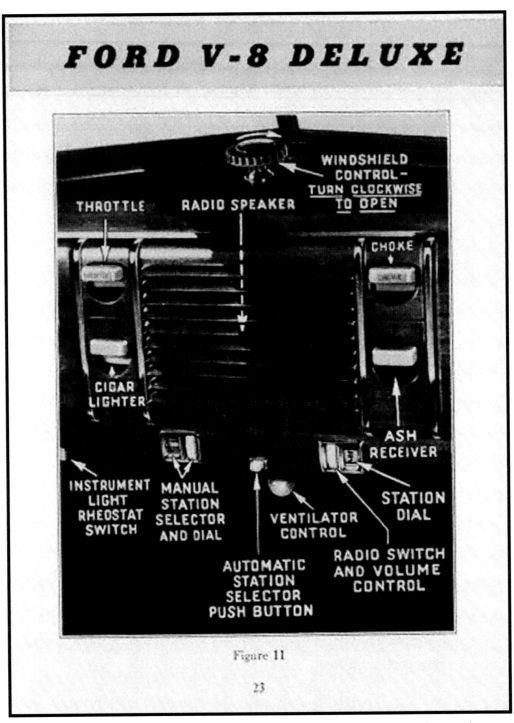

1939 Ford owner's manual

The control knob for the swing out windshield was last used on this 1939 Ford model. Also note the ventilation cowl-top scoop control below the instrument panel.

The new Auto-Lite Hot Air Heater, Model A-20 is claimed by its manufacturers to be the best hot air heater produced regardless of cost, despite its low selling price. Features claimed for this new-type heater are: Easy, quick installation; finger-tip control of register and defroster; heater entirely under hood; free easy upward circulation; instant

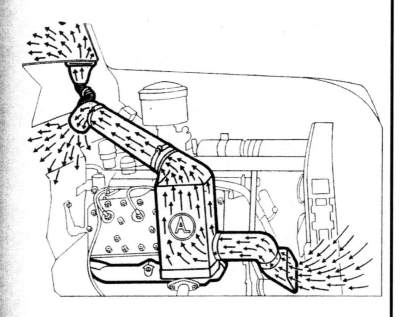

heat; health heat—no fumes; no mechanical parts; quiet operation; permanent defroster; compact installation; easily accessible; quick, positive heat control; low first cost and no maintenance; definite market ready; easy to show and easy to sell; adequate air conditioning; no room lost inside car and no electrical connections.

This Electric Auto-Lite Heater uses the tried and proven method of heating clean, fresh air by passing it around the tubular unit which in turn derives heat from exhaust gases and provides a bountiful supply of healthy heat almost instantly.

Installation is simple and can be made quickly without removing splash pan or cutting off muffler pipes. The Model 20-A, designed for all 85-H.P. Ford V-8 cars, 1937 to 1939 inclusive and for the Mercury is shipped complete with defroster and there are no extras to buy.

Distributed exclusively through jobbers, wholesalers and specialty distributors and sold through Ford dealers, Ford service garages and other retail sales channels where Ford cars are serviced. Write or wire for additional information and also special jobber and dealer discounts to The Electric Auto-Lite Co., Fostoria, Ohio.

Ford Field September 1939

The Electric Auto-Lite Co. offered this exhaust heater with a list price of $12.50.

Ford Field September 1939

Noblitt-Sparks Industries, Inc. offered a new Arvin brand hot water heater with optional double defrosters.

Ford Field   September 1939

Prestone brand antifreeze sold for $2.65 per gallon in 1939.

## Attention Ford Dealers!
# SELL HEATING COMFORT
### Today-tomorrow-and *every* day in the year!

• The best time to sell heaters is when you sell new cars. That's every week day in the year! When a man buys a new car, he's usually feeling "flush". Sell him a Ford Heater *right then and there*... Show him how easy it is to include it in the deferred payments. If he waits, he may buy some other heater and be denied the heating comfort to which he is entitled. Then you have lost a nice profit—a profit you could have had for yourself simply by pointing out the many advantages of *approved* Ford Heater-Defrosters.

• Equip every demonstrator with a heater. Let prospects see how efficiently they work. Owners will appreciate your interest in their Winter Driving Comfort.

## FORD Hot Water and Hot Air HEATERS
### TAILOR-MADE FOR THE FORD—MERCURY—FORD TRUCK
## TWO TYPES TO CHOOSE FROM

**HOT WATER HEATER-DEFROSTER WITH AUTOMATIC, BUILT-IN AIR VENT**

**HOT AIR HEATER-DEFROSTER WITH PRE-HEATED, FILTERED FRESH AIR**

- Two-Speed, Self-Illuminating Switch
- Two Built-in Defrosters
- Separate Defroster Control
- Direct Air Flow
- Indirect Air Flow

- Illuminated Blower Switch
- Two Built-in Defrosters
- Separate Defroster Control
- Convenient Heater Control
- Outlet with Adjustable Heat Deflector

### TO SELL HEATERS—DISPLAY THEM!
You still have time to get handsome, individual display stands for the Ford Hot Air and Hot Water Heater-Defrosters. Ask your Ford Motor Company representative about this free offer, good until September 30, 1939.

**Hot Water Heater-Defroster**
*Manufactured for Ford Motor Company by*
**EATON MANUFACTURING COMPANY**
CLEVELAND, OHIO

**Hot Air Heater-Defroster**
*Manufactured for Ford Motor Company by*
**NOVI EQUIPMENT COMPANY**
NOVI, MICHIGAN

Ford Field   September 1939

Eaton Manufacturing Co. manufactured hot water heaters and defrosters.
Novi Equipment Co. manufactured hot air heaters and defrosters.

Ford Field September 1939

Tropic-Aire Inc. offered this hot water heater with a reversible electric blower.

Ford Field September 1939

The Fulton Co. offered rubber bladed fans and electric frost shields.

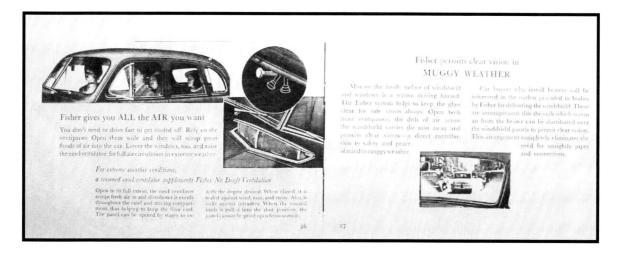

1939 Oldsmobile brochure

The text reads as follows:

**Fisher gives you ALL the AIR you want**
You don't need to drive fast to get cooled off. Rely on the ventipanes. Open them wide and they will scoop great floods of air into the car. Lower the windows, too, and raise the cowl ventilator, for full air circulation in extreme weather.

**For extreme weather conditions, a screened cowl ventilator supplements Fisher No Draft Ventilation.**
Open to its full extent, the cowl ventilator scoops fresh air in and distributes it evenly throughout the cowl and driving compartment, thus helping to keep the floor cool. The panel can be opened by a lever to exactly the degree desired. When closed it is sealed against wind, rain, and snow. Also, it locks against intruders. When the control knob is pulled into the shut position, the panel cannot be pried open from outside.

**Fisher permits clear vision in MUGGY WEATHER**
Mist on the inside surface of windshield and windows is a serious driving hazard.

The Fisher system helps to keep the glass clear for safe vision always. Open both front ventipanes; the drift of air across the windshield carries the mist away and permits clear vision- a direct contribution to safety and peace of mind in muggy weather.
Car buyers who install heaters will be interested in the outlets provided in bodies by Fisher for defrosting the windshield. These are inconspicuous slits thru which warm air from the heater can be distributed over the windshield panels to permit clear vision. This arrangement completely eliminates the need for unsightly pipes and connections.

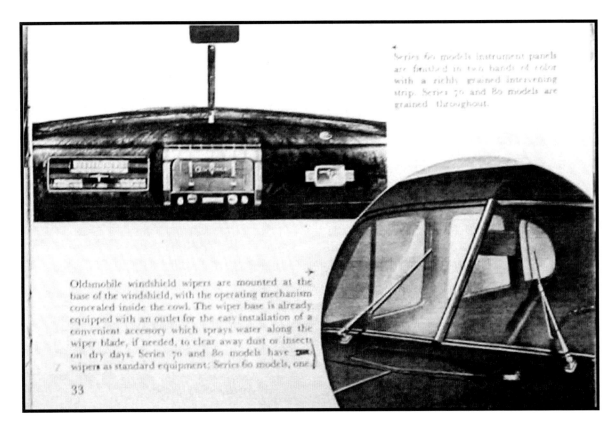

1939 Oldsmobile brochure

Oldsmobile windshield wipers are mounted at the base of the windshield, with the operating mechanism concealed inside the cowl. The wiper base is already equipped with an outlet for easy installation of a convenient accessory which sprays water along the wiper blade, if needed to clear away dust or insects on dry days. Series 70 and 80 models have two wipers as standard equipment. Series 60 models, one.

1940 Hudson brochure

Hudson offered a windshield washer in the 1940 model year.

A **cylinder with radiating flanges** is shown in Fig. 38. When in motion, the current of air blowing against the flanges drives the heat away. This principle is used extensively on motorcycle engines.

The **"copper-cooled"** cylinders are fitted with copper flanges (copper radiates heat quicker) and a draught of air is circulated around cylinders.

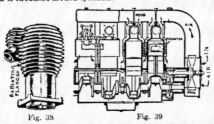

Fig. 38    Fig. 39

A **forced draught air-cooling system** is shown in Fig. 39, formerly used years ago on a prominent make of car. With this system the circulation of air was forced through jackets placed around each cylinder, open at the bottom and top, being connected to a pipe from a centrifugal air blower or fan. The forced air passed the radiator flanges, and out at the bottom.

The Franklin air-cooled engine is a very successful engine for automobile pleasure cars employing the air-cooled method. The six cylinders are $3\frac{1}{4}''$ bore and $4''$ stroke, giving a formula horsepower of 25.3.

By referring to the illustration (Fig. 40), the path of the air is shown, first through the hood, thence over and down through the air jackets. The air is then deflected downwards and out through the flywheel blades.

**Vertical steel fins** are made integral with the individual cylinder casting, by having the iron poured around the strips of steel. Very light aluminum jackets guide the air draught downward from the heads of the cylinders.

Fig. 40. Direct air cooling of the Franklin. The fly wheel is the only moving part of the cooling system. This is termed a "draught system." The later Series 10 car uses a "blast system"; the air being drawn by a fan placed in front.

**Note the vanes in the fly wheel**, which create a suction equal to 2,200 cubic feet every 60 seconds; a continuous flow of air literally wiping the heat away. It is stated that the heat on a Franklin engine is about 350° Fahr. (see Fig. 12, page 107, for Franklin exhaust-heated inlet manifold). This heat is shut off after the engine is warmed up.

The Franklin at one time employed **auxiliary exhaust valves** to assist in dispelling the heat of explosion from the cylinder as rapidly as possible. This method, however, has been discontinued.

Another popular make of air-cooled car is the Holmes.

## HEATING A CAR

There are three methods of heating a car as explained below:

(1) hot water; (2) exhaust gas; (3) hot air.

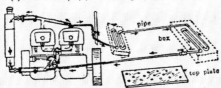

Fig. 41. Hot-water method of heating a car.

**The hot-water method** is shown in Fig. 41. This method can be used only where there is a forced or pump-circulation system. Connections are made with the circulating system at the top of the rear cylinder. The water circulates through the heater, whence it returns to the bottom of the radiator.

The heater is made of regular water pipe, and the housing of aluminum or light cast iron. The floor is cut away, allowing the surface of the heater to be flush with the floor. The top plate, made of aluminum, is then placed over the heater box.

**The exhaust method** for heating is to utilize the exhaust gases instead of water. In this instance the pipes would be connected with the exhaust pipe instead of with the water pipe. Only one side, the inlet, would be connected and an outlet is provided for the emission of the gas.

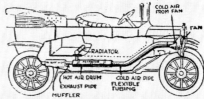

Fig. 42. Hot-air method of heating a car.

**The hot-air method** is shown in Fig. 42. This example illustrates the Brickly (patented) method. The air is taken from the fan through a funnel opening, and a flexible metal hose drives it through a metal jacket 24 to 30 inches long, which covers the "piping hot" exhaust pipe, and warms it thoroughly. It then drives it through a $1\frac{1}{2}$-inch opening in the floor of the car, into a tubular register, along the back edge of the front seat, sending a continuous stream of heated air into the car. (Exhaust gases are not used.)

See also page 1056 for later information on heating a car

Dykes Encyclopedia 1940

Hot air heaters and hot water heaters are described in this article.

## HEATING A CAR

There are four methods of heating a car: (1) by hot water; (2) by hot air; (3) by exhaust gas; (4) by steam. See also page 156.

### Hot-Water Heater[1]

This system derives heat from the water circulating in the engine cooling system. The hot-water heater is mounted on the inside of the dash always as low as possible, and necessarily below the level of the car radiator upper tank. The hot water is forced to circulate through the heater core, which is similar to a radiator core. An electric fan behind the radiator forces the air through the heater core channels into the car. As the air passes through the core, it absorbs the heat from the hot water. The temperature of the hot water determines the temperature of the heated air. The size or area of the heater core determines the quantity of heat available at a given water temperature. The speed of the fan regulates the quantity of heat transferred to car interior and can therefore be used to regulate the temperature within the car. Adjustable deflectors on heater serve the same purpose.

Electric connections: There are two leads from the fan motor. One is grounded, and the other is connected to the heater switch. The other terminal of heater switch is connected to ammeter terminal opposite the battery connection, or to the coil side of the ignition switch. If the latter connection is made, the heater will always be turned off when the ignition is turned off. Some heaters are grounded internally but have two leads. Be sure to connect the one marked "Bat" to the live side of the ammeter.

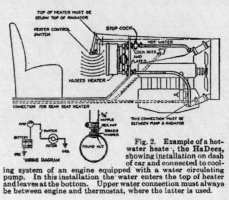

Fig. 2. Example of a hot-water heater, the HaDees, showing installation on dash of car and connected to cooling system of an engine equipped with a water circulating pump. In this installation the water enters the top of heater and leaves at the bottom. Upper water connection must always be between engine and thermostat, where the latter is used.

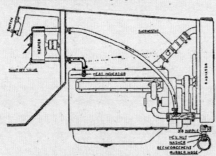

Fig. 1. Example of a hot-water heater, the Tropic Aire Senior model shown mounted on the dash of car with connections to the hot-water circulation system of an engine equipped with a water circulating pump. In this installation the water enters the bottom of heater and leaves at the top.

Where engines are not equipped with water pumps and use thermo-syphon circulation, the heater is placed as low as foot room will allow and the inlet is connected from upper hose to top of heater and outlet to lower hose and at bottom of heater.

Thermostat: Most engines run between 160° and 180° Fahrenheit, on the average. In the summer they may get as high as 180° and may get as low as 160°. In the winter the water temperature may get considerably below 160°. In such cases, to decrease the time required to warm up engine, many cars are equipped with thermostats, but they may not be suitable for a hot-water heating system.

There are three types of thermostats: standard, high reading, and adjustable. With the standard type, the temperature range of water is between about 148° to 155° F., which is sufficiently high for good heater performance but not high enough to cause undue evaporation of alcohol anti-freeze solutions. (Alcohol anti-freeze solutions can be run at a maximum temperature of 160° F. without undue loss.) The temperature range of water on the high-reading type is between 160° and 175° F.; and with the adjustable type, the temperature may be varied anywhere between 148° and 180°.

On cars in which a water-line thermostat is installed, the upper hose connection to the heater should be made below the thermostat.

When a special thermostat is installed, remove the equipment thermostat if there is one.

On all "V" type engines, two thermostats should be installed.

Air vents are provided at top of heater tanks. When filling radiator with a considerable quantity of water (as when the car has been drained), the vent should be opened to permit the trapped air to escape until water flows out of it. If the entire car is drained to prevent freezing, be sure to open air vent so that the heater will drain.

### Steam Heating[1]

With this system heat is utilized from the exhaust to generate steam. Temperatures in the exhaust pipes range from 350° to over 1,400° F. The parts of this system are shown in Fig. 3.

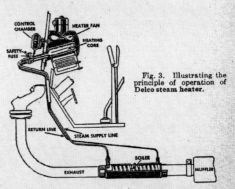

Fig. 3. Illustrating the principle of operation of Delco steam heater.

Operation: The operation of the system is quite similar to that of the steam house-heating system. One ounce of water is placed in the system. The system is filled by disconnecting the return line at the radiator. This water is used over and over again. Gravity carries this water through the return line to the boiler located around the exhaust pipe of the car. Immediately after the engine is started, the boiler is surrounded with exhaust heat having an average temperature range of 900° F. This intense heat quickly converts the 1 ounce of water (or ice) into steam. Steam rises through the steam supply line to the heating core, from which heat is radiated and distributed in the car from the heater fan. After the steam is condensed into water in the radiator, it flows down the return line to the boiler.

Steam pressure is automatically maintained at about 100 pounds per square inch by the patented control chamber, which governs the amount of water in circulation. For this reason, all threaded connections must be very tight at all times. The purpose of the safety fuse is to protect the system from abnormally high pressure, which might be caused by having too much water in the system or by clogged lines. If the pressure exceeds about 650 pounds per square inch, a thin metal disc on the fuse will be ruptured, permitting all of the steam to escape.

---

[1] Examples are 1935 models. See also pages 649-54 for other accessories.

Dykes Encyclopedia 1940

This article describes the Delco steam heater.

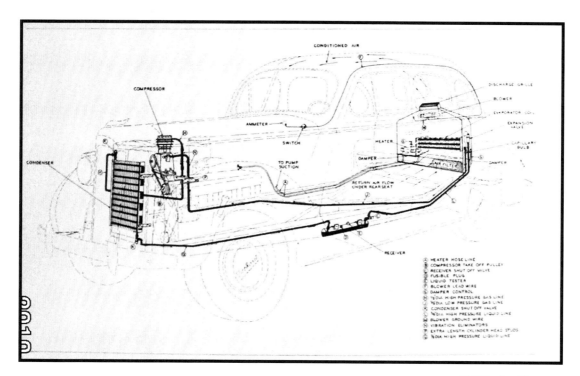

1940 Packard image

The first air-conditioned production vehicle. The retail price was reported to be $275.

Image taken Circa 1993

A 1940 Packard fitted with factory air conditioner.

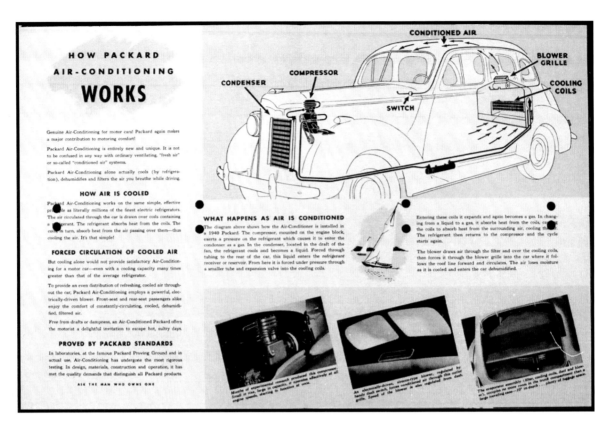

1940 Packard brochure

The 1940 Packard air conditioner system.

Both Images Circa 1993

1940 Packard air conditioner control and compressor.
The compressor had no clutch. The drive belt was removed for winter driving.

NEW YORK JOURNAL AND AMERICAN • • • • WEDNESDAY, JULY 24, 1940

# New Air-Cooled Packard Foils Humidity on Tour

**New Device Like Home Refrigerator**

By R. M. VANDIVERT.

"Tempering the wind to the shorn lamb," often referred to as one of the great phenomena provided by an all-wise providence, has its parallel in motor car luxury made available in current Packard enclosed sedan models.

Packard engineers, profiting by the advances in house refrigeration and home air conditioning, have applied these developments to motor car comfort and created weather conditioning to meet all seasons and all outside temperatures.

Recently the writer took a four-day trip through New York State,

The seven-passenger Packard super-eight, equipped with air-cooling equipment, which was used on a recent test trip through New York and New England.

This simple switch, incorporated in the instrument panel, governs the speed of the blower which controls the volume of air. This is regulated by a dial rheostat.

The compressor, small in size but large in capacity, operates effectively at all engine speeds.

New England and New Jersey, during which the variable weather provided an opportunity to give the "weather-conditioner" a thorough try-out. Not only did we ride in air-cooled comfort during two humid unpleasantly warm days, but we also encountered a period of cold, clammy rain when slightly warmed, dehumified air added greatly to the creature comfort of "down east" touring.

**SIMPLE SYSTEM.**

The transition from air-cooled comfort to equal comfort with warmed air is simplicity itself. The same air circulatory system provides car cooling or car heating, as one may desire, by a simple adjustment of damper controls located in the trunk compartment of the car, an operation requiring no particular skill and but a few seconds of time.

For hot weather driving the car is cooled by mechanical refrigeration. The cooling principal is the same as that used in household refrigerators. It is simply a matter of circulating air over coils which contain a refrigerant. The refrigerant absorbs heat from the coils and a coils absorb heat from the air, thus cooling the air, which is dehumidified as it is cooled.

The capacity of the Packard cooling system at high touring speeds is rated as equivalent to the cooling effect of placing one and one half to two tons of ice in the car to be melted in twenty-four hours.

The evaporator assembly, which includes filter, cooling coils, heater duct and blower, is located at rear of the trunk compartment leaving plenty of room for ordinary travel luggage.

**WORKS WITH ENGINE.**

The cooling system functions emmediately when the car engine is started. It is regulated by a simple switch attached to the instrument panel within easy reach of the driver, which controls the amount of cooled air desired. With the windows closed, the air effectively filtered to remove dust or pollen as it passes through the air-conditioner, and regulated by the speed of the blower, passes through a duct located behind the rear seat, circulates along the roof to the front of the car and returns to the rear along the car floor. This flow of air effectually results in the permeation of conditioned air throughout the entire car.

To heat the car, the dampers are changed and air heated for circulation by passing through a score in which hot water is circulated.

**COMFORTABLE RIDE.**

The demonstrator we drove was a Packard Super-eight seven passenger sedan, big, roomy, luxurious but at the same time easy to handle at touring speeds on the highway and in traffic. During the trip we drove ultra-modern highways and back-country roads and all were negotiated with a maximum of comfort. Even in steep mountainous country in the Berkshires, the Catskills, the Ramapo and other ranges we crossed, the car took everything in high without effort.

In spite of the fact that the cooling unit was kept in almost constant use in much of this mountain travel, there was no perceptible loss of motive power in handling the car nor was there any noticeable extra consumption of gasoline. Considering the size of the car, the roads traveled, and the touring speeds maintained, the consumption of fuel was remarkably low during the entire trip.

S-P New York

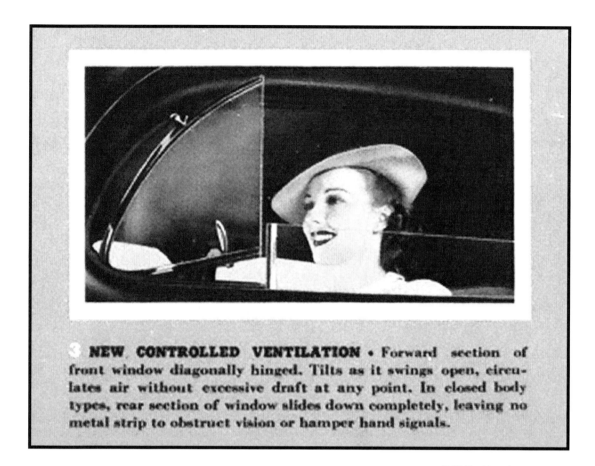

Ford added ventipanes to its vehicles in 1940

1940 Ford brochure

A 1940 Cadillac fitted with a sunshade over the windshield

Richard Kughn Collection

Stahls Collection

A 1940 Ford fitted with a hot air heater.

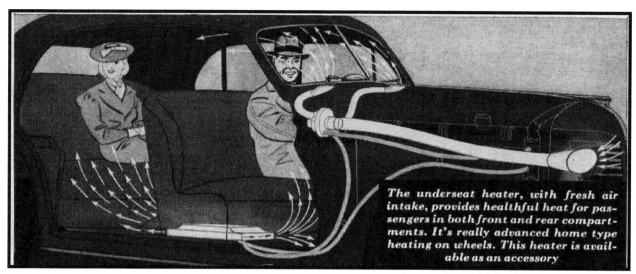

*The underseat heater, with fresh air intake, provides healthful heat for passengers in both front and rear compartments. It's really advanced home type heating on wheels. This heater is available as an accessory*

1940 Buick brochure

### Car Heaters and Defrosters

In designing heaters for the 1940 cars, every conceivable type of weather condition has been considered, and the best possible location in every body model has been studied for maximum benefit to all passengers. The result is the underseat heater—the most important heater development for 1940.

The primary advantage in this heater is that the rear compartment floor, which has been, with the dash type heaters, practically void of heat, can now be heated to the exact degree desired by the rear compartment passengers. The other important advantage is—the air expelled from the heater, decidedly warms the lower portion of the car first and subsequently raises to the higher levels, resulting in heating the passengers' feet first and the breathing air last as scientifically it should.

With previous types of heaters the hot air has been expelled mostly in the face of the passengers, thus causing the breathing air to be much warmer than the air contacting other parts of the body. In the cases of night driving and like circumstances this tended to a drowsiness on the part of drivers which is undoubtedly a safety hazard. Results of tests with the dash type heater have indicated that in warming the rear compartment to any degree of comfort, the front compartment is made unbearably hot. With the underseat heater a constant, uniform temperature can be obtained throughout all parts of the car interior.

With this type of heater there is available, a dash mounted defroster, consisting of a fan and core, to adequately defrost, de-ice and de-humidify the windshield.

The difference in this type of heating and the previous dash type might be likened to the difference between heating a room with an open fireplace and scientifically heating a room with a modern balanced heat distribution system.

For smaller cars, coupes, etc., the dash mounted heater is available. The heater construction is similar to previous types with defroster mounted on top, and will adequately serve this type of car.

For cars in extremely cold territories a combination of the underseat heater and dash mounted heater may be used. This entails the installation of the underseat heater combined with a dash heater and defroster unit. A car so equipped will be comfortable in the most extreme sub-zero weather.

### Fresh Air Inlet

A fresh air inlet is provided as optional equipment on the underseat heater, defroster equipped, cars. This inlet consists of a flexible hose which takes the fresh air in at the radiator grille and expels it through the defroster unit on the dash. The amount of fresh air taken in is determined by the position of a valve controlled by a lever on the instrument panel. The air is expelled onto the windshield and into the car either by the pressure of the air-stream, or, if this is not sufficient, by operating the defroster. This system may be operated in winter or summer and the temperature or amount of fresh air controlled by the valve. In conjunction with the under-seat heater it provides a means of balancing the temperature in the car by means of fresh circulated air, or, in hot weather, of introducing fresh air into the car.

1940 Buick engineering document

A hot water heater mounted underseat with a fresh air inlet and blower.

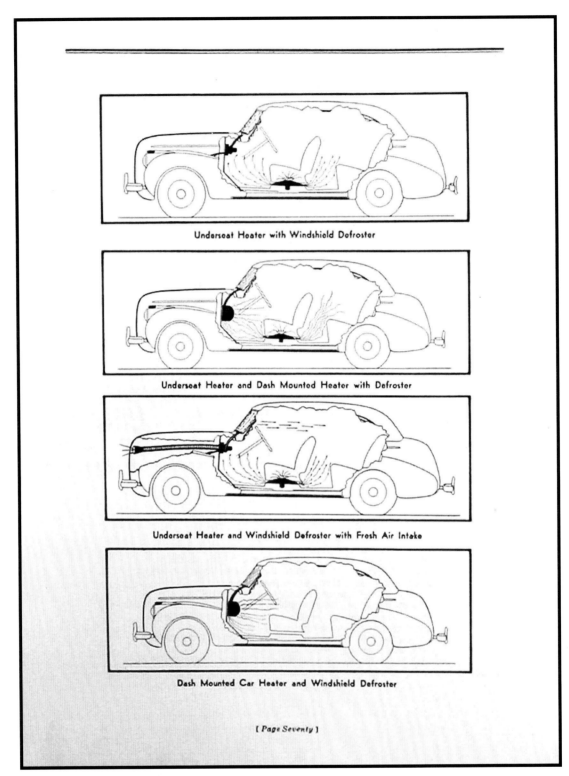

1940 Buick engineering document

Four alternative heater configurations were offered on this 1940 Buick.

Circa 1940 brochure

Stewart Warner offered this South Wind brand gasoline heater.

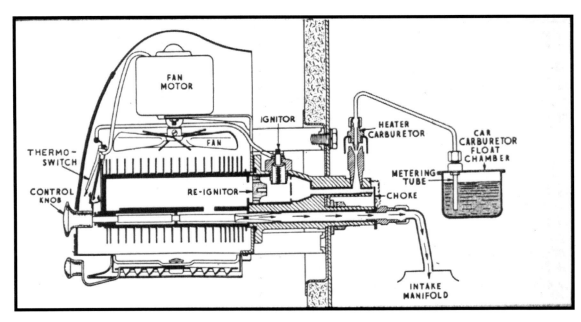

Circa 1940 Stewart Warner South Wind gasoline powered heater.

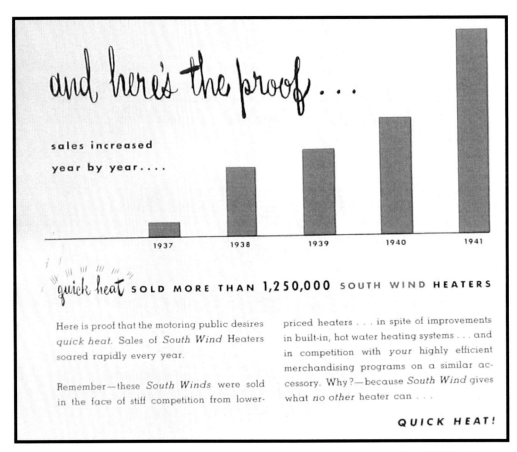

Circa 1941 brochure

This heater was a best seller for many years.

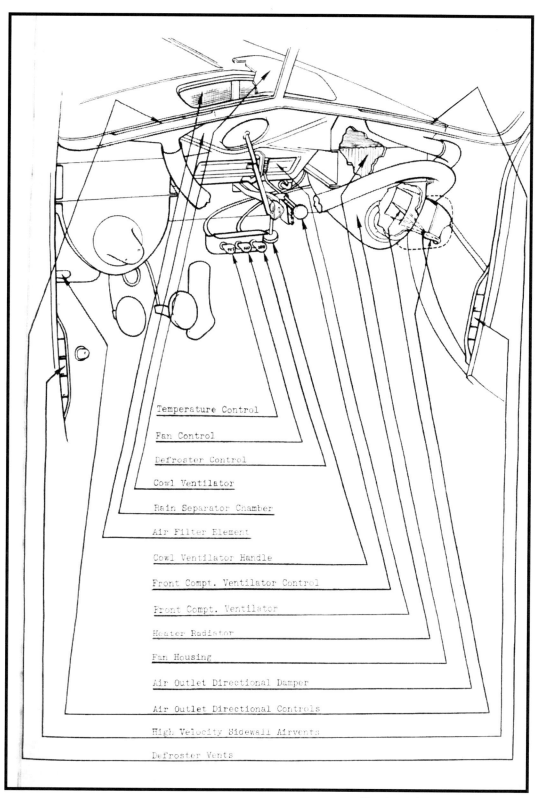

Chrysler engineering document

The 1940 Chrysler Corporation all new heater and ventilation system.

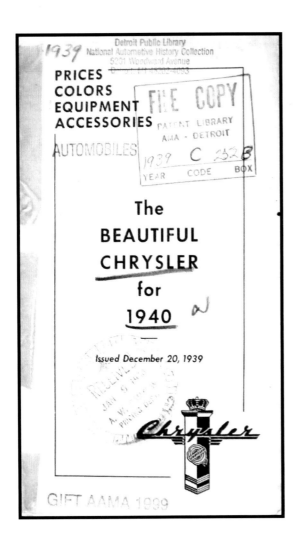

1940 Chrysler brochure

Note the Kool Kooshion listed along with several heater alternatives and attachments.

1941 Packard brochure

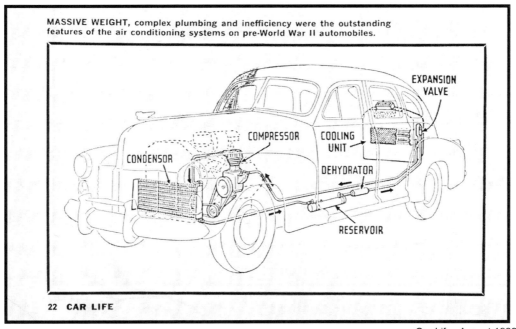

Car Life August 1966

The 1941 Cadillac air conditioner system.

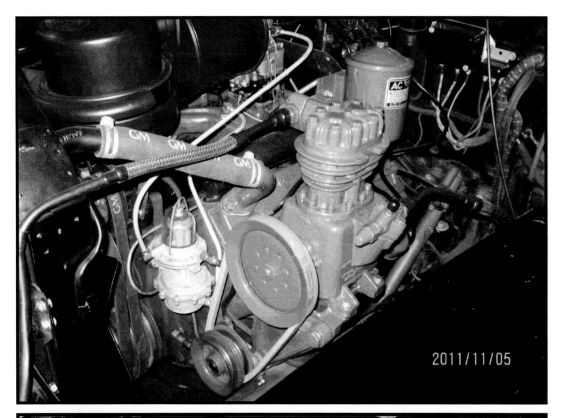

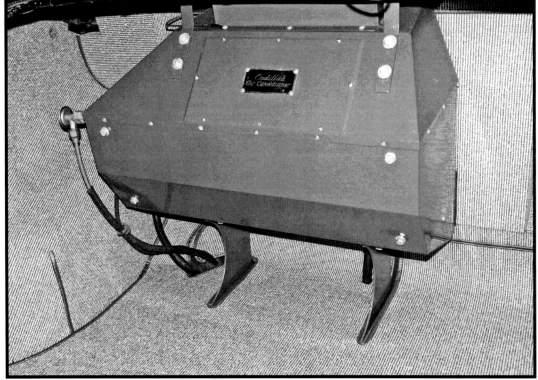

Lower image courtesy Richard Kughn Collection

The 1941 Cadillac air conditioner compressor and evaporator.

Both images courtesy Richard Kughn Collection

A 1941 Cadillac fitted with factory air conditioner. The compressor is shown above.

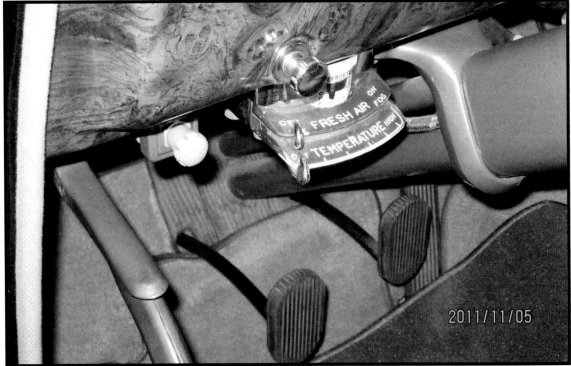

Richard Kughn Collection

A1941 Cadillac air conditioner rear package seat outlet register and instrument panel control

## AUTOMATIC HEATING SYSTEM

For 1941 Cadillac is proud to announce the most outstanding heating system ever developed for automobiles, an exclusive development of Cadillac engineers, and available only on 1941 Cadillacs.

For the first time it is possible to set the heat control in your car at the beginning of winter for the temperature you desire, just as you set the thermostat in your home. From then on the Cadillac Automatic Heating System will heat both front and rear compartments equally, provide fresh air ventilation, and defrost the windshield in normal winter weather, all automatically, without having to touch a single control.

This new system is divided into three basic functions, automatic temperature controlled heating, ventilating and defrosting.

Automatic temperature controlled heating is accomplished by two heaters underneath the front seat, thermostatically controlled both as to water flow and fan speed. The Automatic Heating System is connected into the ignition lock so that when the ignition key is turned on the heater is automatically in operation. Similarly when the ignition is off the heater also is off. When the car is started, and the water in the cooling system is cold, both fans are completely shut off preventing any blast of cold air on the passengers.

After the water in the cooling system is heated both fans are automatically turned on at high speed. They continue at high speed until the temperature reaches the point set on the temperature control, when the fans go to slow speed and water circulation in the heater is restricted.

If the temperature continues to go higher than that set by the control, the water flow is further restricted. If the temperature drops, however, water circulation is increased and the fans are returned to

14

1941 Cadillac accessory brochure

The 1941 Cadillac offered an automatic temperature control heater system.

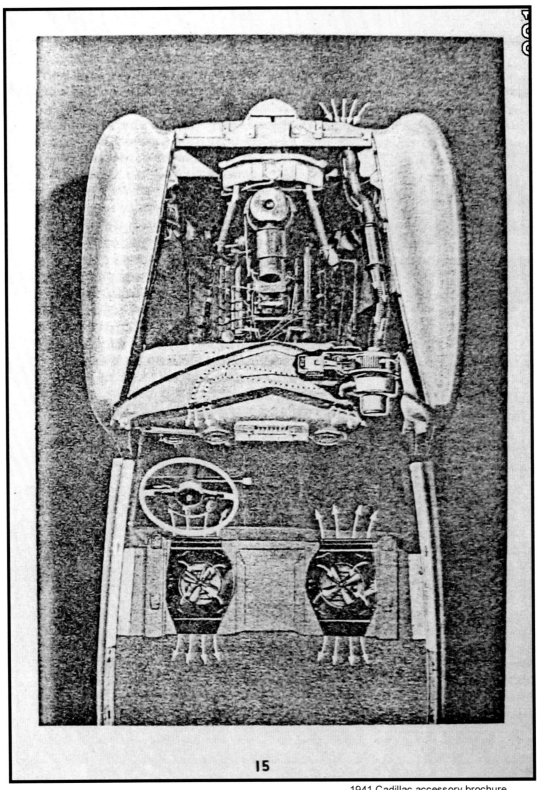

1941 Cadillac accessory brochure

The 1941 Cadillac offered an automatic temperature control heater system. Note the dual underseat heaters.

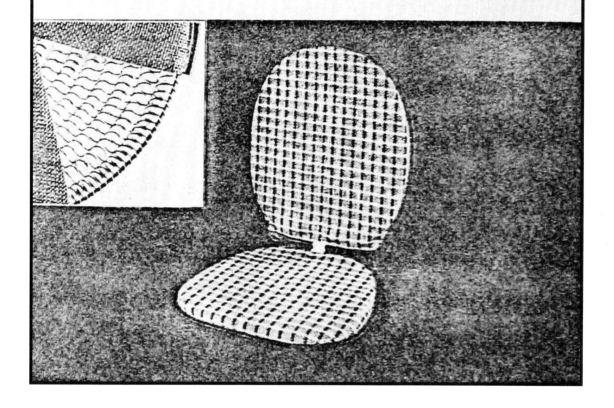

1941 Cadillac accessory brochure

The cool cushion was constructed of coil springs and a woven rice paper cover.

Images circa 1993

A 1941 Buick fitted with a window mounted evaporative cooler.

# MISCELLANEOUS

## CLIMATIZER

The Climatizer is a built-in, fresh air heating and ventilating system which draws air from an adjustable ventilator opening on the left side of the cowl, passes it through built-in air ducts to the filter where dust and dirt are removed, heats it if desired in a core below the front seat, and then (by a motor-driven fan) spreads the fresh air uniformly throughout the car interior. The defroster unit is a separate fan blower and heater coil, mounted on the dash behind the instrument board. It forces heated air through two flexible tubes and built-in air ducts to remove condensation and frost from the windshields.

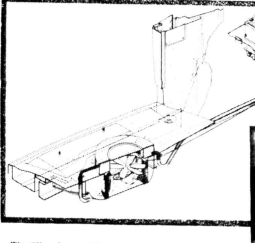

The Climatizer and Defroster controls are located on a panel just below the center of the instrument board. The left hand switch, with identification FRESH AIR stamped on the panel, regulates the Climatizer motor which controls the amount of fresh air delivered to the car interior through the Climatizer.

The center control, marked HEAT on the panel, regulates the quantity of hot water which circulates through the Climatizer heater core. When the control is pushed in, the water is shut off, and the flow is increased as the control button is pulled outward.

The right hand control, marked DEFROSTER on the panel, controls the defroster motor. This control operates in the same manner as the fresh air control.

The small conveniently located handle on the left side of the car controls the air intake to the Climatizer. Turn the handle clockwise to open the air intake. With the handle down, the intake is half open. When the handle is upward, the intake is either closed or wide open. The AIR INTAKE must be opened in order to obtain heat.

The only service attention that will ordinarily be required for the system is that of cleaning the filter element. Take off the bottom plate which is attached to the heater housing by screws. The filter is held in position on the heater core and motor support by two coil springs which hook into the shell of the filter element. Remove the filter element and wash it in clean gasoline. Then blow it dry with air and saturate it with clean engine oil, permitting the excess to drain. Place the filter element in position and hook up the two coil springs. Replace the lower cover.

The frequency with which the filter needs servicing is governed entirely by operating and atmospheric conditions. If the car is operated on paved roads, it should not be necessary to clean the filter more than three or four times a year. However, if the car is operated under dusty conditions, especially when the system is used both for heating in winter and ventilation in summer, the filter will need more frequent cleaning.

To drain the cooling system completely as for car storage in freezing weather, drain the Climatizer also. To do this, remove the hose from the cylinder head fitting and hold the open end of the hose below the heater level until the heater has drained.

202

1941 Studebaker shop manual

The Studebaker Climatizer system was mounted under the driver's seat.

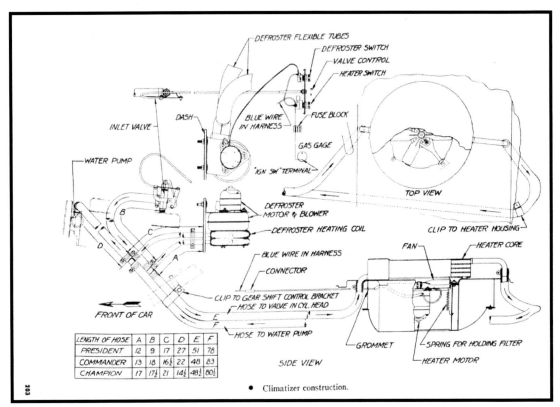

1941 Studebaker shop manual

An engineering drawing of the Studebaker Climatizer system.

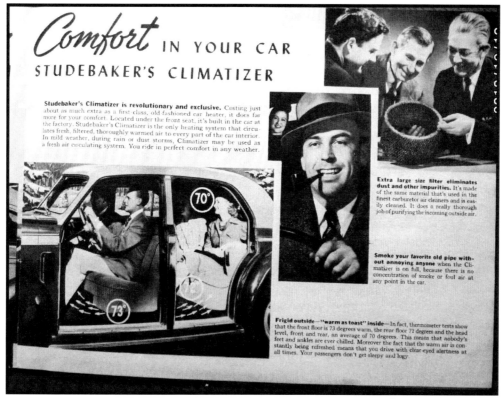

1941 Studebaker brochure

The Studebaker Climatizer system featured an air filter.

Motor November 1941

E.A. Laboratories, Inc. offered two models of their hot water heaters.

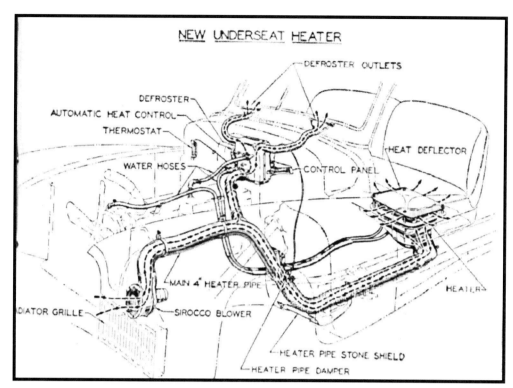

Motor Service October 1941

Pontiac has developed a heating and ventilating system available at extra cost that meets *every* comfort requirement. Just to cite a few of its advantages: it delivers four times more heat than last year's heater, it maintains car temperature at the desired level, it de-ices as well as defrosts the windshield, it eliminates steaming and fogging of windows, and can even be used to cool off the car in hot weather!

The fundamental feature of this sensational new unit is a big blower and intake which forces the air into the car at a rate that is faster than it can leak out. Consequently, warm air floods the interior of the car giving almost perfect heat distribution without drafts or air currents.

An important health factor is that only fresh outside air is used and the air in the car is completely changed every two or three minutes depending on the speed of the car. Used air is *not* recirculated as with ordinary heaters.

Visibility and safety is greatly improved because all steaming and clouding of windows is eliminated. Humidity is kept below the dew point by the heated fresh air except under extraordinary conditions.

The heater delivers four times more heat than last year's unit and therefore warms up the car much quicker after the engine is warm. A thermostatic control maintains the flow of heat at the desired temperature.

A defroster unit is an integral part of this new heater. When control knob is set at "Normal," cool outside air is brought in at breathing level. This keeps the windshield cold to prevent snow from melting and freezing on the windshield.

Any frost or ice that may form on the windshield while the car is not in use can be melted by turning knob to "De-Ice" position. This delivers enough warm air to the windshield to take care of average defrosting conditions. In extreme cases, turn off Heater Air completely and you send all the air through the defroster to melt frost and ice in a hurry.

Although this only partially describes the functions of Pontiac's New Underseat Heater and Ventilator, you can see that it offers many advantages over former heater-defroster units.

1942 Pontiac brochure

This 1942 Pontiac Venti Heat hot water heater had a fresh air inlet and a defroster system.

1942 Pontiac brochure both images

Note the fuel pump booster for the windshield wipers, windshield washers, rear window wiper and venetian blind sunshade.

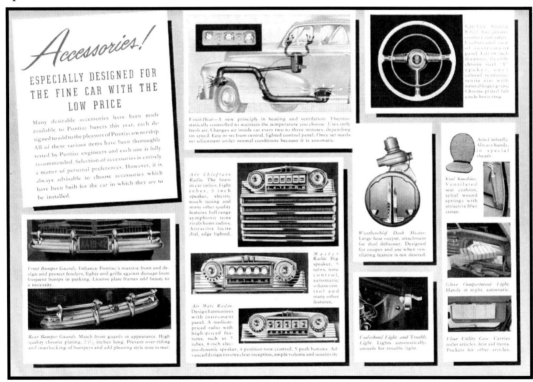

Note the Kool Kooshion- the forerunner of ventilated and cooled seats. Also note the low cost Weatherchief dash heater for coupes and sedans without the fresh air heater.

### OLDSMOBILE Condition-air

## Heating and Ventilating System
## Pressure Circulates Heated Filtered Air

The last word in motor car heating and ventilation. Fresh, outside air is forced into the Condition-Air unit, where it is filtered, heated to desired temperature, and uniformly distributed throughout the car. At slow speeds, a large auxiliary fan maintains fresh air circulation. Temperature is thermostatically controlled and is maintained by varying the water flow through the heater core, rather than by fan speed. Powerful defrosting fan keeps windshield free from frost and mist.

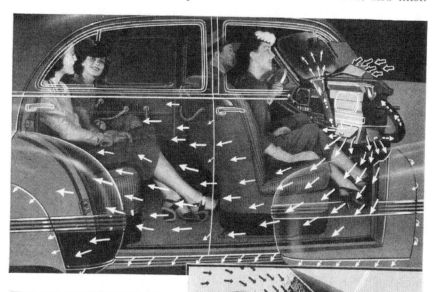

**OTHER QUALITY HEATERS**
*De Luxe Dual Flow Heater* provides exceptional heating capacity and defroster performance. A variable speed, reversible motor permits either direct or indirect heating. *Underseat Heater and Defroster* operates efficiently to give both front and rear seat passengers abundant warmth. Installed under front seat. Defroster is separate installation.

1942 Oldsmobile brochure

Condition-Air was the name used for this Oldsmobile heater and ventilation system.

1946 Studebaker brochure

The Studebaker air inlet for the Climatizer hot water heater was thru the side vent.

1946 Studebaker brochure

The Studebaker Climatizer hot water heater system included the instrument panel control, an accurately balanced blower, a cleanable air filter and a sturdy copper brass heater core. The separate defroster sub system consisted of a blower, a hot water heat exchanger and built in defroster ducts.

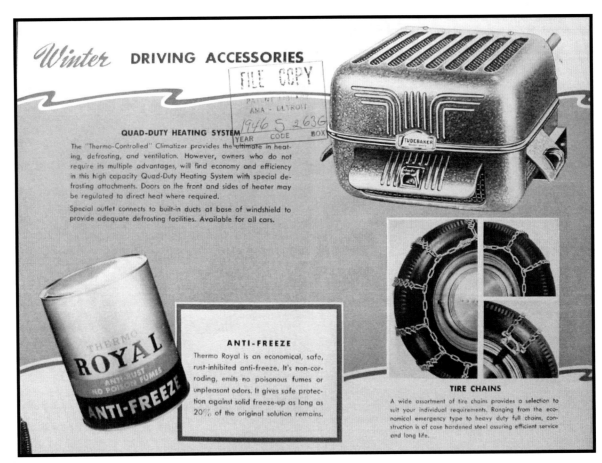

1946 Studebaker brochure

A low cost recalculating hot water heater called the Quad Duty Heating System was offered for vehicles not equipped with the deluxe Climatizer fresh air system. Defrosters were an extra option with the Quad Duty heater.
Note the Studebaker antifreeze labeled Thermo Royal.
Tire chains were also offered in a wide assortment as genuine accessories. Tire chains were used to drive in deep snow and icy roads.

Circa 1946 brochure

Motorola offered two models of its gasoline-powered heater. One for underseat mounting the other for cowl mounting.

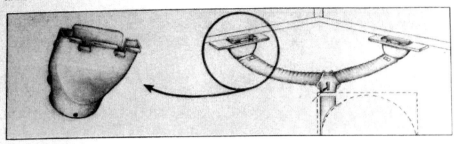

## TROPIC-AIRE

## Universal Windshield Defroster Attachments

*SNAP IN TYPE FOR USE WITH THE
TROPIC-AIRE 1947-1948 HEATERS*

Especially designed to fit the new larger defroster outlet connection on all 1947-48 Tropic-Aire dash model heaters. The snap-in metal nozzles, large size reinforced flexible tubing and new "Y" outlet attachment (with butterfly shut-off) permit an unrestricted flow of hot air from the heater to the windshield for rapid defrosting.

This Universal defroster attachment will fit all cars with the exception of the Chrysler, DeSoto and Dodge automobiles, years 1941-1947. These cars are equipped with built in ducts and require only a piece of flexible tubing and a defroster outlet attachment. Now, only two models are required for stock to fit all models of passenger cars, a tremendous saving in investment to you.

EFFECTIVE NOVEMBER 1, 1947

| To fit openings below the windshield on the following: | MODEL | LIST PRICE |
|---|---|---|
| For all cars, except* | DU5000 | 5.00 |
| *Chrysler<br>*De Soto  1947 to 1941<br>*Dodge | D2842 | 2.00 |

## TROPIC-AIRE, INCORPORATED

4501 WEST AUGUSTA BOULEVARD          CHICAGO 51, ILLINOIS

TA 699

November 1947 brochure

Tropic-Aire, Inc. listed two models of windshield defroster attachments. Note that the 1941-1947 Chrysler vehicles were factory-fitted with defroster ducts.

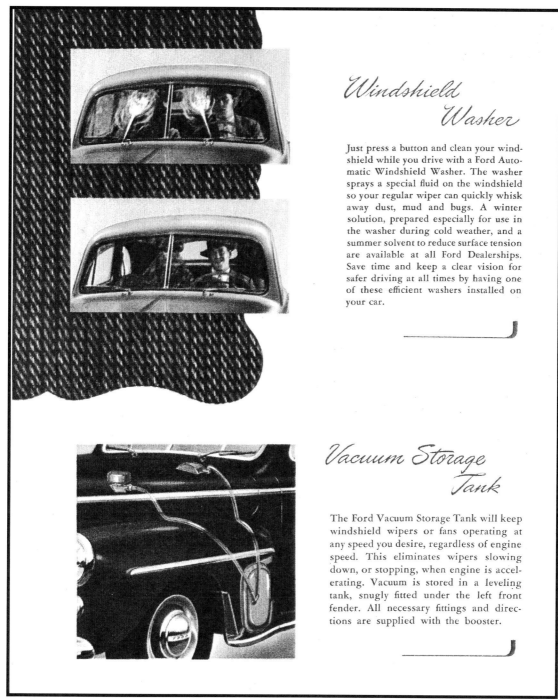

1947 Ford accessory brochure

Ford offered windshield washers and a vacuum storage tank. The tank was used to "keep windshield wipers or fans operating at any speed you desire, regardless of engine speed. This eliminates wipers slowing down, or stopping, when engine is accelerating".

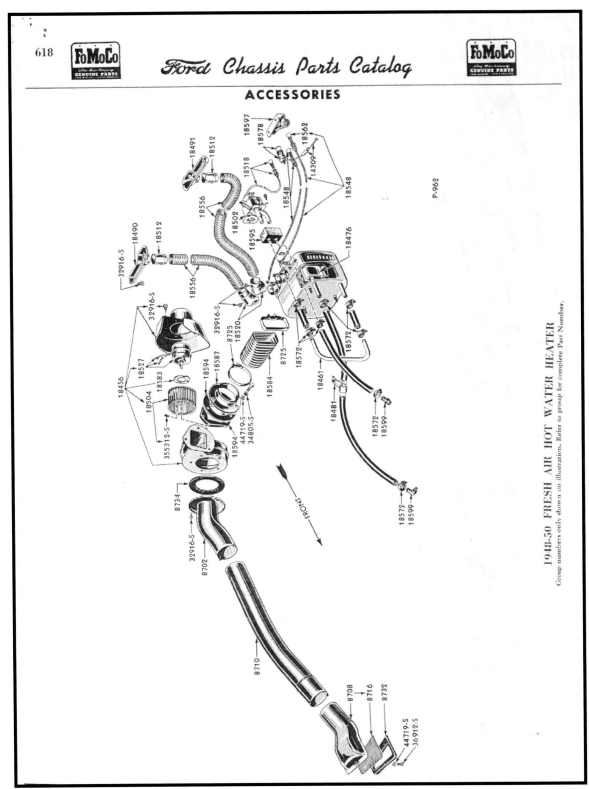

Ford Truck Parts Manual 1948-1950

Ford introduced its first fresh air hot water heater with thermostatically controlled water valve on the 1948 Ford pickup.

# South Wind
## MODEL 977 DELUXE HEATER INSTALLATION SUGGESTIONS

This installation sheet (Packing Material No. PM-5016) has been prepared to assist the service man in his installation problems relative to the 977-A Heater as well as to approach standardization on as many installations as possible.

The front side of this sheet is divided into sections which are designated by the circled letters A, B, C, etc. These sections contain details of passenger car and truck installations. On the back side of the sheet are photographs showing the heater position in eight 1946 passenger cars and trucks.

Section "A" consists of an installation chart containing duct measurements, fuel line sizes and locations. Section "B" contains precautionary notes and instructions on testing the installation. Sections "C" and "G" provide pictures and lists of the parts contained in an installation. Sections "D," "E," and "F" illustrate the types of mounting. Section "H" shows how to make the two types of fuel fitting installations. Section "I" describes the electrical connections. On the back side of the sheet under each photograph is a summary of installation instructions for that specific vehicle.

## INSTALLATION CHART

| AUTOMOBILES | Intake Duct Length* | Outlet Duct Length* | Exhaust Length* | Auto Fuel Line Size | Location of Heater | Location of Hole in Fire Wall | Location of Cold Air Intake |
|---|---|---|---|---|---|---|---|
| Plymouth | 30" | 15" | 12" | 5/16" | R.H. | R.H. | Inside Vehicle Grille |
| Dodge | 30" | 15" | 12" | 5/16" | R.H. | R.H. | Inside Vehicle Grille |
| DeSoto | 30" | 15" | 12" | 5/16" | R.H. | R.H. | Inside Vehicle Grille |
| Chrysler | 30" | 15" | 12" | 5/16" | R.H. | R.H. | Inside Vehicle Grille |
| Pontiac | 36" | 30" | 8" | 5/16" | R.H. | R.H. | Inside Vehicle Grille |
| Oldsmobile | 30" | 30" | 12" | 5/16" | R.H. | R.H. | Inside Vehicle Grille |
| Chevrolet | 36" | 36" | 24" | 5/16" | L.H. | L.H. | Inside Vehicle Grille |
| Studebaker (Champion) | | | | 1/4" | R.H. | R.H. | |
| Ford | | | | 1/4" | | Center | Inside Vehicle Grille |
| **TRUCKS** | | | | | | | |
| Dodge | 40" | 24" | 24" | 5/16" | Tie Rod | R.H. | Inside Vehicle Grille |
| G.M.C. | 40" | 6" | 24" | 5/16" | Tie Rod | R.H. | |
| International | 36" | 12" | 12" | 5/16" | R.H. | R.H. | |
| Chevrolet | 36" | 36" | 24" | 5/16" | Tie Rod | R.H. | Inside Vehicle Grille |
| Federal | 36" | 6" | 24" | 5/16" | Tie Rod | Center | |

*Approximate Lengths.

1948 South Wind gasoline heater brochure

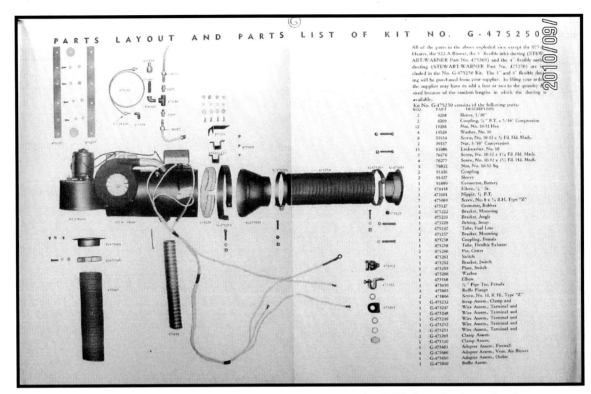

1948 South Wind gasoline heater brochure

## FUEL FITTING INSTALLATION

Figs. 1 and 2 illustrate the Horizontal and Vertical type of installation.

Install the fuel fittings in the fuel line downstream of the filter and on the discharge side of the fuel pump. To prevent vapor locking of the heater, insert the "T" connection close to the carburetor and direct the leg *down* so that any vapor in the fuel will rise into the carburetor line. An elbow is used in the "Vertical Fuel Line" to direct the "T" down. A vibration loop must be placed in the 1/8" heater fuel line.

To install the fuel fittings, select and mark the best location on the car fuel line. Disconnect and remove the fuel line. Using a tube cutter or hacksaw, cut out a 2⅛" section at the selected location for the insertion of the fuel fittings.

The factory furnishes fittings for 5/16" fuel lines. Ford and Studebaker vehicles have ¼" fuel lines. For these two installations, the installer will have to furnish the following parts: (2) ⅛" pipe x ¼" Compression coupling, (2) ¼" Compression sleeves, (2) ¼" Compression nut.

## WIRING AND COLOR CODE

This diagram shows only the wires installed by the service man. The wires are color coded to simplify installation. In the diagram the dotted lines represent No. 18 wire or larger and the solid lines No. 14 wire or larger. The circled letters are the key to the color code chart below.

Before the 4" outlet duct is attached to the firewall adapter, connect the quick disconnects on the wiring harness to the quick disconnects on the short lengths of color coded wires attached to the relay. The free end of the harness *is* threaded through the firewall adapter grommet, and enters the compartment through the firewall adapter. Then thread the harness through the grommet on the grille. Complete the connections as shown in the diagram and chart.

### WIRE COLOR CODE

| PART | COLOR | LENGTH | ELECTRICAL CONNECTION |
|---|---|---|---|
| 91880 | Black | 33" | Dead side of Automobile Ignition Switch to Heater Control Switch |
| G-475247 | Black-Yellow Tracer | 48" | Thermostat to No. 6 Terminal on Relay |
| G-475248 | Black-Yellow Tracer | 24" | "Heat" Terminal of Heater Control Switch to Thermostat |
| G-475249 | Black | 60" | "Heat" Terminal of Heater Control Switch to No. 1 Terminal on Relay |
| G-475252 | Black-Blue Tracer | 60" | "Fan" Terminal of Heater Control Switch to No. 5 Terminal on Relay |
| G-475253 | Black-Red Tracer | 38" | Battery to No. 4 Terminal on Relay |

1948 South Wind gasoline heater brochure

1948 brochure

The Anderson Co. offered curved windshield wiper blades in 1948.

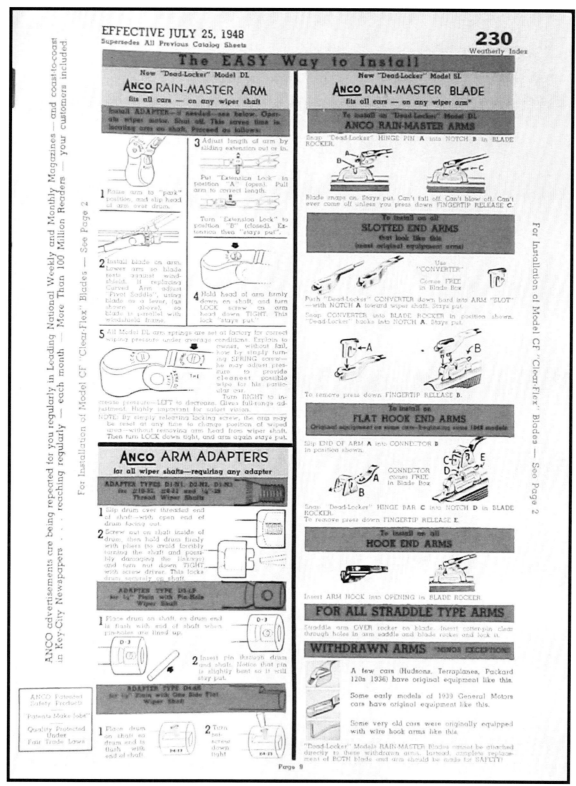

1948 brochure

This Anderson Co. brochure shows the numerous fittings used for windshield wipers in the era.

1948 brochure

American Bosch Corp. offered rear window wipers for retrofit to many models of 1942-1948 vehicles.

## Application List
### REAR WINDOW PASSENGER CAR INSTALLATIONS FOR AMERICAN BOSCH ELECTRIC WINDSHIELD WIPERS

REVISED MARCH, 1948

| MAKE OF CAR | YEAR | MODEL OF CAR | MODEL WINDSHIELD WIPER |
|---|---|---|---|
| Buick | 1946 | Mdls. 50 & 70 Super & Roadmaster | 2 JC |
| | 1947-48 | 2 door Model 46S | 8 JC |
| | 1947-48 | 2 door Models 56S, 76S; 4 door Model 41 | 2 JC |
| | 1947-48 | 4 door Models 51, 71 | 2 JC |
| Chevrolet | 1942-48 | 2-4 door Special Deluxe, Master Deluxe, Stylemaster & Fleetmaster Sedans | 8 JC |
| | 1942-48 | 2 door Fleetline Sedans | 10 HC |
| | 1942-48 | 4 door Fleetline Sedans | 9 JC |
| | 1948 | Stylemaster & Fleetmaster Club & Business Coupe | 26 IC |
| Chrysler | 1942-48 | 2-4 door Sedans | 8 NC |
| De Soto | 1942-48 | 2-4 door Sedans | 8 NC |
| Dodge | 1946-48 | 2-4 door Sedans | 8 HC |
| Ford | 1942-48 | 2-4 door Sedans | 8 HC |
| Frazer | 1947-48 | 4 door Sedan & Manhattan | 9 JC |
| Hudson | 1946-48 | 2-4 door Super 6 & Super 8 Sedans | 8 HC |
| Kaiser | 1947-48 | Special 4 door Sedans | 9 JC |
| Lincoln | 1946-48 | 4 door Sedans | 8 HC |
| Mercury | 1942-48 | 2-4 door Sedans | 8 HC |
| Nash | 1947-48 | Mdls. 600 & Ambassador 4 door trunk Sedan & 2 door Brougham | 2 IC |
| | 1948 | Deluxe Business Coupe | 2 IC |
| Oldsmobile | 1946-47 | Mdls. 66 Club Sedan, 76-78 Club and 4 door Sedans | 10 HC |
| | 1946-47 | Mdls. 66 Club Coupe & 4 door Sedans | 10 HC |
| | 1948 | Model 66 2 door Sedan | 10 HC |
| | 1946-47 | Mdl. 98 4 door Sedan | 2 JC |
| | 1947 | Mdl. 98 2 door Sedan | 9 JC |
| Packard | 1946-47 | 2 door Deluxe Clipper | 9 NC |
| | 1946-47 | 4 door Deluxe & Super Clipper | 10 HC |
| | 1948 | 2-4 door Deluxe, Super & Custom Super | 8 HC |
| Plymouth | 1942-48 | 2-4 door Sedans | 8 IC |
| Pontiac | 1946-47 | 2-4 door Torpedo Sedan | 8 JC |
| | 1946-47 | 4 door Streamliner Sedan & Torpedo Sedan Coupe | 10 HC |
| | 1948 | 2 door Streamliner Sedan | 10 HC |

## *It's Easy to See..* REAR-VIEW SAFETY PAYS!

**SOLD BY**

1948 brochure

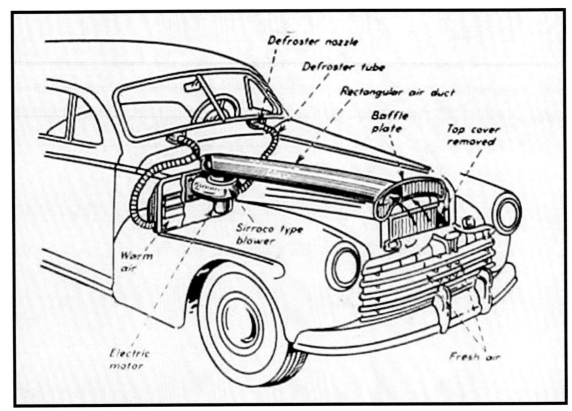

Image from paper by F.A. Ryder *Modern Developments in Vehicle Heating* 1949

1947 Ford fresh air heater

Vehicle Owner Larry & Maureen Wolohon

1948 Ford fitted with fresh air heater (4 images).

Image of new heater core and distribution duct courtesy of lilroesgrey on eBay

1948 Packard washer nozzle.

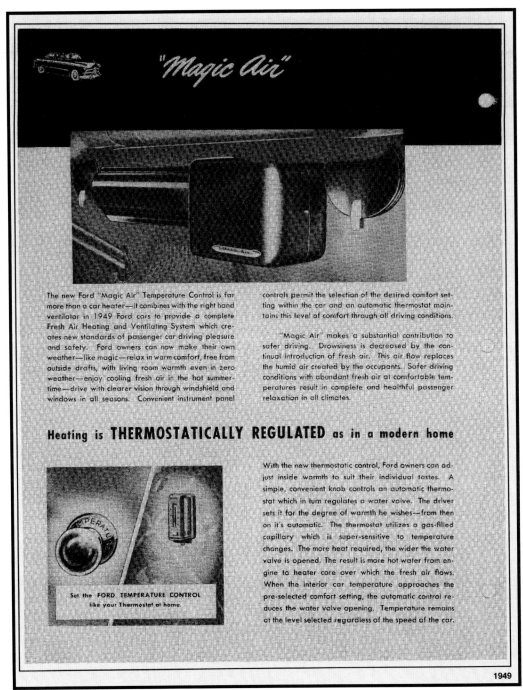

1949 Ford brochure

Ford offered a thermostatically controlled fresh air heater on its new 1949 model. The deluxe system was called Magic Air.

1949 Ford brochure

The instrument panel heater control was made up of four spherical chrome-plated knobs on the 1949 Fords.

1949 Ford brochure

This is a list of parts for the 1949 Ford Magic Air heater.

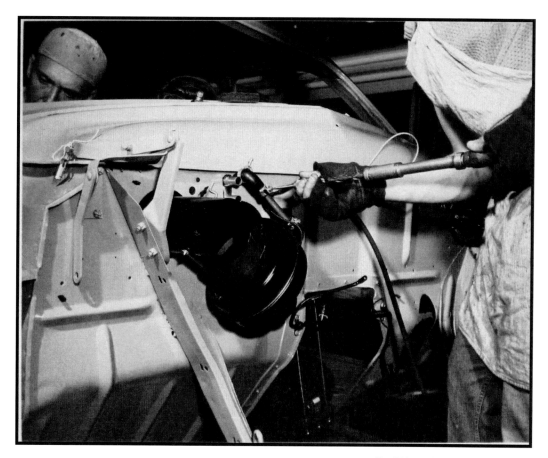

Ford Motor Company image

Note the heater control dangling thru the tunnel opening during installation of the heater hose clamp on a 1949 Ford.

1949 Ford brochure

Ford offered a hot water recirculation heater as a low cost alternative to Magic Air.

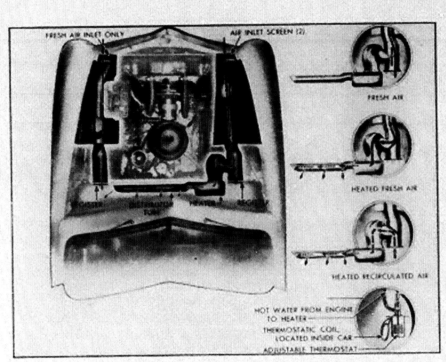

**Built-in Ventilating System and Ford Fresh Air Heater.**

# VENTILATION SYSTEM

Ford cars for 1949 are equipped with a new type of interior ventilating system.

Screened openings at the front of the car, between the radiator core and fenders, admit fresh air to two large diameter ducts leading to the driver's compartment. The outlet end of each of these ducts is provided with a hinged register. Each duct also has a control valve operated by means of a Bowden wire control on each side of the steering column. The control knob at the right functions to set its valve at any one of three positions, as shown in the illustration: (a) to admit fresh air directly into the body—(b) to direct fresh air through the heater fan—(c) to re-circulate heated air in the body.

The Ford fresh air heater, when installed as an integral part of the ventilation system, is equipped with an adjustable thermostat to regulate the flow of hot water to the heater.

Controls for turning the heater on and off, and also for adjusting the thermostat to provide the desired temperature, are located on the instrument panel, together with the defroster control.

The thermostatic heat control valve is separately mounted just above the heater, and may be removed as a unit for service or replacement.

1949 Ford brochure

The 1949 Ford ventilation system used two ducts in place of the previous cowl-top scoop.

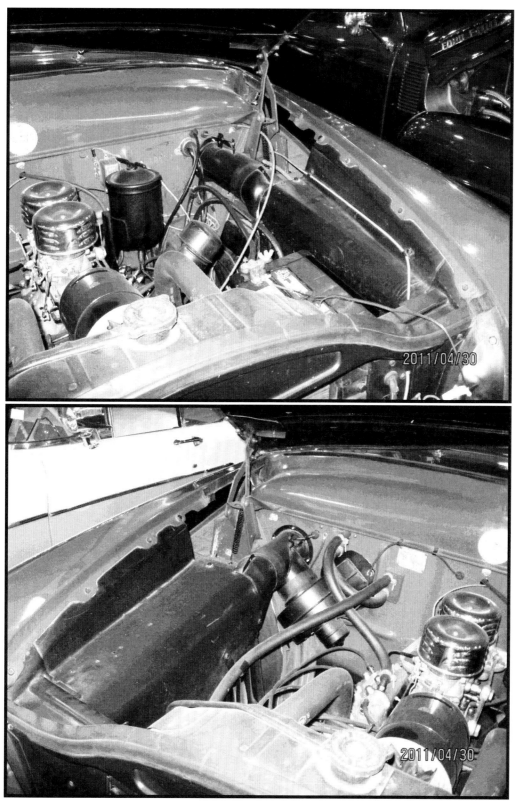

Ed Meurer Collection

The 1949 Ford ventilation and heater air inlet system.

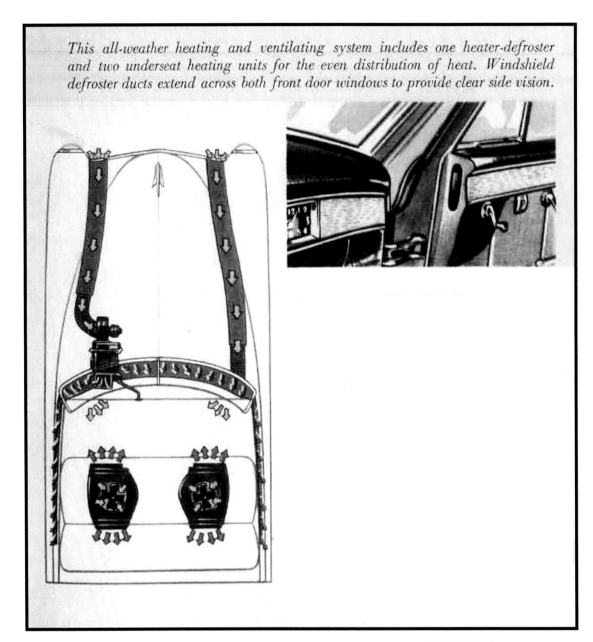

1949 Cadillac brochure

The 1949 Cadillac was fitted with two underseat heaters and side window de-misters.

### Luxury Lap Robe

*126*—For people who want the best in car appointments. Pure virgin wool. Soft, suede-like finish in solid colors with harmonizing checked linings. Reversible. Deep, roomy, set-in pockets. Tuck-in flap holds robe securely. Colors: Apple

*Lap robe—ideal for picnics, football games, or name the occasion*

| 127 | 1949 MERCURY ACCESSORIES |

Blossom Cream with brown and white check; maroon with blue and white check; and light green with brown and white check. A choice of smart monograms is available (at extra cost) to add a personal touch.

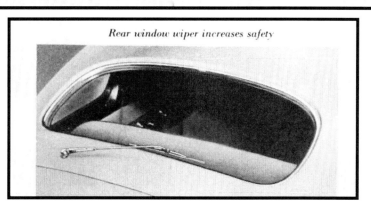

1949 Mercury Data Book 3 images

The 1949 Mercury offered monogrammed lap robes and a rear window wiper.

Image by Adrian Pingstone wikipedia

A 1950 MG fitted with dual wipers and a cross-link with idler.

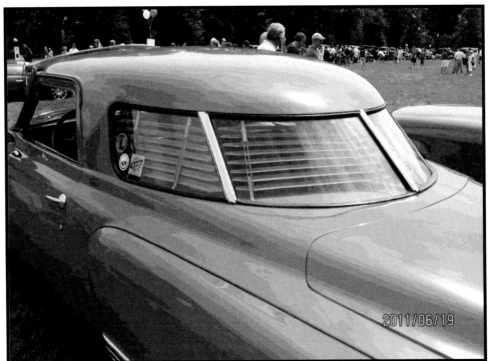

Vehicle Owners Harvey and Julie Snitzer

A 1950 Studebaker fitted with genuine Studebaker accessory window blinds.

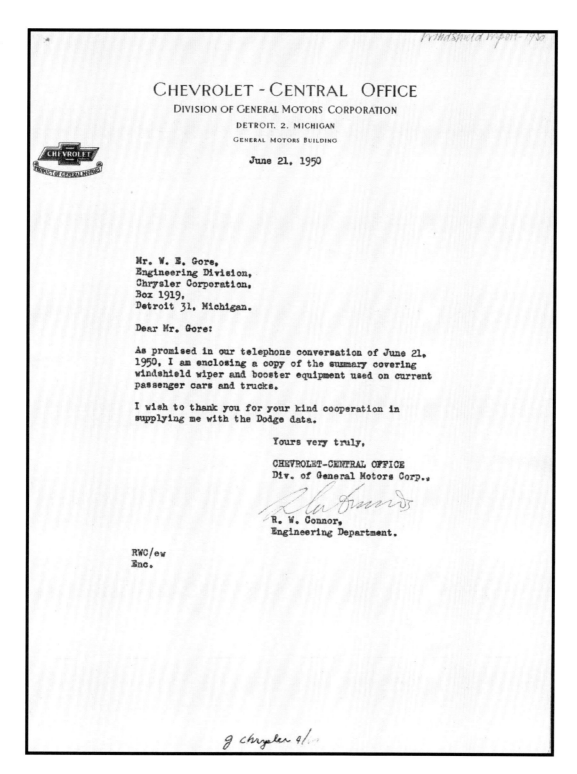

1950 Chevrolet engineering letter

Letter from R. W. Connor to W. E. Gore listing windshield wiper motors used in the 1950 era on many North American cars and trucks.

## WINDSHIELD WIPER AND BOOSTER EQUIPMENT SUMMARY

**PASSENGER CARS**

| MAKE | TYPE OF WIPER | BOOSTER EQUIPMENT (Vacuum pump unless otherwise noted) |
|---|---|---|
| Buick | Dual vacuum | Standard |
| Cadillac | Dual vacuum | Standard |
| Chevrolet | Dual vacuum | Extra cost - Factory installed |
| Chrysler | Dual electric | None required |
| De Soto | Dual electric | None required |
| Dodge * | Dual electric | None required |
| * Wayfarer | Dual vacuum | Extra cost - Factory installed |
| * Meadowbrook | | |
| Ford 6 | Dual vacuum | None available |
| 8 | Dual vacuum | Extra cost - Factory installed |
| Frazer | Dual vacuum | Standard |
| Hudson | Dual vacuum | Extra cost - Factory installed |
| Kaiser | Dual vacuum | Standard |
| Lincoln | Dual vacuum | Standard |
| Mercury | Dual vacuum | Standard |
| Nash | Dual vacuum | Standard with overdrive or hydramatic |
| | Dual vacuum | None available with standard transmission |
| Oldsmobile | Dual vacuum | Standard |
| Packard | Dual vacuum | Standard |
| Plymouth | Dual vacuum | Standard (Extra cost w&s) |
| Pontiac | Dual vacuum | Standard |
| Studebaker | Dual vacuum | Standard on Commander |
| | | Extra cost - Dealer installed on Champion |

**TRUCKS**

| MAKE | TYPE OF WIPER | BOOSTER EQUIPMENT (Vacuum pump unless otherwise noted) |
|---|---|---|
| Chevrolet | Dual vacuum | Extra cost - Factory installed |
| Dodge 1/2 to 2-1/2 | Dual vacuum | Extra cost electric - Factory installed |
| 2-3/4 up | Dual electric | None required |
| Ford F1-F6 with 6 cyl. engine | Dual vacuum | Extra cost electric auxiliary pump - Dealer installed |
| Ford F1-F8 with 8 cyl. engine | Dual vacuum | Standard |
| GM Truck & Coach | Dual vacuum | Extra cost - Factory installed |
| International | Dual vacuum | Extra cost - Dealer installed. Electric dual wipers also available as extra cost option. |
| Studebaker | Dual vacuum | Extra cost - Factory installed. Electric booster pump also available as extra cost option. |

6-20-50

1950 Chevrolet engineering letter

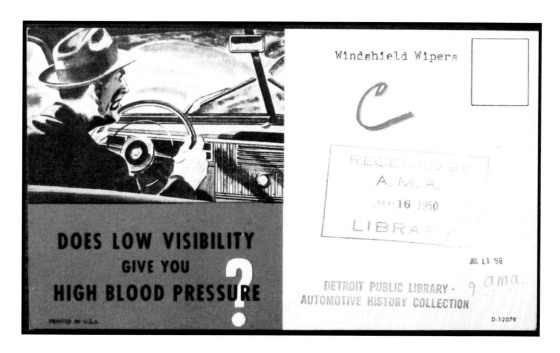

1950 Chrysler MoPar post card

MoPar offered a foot operated windshield washer in 1950.

Vehicle Owner Bob Elton

A 1950 Hudson vacuum motor and cable windshield wiper system.

Stahls Collection

The 1951 Hudson heater system. Note the defroster hoses and the temperature control valve in the upper right. The valve was controlled with the Bowden cable. Because the heater was clearly visible from the driver's seat, it was finished as a Class A surface and contained a Hudson Weather Control logo.

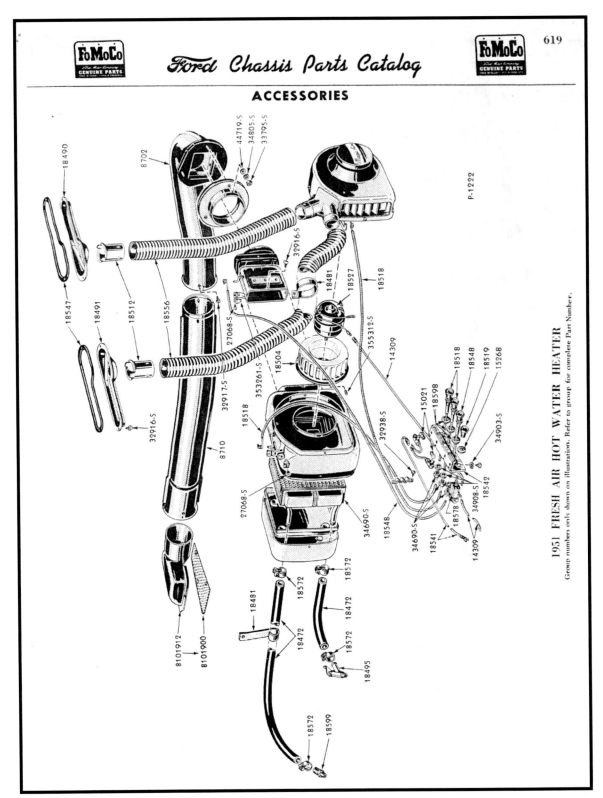

Ford Truck Parts Manual 1951

Ford offered this fresh air hot water heater with a blend-air door. This system eliminated the need for a water valve.

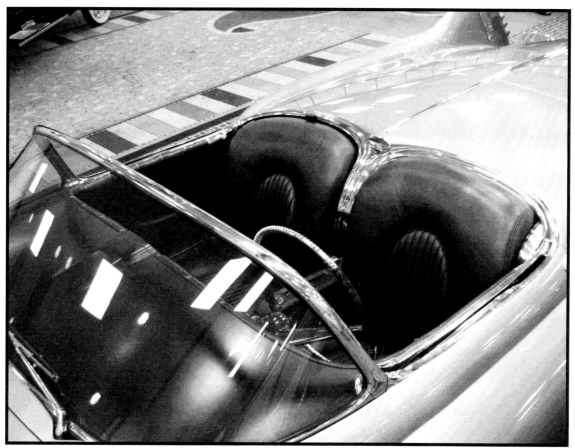

Images courtesy of Mike and Debbie Rowand

The 1951 Buick Le Sabre concept vehicle was fitted with heated seats.

Image courtesy of Automobile Quarterly

Image taken at the Buick Gallery & Research Center

The 1951 Buick XP 300 concept vehicle was fitted with four windshield wipers.

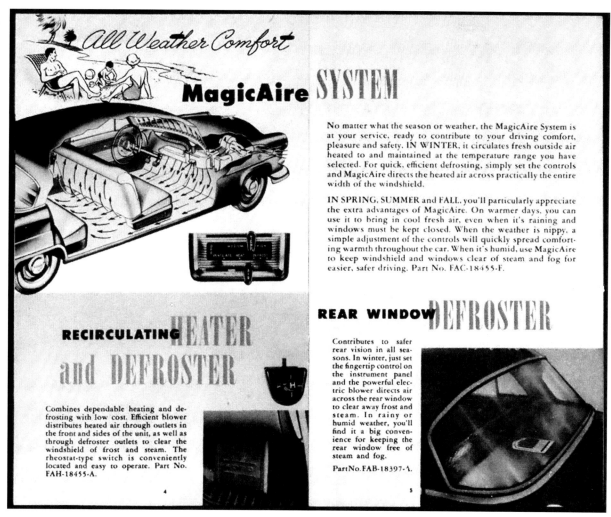

1952 Ford brochure

This 1952 Ford brochure shows an early instrument panel heater control that used slide levers rather than push-pull knobs. A rear window defroster was offered on this vehicle.

1952 brochure

American Bosch Corp. offered this electric windshield wiper motor with 1 or 2 speeds.

*An air conditioner of this type is easy to install, reduces heat by 20°*

*To alleviate that hot, sticky ride this porous cushion is just the thing*

Motor Trend  August 1952

Early devices to ease the discomfort of hot weather driving were the window hang-on evaporative cooler and the cool cushion/Kool Kooshion.

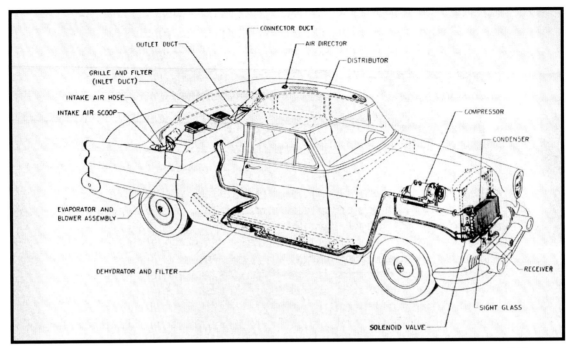

Motor January 1953

The 1953 Buick air conditioner system.

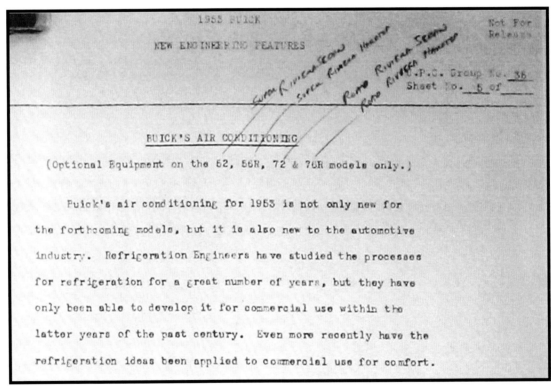

1953 Buick New Engineering Features

The 1953 Buick air conditioner system.

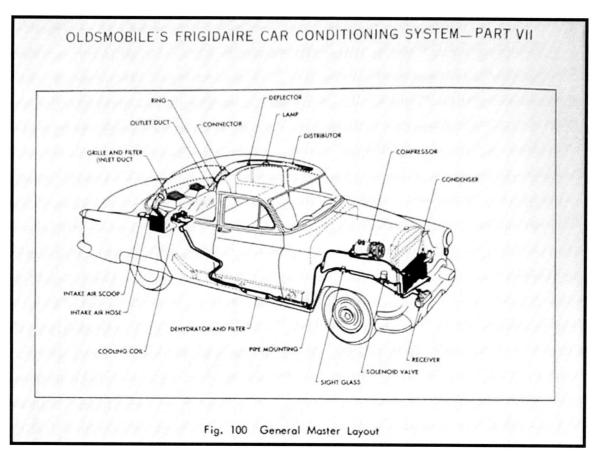

1953 Oldsmobile service manual

The 1953 Oldsmobile air conditioner system.

## A WORD TO ALL OLDSMOBILE SERVICE MEN

With the introduction of Oldsmobile's Frigidaire car conditioning a new field of service has been opened to all Oldsmobile service men.

Air conditioning as it pertains to our car is so new, that you service men are faced with a new experience in learning how to service it.

In entering this new field of service, you are going to hear strange words and new words as well as terms that are entirely foreign to you. You will be required to work with tools and instruments of which you do not have the "feel". Yet, if you will study this manual you will have a pretty good idea of how Oldsmobile's car conditioning system is serviced.

Because so little is known of air conditioning by the average automobile mechanic we have tried to bring you a simple story of the basic principles and fundamentals of refrigeration and air conditioning.

We have for your benefit taken certain liberties to eliminate complicated formulas and unfamiliar scientific phrases to bring you this simple story. We believe you will realize that further study of refrigeration text books should be made before a service man can be considered a complete craftsman who knows the WHY as well as the HOW of refrigeration.

I hope you will find this simple story of refrigeration interesting and educational.

*E. E. Kohl*
E. E. Kohl
General Service Manager

1953 Oldsmobile service manual

Letter from E.E. Kohl to all Oldsmobile service men regarding air conditioning.

# CHAPTER II
# LOCATION AND FUNCTION OF UNITS

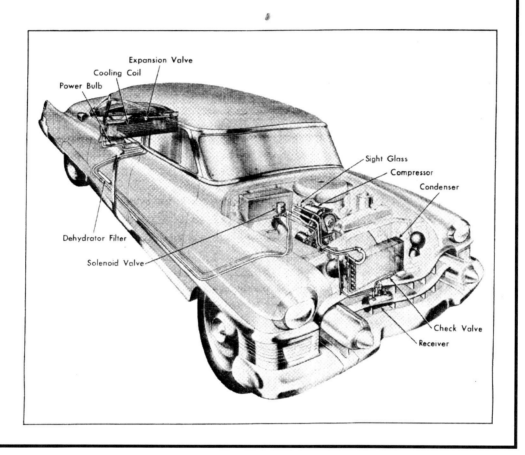

1953 Cadillac service manual

The 1953 Cadillac air conditioner system.

# CONTROL PANEL

**SWITCH-"ON"** provides full cooling

**SWITCH in "VENT"** position turns on **BLOWERS** only

As you can see here, the control panel is relatively simple, both in its construction and in the operation of it. The toggle switch has only 3 positions -- VENT, ON, and OFF. The control <u>lever</u> controls the temperature to the degree preferred by the driver. To make the temperature cooler, it is only necessary to move the lever to the right. The far right position of the lever provides maximum coolness -- with the switch in the ON position, of course.

1953 Cadillac service manual

The 1953 Cadillac instrument panel air conditioner control.

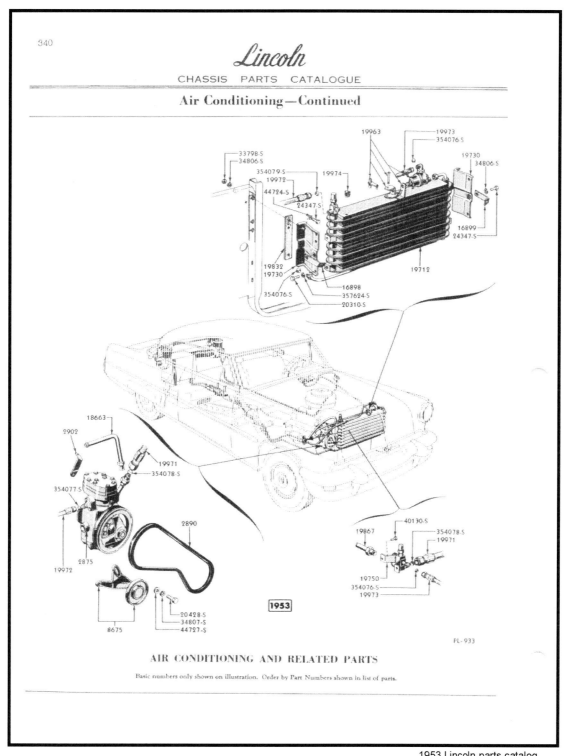

The 1953 Lincoln air conditioner system.

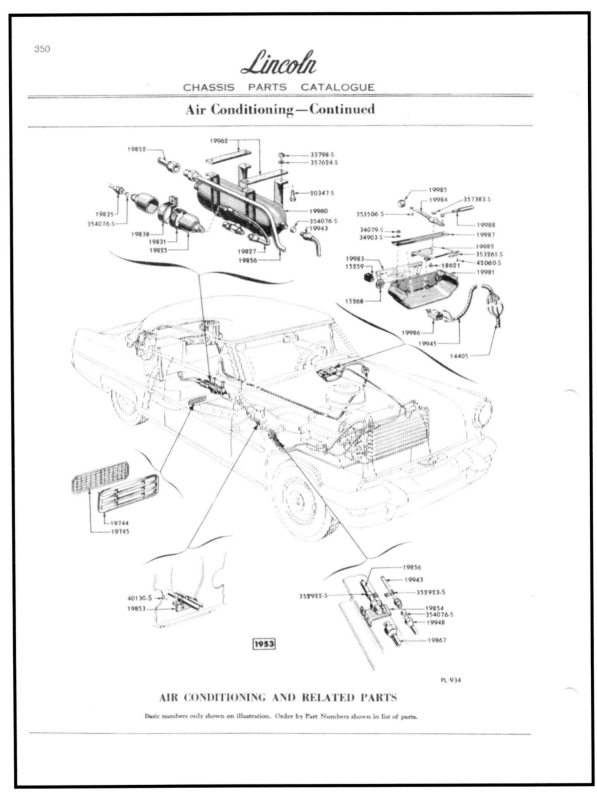

The 1953 Lincoln air conditioner system.

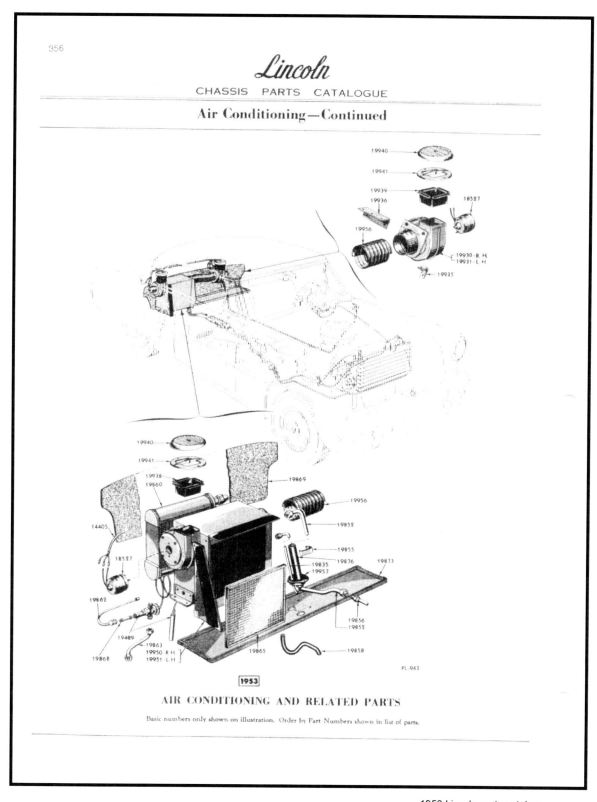

The 1953 Lincoln air conditioner system.

1953 Lincoln parts catalog

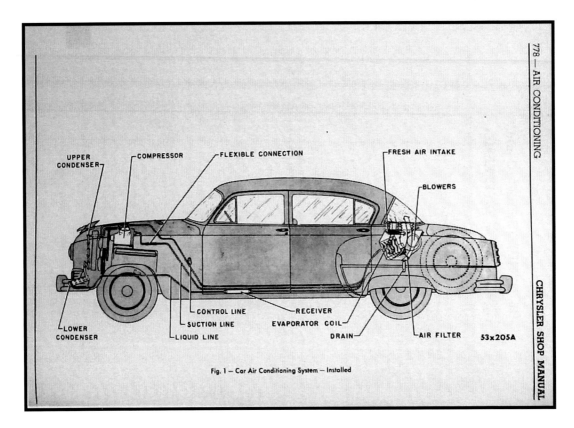

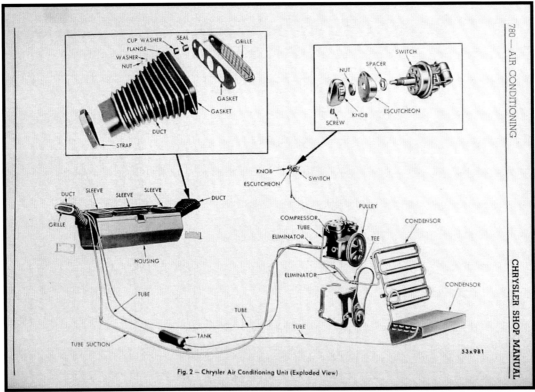

1953 Chrysler service manual

The 1953 Chrysler air conditioner system.

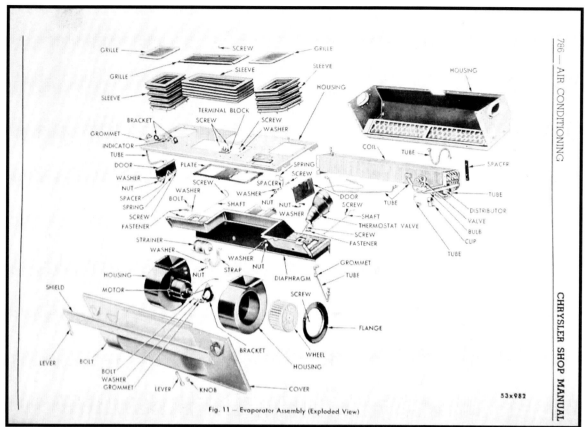

1953 Chrysler service manual

The 1953 Chrysler air conditioner system.

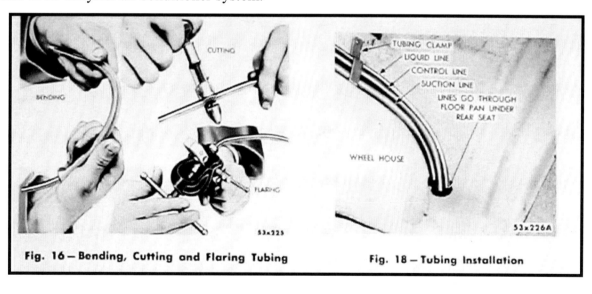

1954 Dodge service manual

Flare fittings were used on the 1953 refrigeration tubing connections.

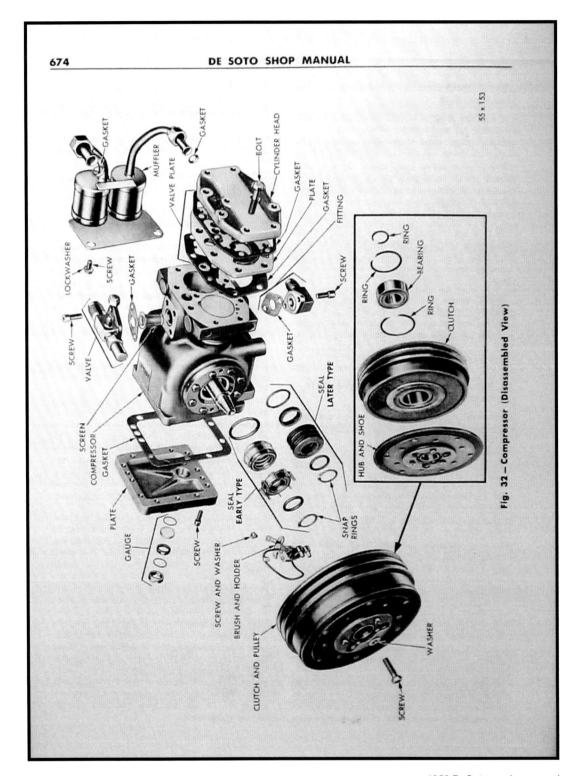

1953 DeSoto service manual

The 1953 DeSoto air conditioner compressor with electromagnetic clutch

# AIR CONDITIONING

## GENERAL INFORMATION

The Chrysler Airtemp Air Conditioning System, as shown in (Figure 1) is a mechanical vapor compression refrigeration system designed by Chrysler engineers for cooling the passenger compartment of a motor vehicle.

In cooling the passenger compartment of the automobile, Chrysler Corporation now offers to owners of Dodge Cars the highest capacity air conditioning system yet disclosed for use in passenger automobiles. Simple to operate, the new system delivers fresh, cool dehumidified air a few minutes after starting, and creates a comfortable atmosphere within the car even when moving through slow traffic in summer weather.

Cooling of the automobile by means of refrigeration represents a development of considerable interest to both the automobile and the air conditioning fields.

Chrysler Corporation research on car air conditioning began in 1939 with the formation of a team of automotive engineers and specialists from Chrysler's Airtemp Division, one of the Country's leading manufacturers of heating and air conditioning equipment.

Their combined efforts have resulted in the present highly-successful air conditioning system for passenger cars.

Foremost among many features of the system is the provision for introducing fresh air into the car, thus preventing the staleness associated with automobile air conditioners that merely recirculate the same air. With the Chrysler Airtemp Air Conditioning System both fresh and recirculated air are filtered before they reach the car interior. Sensitive automatic controls guard the occupants against chilliness once the temperature has been lowered to a comfortable level. On humid days the system adds much to passenger comfort by removing excess moisture from the air.

Installation of the unit is neat and attractive, employing little-used sections of the car to house the major components of the unit. Except for the external air vents, its presence is hardly noticeable to the casual observer. A single switch on the instrument panel is the only control with which the driver is concerned.

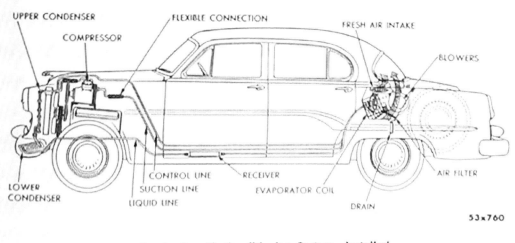

Fig. 1 — Car Air Conditioning System — Installed

1954 Dodge service manual

A description of the 1954 Dodge air conditioner system.

**4**            **DODGE SHOP MANUAL**

Passenger comfort is greatly enhanced in air conditioned cars by the continuous introduction of fresh air, amounting to about 25% of the total volume being circulated. This prevents staleness, and mildly pressurizes the car interior so that air will leak *outward* carrying off smoke, carbon dioxide and other odors. At the same time, the inward seepage of hot dusty air is prevented. Both fresh and recirculated air in the Airtemp Air Conditioning System are filtered after cooling. (Some early models were not equipped with a filter, while later models are filter equipped.)

# SPECIFICATIONS

## COMPRESSOR

| | |
|---|---|
| Location | On right bank cylinder |
| Type | V-type 4 cylinder |
| Bore | 1½ inch |
| Stroke | 1⅜ inch |
| Valve | Reed Type Valve |
| Speed | 625 rpm at 25 mph |
| Oil Capacity (Refrigeration Oil MS) | 32 ounces |
| Weight | 75 pounds |

## CONDENSER

| | |
|---|---|
| Location | Front of car |
| Type | (Upper) Steel tubing spiral steel fins |
| | (Lower) Copper tubing aluminum fins |
| Weight | (Upper) 10 pounds |
| | (Lower) 20 pounds |

## RECEIVER

| | |
|---|---|
| Type | Cylindrical Steel Container |
| Location | Right side of frame |
| Refrigerant | Freon 22 |
| Total Charge | 5.0 pounds |

## EVAPORATOR

| | |
|---|---|
| Location | Luggage Compartment |
| Length | 43 inches |
| Depth | 9 to 13 inches |
| Height | 11 inches |
| Weight | 70 pounds |
| Space Occupies | 3½ cubic feet |

## BLOWERS

| | |
|---|---|
| Type | Centrifugal |
| Location | In evaporator unit |
| Capacity | 400 cubic feet of air at high speed |
| Current Draw | Approximately 20 amps |

1954 Dodge service manual

Specifications of the air conditioner components used on the 1954 Dodge.

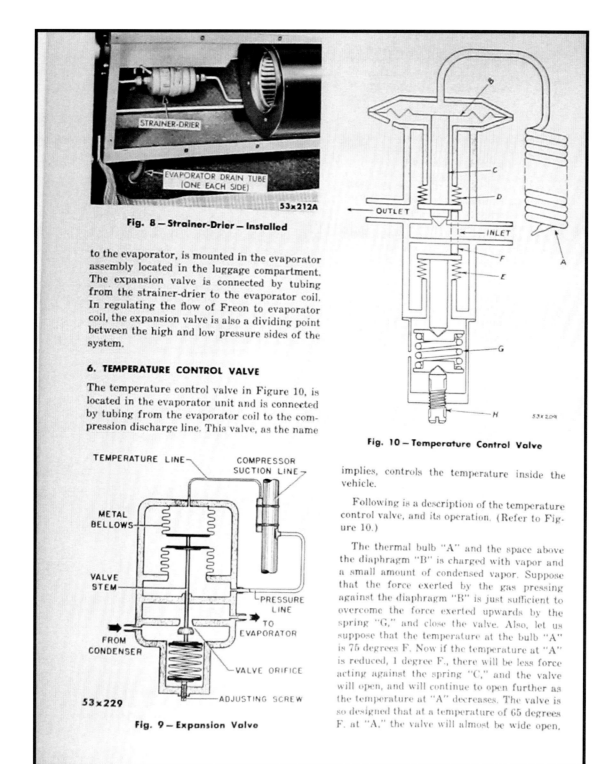

Fig. 8 – Strainer-Drier – Installed

Fig. 9 – Expansion Valve

Fig. 10 – Temperature Control Valve

to the evaporator, is mounted in the evaporator assembly located in the luggage compartment. The expansion valve is connected by tubing from the strainer-drier to the evaporator coil. In regulating the flow of Freon to evaporator coil, the expansion valve is also a dividing point between the high and low pressure sides of the system.

## 6. TEMPERATURE CONTROL VALVE

The temperature control valve in Figure 10, is located in the evaporator unit and is connected by tubing from the evaporator coil to the compression discharge line. This valve, as the name implies, controls the temperature inside the vehicle.

Following is a description of the temperature control valve, and its operation. (Refer to Figure 10.)

The thermal bulb "A" and the space above the diaphragm "B" is charged with vapor and a small amount of condensed vapor. Suppose that the force exerted by the gas pressing against the diaphragm "B" is just sufficient to overcome the force exerted upwards by the spring "G," and close the valve. Also, let us suppose that the temperature at the bulb "A" is 75 degrees F. Now if the temperature at "A" is reduced, 1 degree F., there will be less force acting against the spring "C," and the valve will open, and will continue to open further as the temperature at "A" decreases. The valve is so designed that at a temperature of 65 degrees F. at "A," the valve will almost be wide open.

1954 Dodge service manual

Air conditioner expansion valve and temperature control valve.

# SERVICING AIR CONDITIONING SYSTEM

## 12. PRECAUTIONS IN HANDLING FREON

**CAUTION**—Do Not:

### Expose Eyes to the Liquid

When working around a refrigerating system liquid refrigerant may spill and hit your face. If the eyes are protected with goggles (Tool C-3355) or glasses no serious damage may result. If a splash of refrigerant does hit the eyes, the eyes must not be rubbed. Freon 22 is at least—40 degrees F.—a temperature that the eyes cannot normally withstand. Apply cold water immediately to the area of the eye to gradually get the temperature above freezing point. Use of an antiseptic oil is helpful as a protective film over the eye ball until medical aid can be obtained.

**CAUTION**—Do Not:

### Discharge in Areas Where an Open Flame is Exposed

Discharging large quantities of Freon 22 in an average size work room can usually be done safely as the vapor will produce no ill effect. However, this should never be done if the area contains an open flame such as torch or gas heater. Freon 22 normally is non-poisonous. Concentration of the gas in live flame will produce a poisonous gas. Splashing Freon 22 on bright metal or chrome should be avoided as the gas will tarnish bright metal.

**CAUTION**—Do Not:

### Leave Charging Drum Uncapped

The drum should never be exposed to radiant heat as the resultant pressure from such heat may cause the safety plug to release or the drum to explode. In charging the system it may be necessary to heat the drum to raise the drum pressure higher than the pressure in the system. Use a pail of hot water no hotter than you can put your hand into, or a heated rag wrapped around the drum is all the heat required. If the occasion arises for refilling a small charging drum from a larger drum, never fill the drum completely with liquid. Space should be left in the filled drum for expansion. If the drum is completely filled, and the temperature was increased, extreme high pressure would result. Never transport the charging drum in the passenger compartment; this can be done more safely in the trunk compartment.

## 13. PRECAUTION IN HANDLING TUBING

### Cleanliness During Storage and Installation

It is of the utmost importance that refrigeration tubing be kept clean and dry. Refrigeration tubing is ordinarily purchased in rolls with the ends crimped and sealed air tight. When tubing is used from a roll, reseal the unused portion of the roll to prevent dirt and moisture from getting inside. A piece of tubing that has been cut, flared and prepared for installation should have the ends sealed until the actual installation is being made.

### Cutting and Flaring

Tubing should always be cut with a regular tube cutter (Special Tool C-873) (Figure 16), and never with a hack saw. After cutting the tube, eliminate all burrs and provide a good seating area when the tube is flared. The tube should be double flared with Special Tool C-804. Always inspect a flare before installation to determine if there are any cracks or blemishes on the flare that would cause a possible leak. It should be emphasized that the retention of Freon in a system requires flare or brazed joints of the highest quality.

When repairing a flare joint, it is advisable to reface the seat of the male fitting with special tool C-3366. This is very important to achieve a leak free joint.

### Tightening Flare Joints

Always use a second wrench on both the male fitting and the nut when tightening a flare joint (Figure 17). This will prevent the tube from twisting or cracking.

*NOTE*

*If a flare joint leaks when tested, try tightening the nut. If the leak still persists, it is probably due to a faulty joint, and the male fitting may have to be refaced and a new flare made on the tube.*

### Securing of Tubing

Tubing that is left free to vibrate and move about excessively will soon harden the area of the tube at the flare section so that it may be-

1954 Dodge service manual

Freon handling instructions.

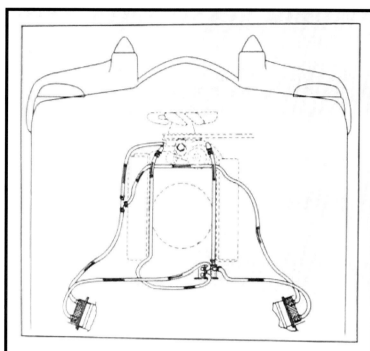

Fig. 16-B-1 Water Flow

## (2) Air Flow

The flow of air through the heater and defroster system is illustrated in Fig. 16B-2. Outside air is drawn in through the cowl air scoop.

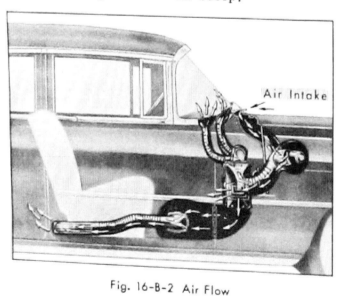

Fig. 16-B-2 Air Flow

1953 Cadillac (1954 service manual shown)

Cadillac offered dual heaters in the 1953 model year.

327

1954 Pontiac brochure

The air conditioner evaporator and blower assembly was mounted in the engine compartment rather than in the trunk on the 1954 Pontiac.

Motor Trend   August 1954

The 1954 Nash had the evaporator mounted in the engine compartment rather than the trunk.

**THE IDLER** blower-type car cooler is just the thing for hot-weather driving comfort. Adjustable legs position the air conditioner on the front floor of your car. Installation is simple whether you use it in your car, trailer or home. Price, $89.50, f.o.b. For further information, write to the Jumbo Equipment Co., 1012 S. Los Angeles St., Los Angeles 15, Calif.

Motor Trend July 1954

Note this device was similar to the window mounted evaporative coolers. It was not a refrigeration device.

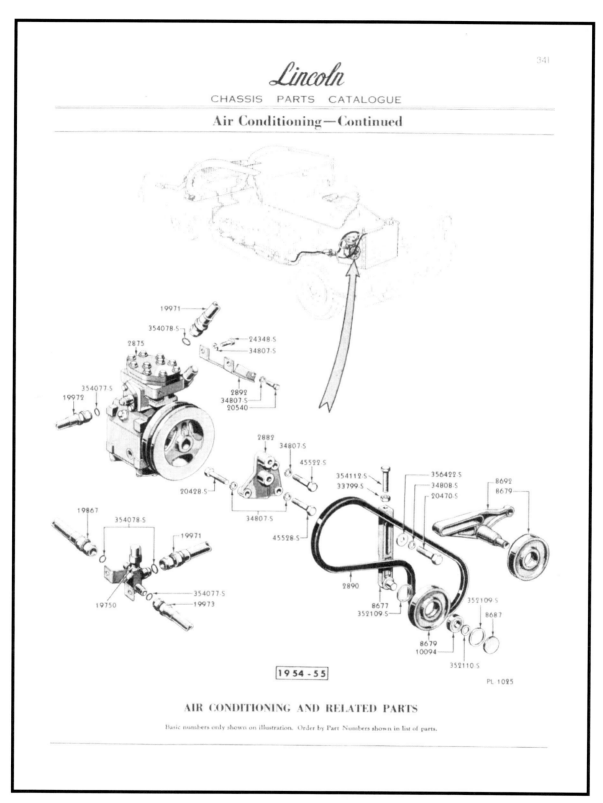

The 1954-55 Lincoln air conditioner system.

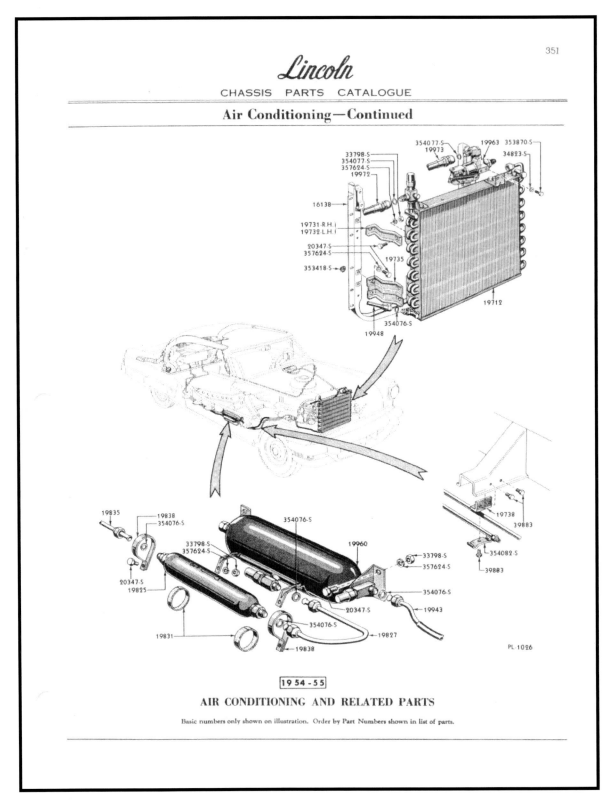

The 1954-55 Lincoln air conditioner system.

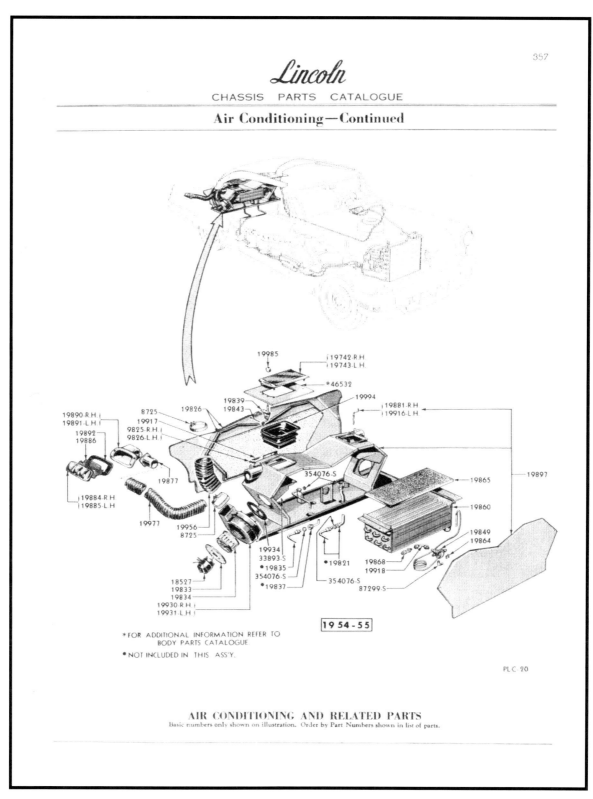

The 1954-55 Lincoln air conditioner system.

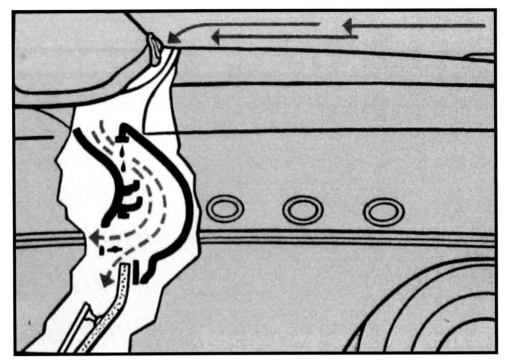

1954 Buick ventilation brochure

The 1954 Buick ventilation air intake was the decorative lower windshield moulding.

Motor Trend March 1954

The 1954 Cadillac featured front doors that housed the rear seat heat ducts

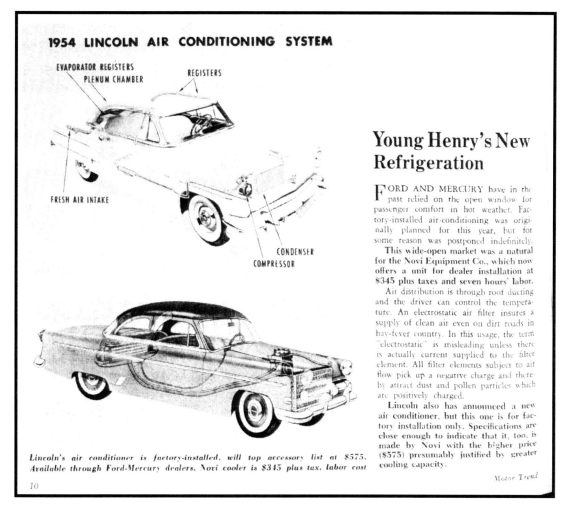

Motor Trend August 1954

Ford cars in 1954 had the air conditioners installed as aftermarket units. The Lincoln used factory-installed units. Both Ford and Lincoln air conditioners were manufactured by the Novi Equipment Co.

1954 Packard brochure

Note that Packard released the first air-conditioned vehicle in 1940 and waited until 1954 to release the next one. This new vehicle also featured a dual heater system.

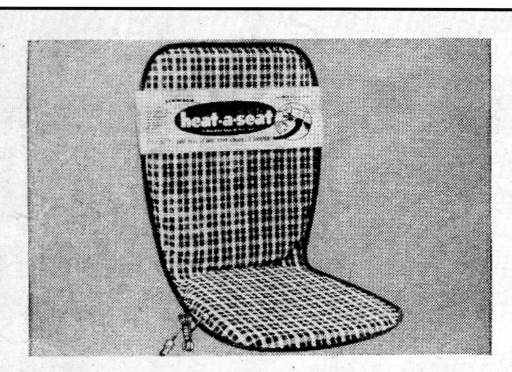

Why huddle on a cold car seat this winter? Enjoy cozy warmth with Heat-A-Seat. Plugs into cigarette lighter and warms instantly. Un-plugged, it's a comfortable air-vented summer cushion. Triple insulated with rubberized silicone Fiberglas; draws less than 4 amps. 100% safety approved; unconditionally guaranteed. Covered in attractive plaid plastic-coated fabric. Red, Blue, or Green. Complete with U.L. approved wire connection. Only $5.95. Item #7.

1954 Newhouse Automotive Ind.

A 1954 aftermarket heated seat called the Heat-A-Seat. Newhouse Automotive Ind. offered it for $5.95.

A 1954 vacuum-powered fan sold by Leroy Auto Sales Co.

A 1954 aftermarket set of venetian blinds sold by the J C Whitney Co.

Vehicle Owner Philip Fischer (2 images)

A 1954 Buick with a desert water bag to keep the water cool during hot weather driving.

A 1954 Buick windshield washer jar.

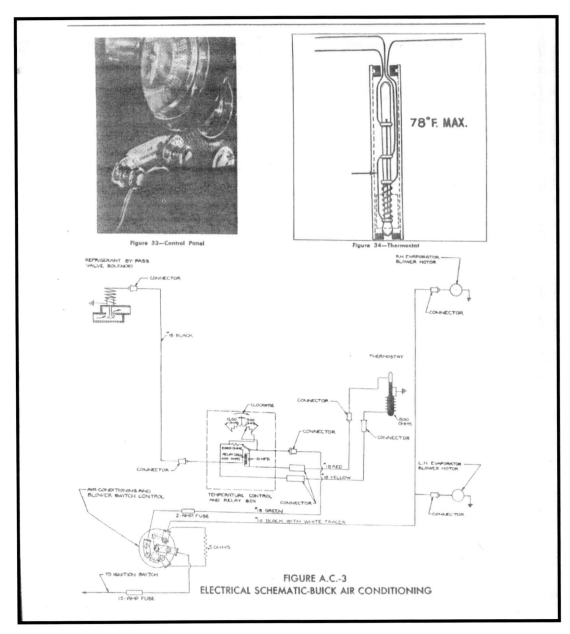

The electrical schematic for the 1954 Buick air conditioner.

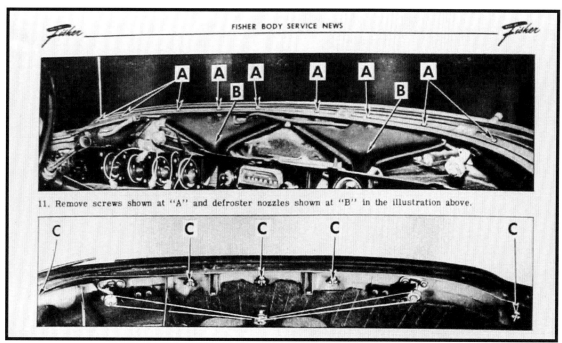

1954 Buick Fisher Body Service News

The 1954 Buick used molded plastic defroster nozzles. It also used cable driven windshield wipers powered by a vacuum motor.

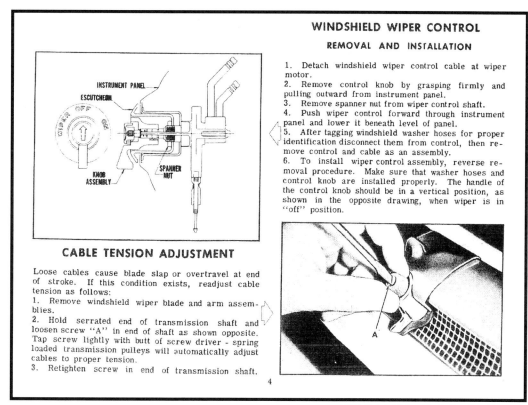

1954 Buick Fisher Body Service News

The 1954 Buick windshield wiper control and cable adjustment procedure.

Image courtesy of The Auto Collections, Imperial Palace Hotel & Casino

A 1954 Cadillac fitted with factory air conditioning. Note the thickness of the condenser.

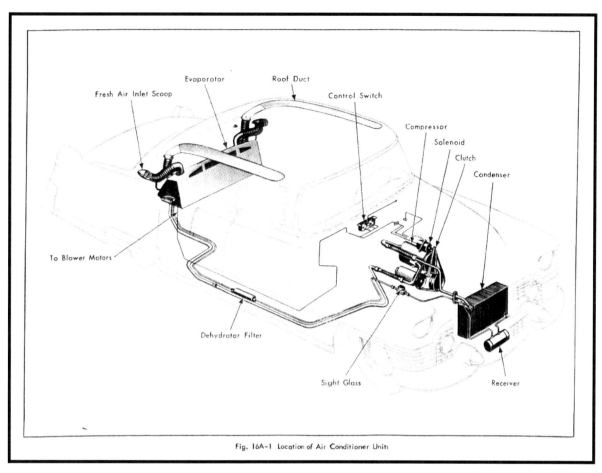

1954 Cadillac service manual

The 1954 Cadillac air conditioner compressor and unique solenoid actuated clutch

Fig. 16A-2 Clutch Solenoid

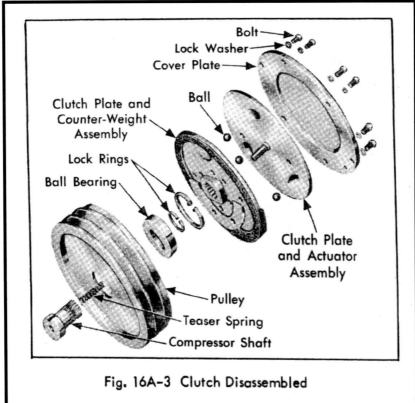

Fig. 16A-3 Clutch Disassembled

1954 Cadillac service manual

The 1954 Cadillac unique solenoid actuated clutch.

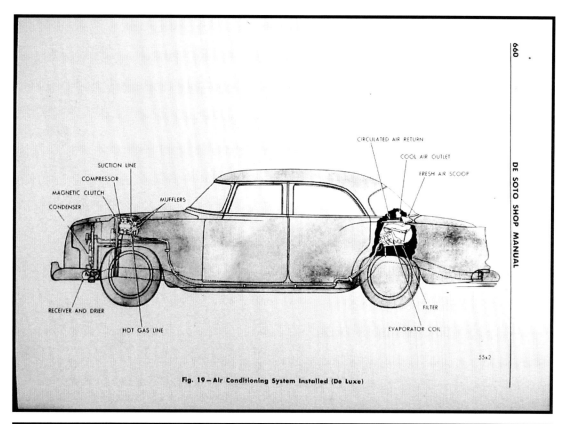
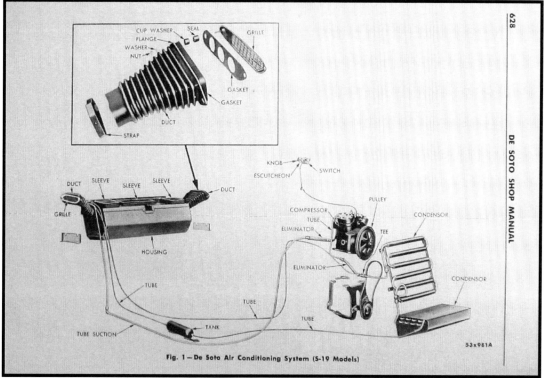

1954 DeSoto service manual

The 1954 DeSoto air conditioning system. Note the two condensers.

345

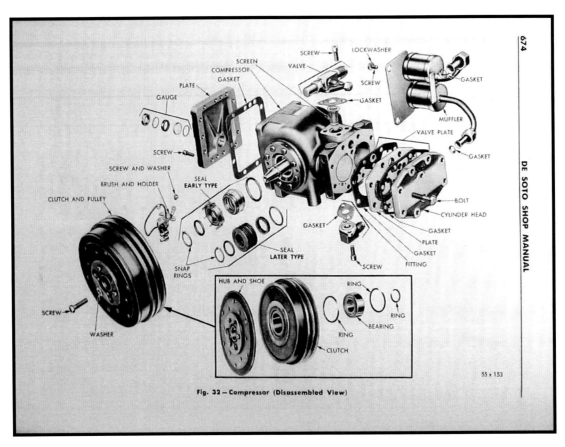

1954 DeSoto service manual

Chrysler Corporation released two air conditioner systems for the 1954 model year. The standard system used R-22 refrigerant and had no magnetic clutch. The deluxe system used R-12 refrigerant and had a magnetic clutch similar to those still in use.

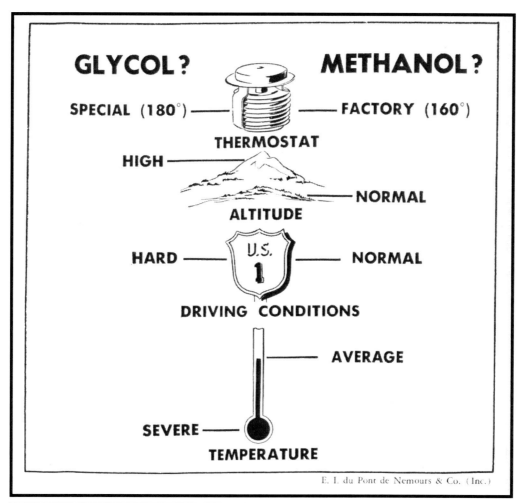

Motor Trend November 1954

In 1954, both alcohol and glycol based antifreeze products were being sold. These charts show the comparison of the two types. The upper chart is by E.I. du Pont de Nemours & Co.

Vizo brand windshield washer antifreeze.

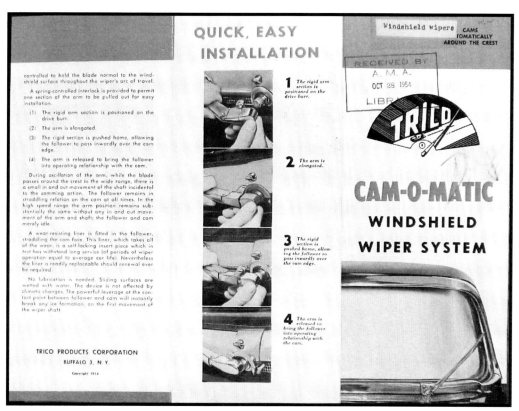

1954 brochure

Trico Products Corp. developed the Cam-O-Matic windshield wiper for curved glass.

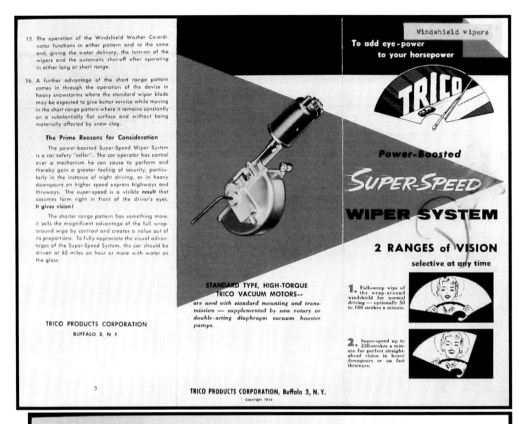

1954 brochure

Trico Products Corp. offered the Power-Boosted Super Speed wiper system for use during fast driving conditions. It featured faster wipes in a narrower arc.

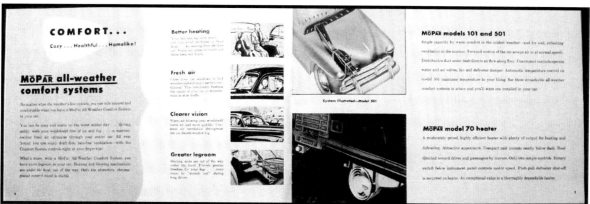

1954 brochure

The MoPar brand offered three heaters for 1954 Chrysler vehicles: a fresh air heater system with manual temperature control, a fresh air heater system with automatic temperature control and a low cost recirculation heater.

JC Whitney offered this door cooler screen in 1955.

A DUAL PURPOSE is claimed for this air conditioning unit, which is made to fit any make car, new or old. Called the Smiley Air Conditioner, the unit is designed to remove the impurities suspended in the air in your car, in addition to drawing in fresh washed and cooled air from the outside. It changes the air 3½ times every minute. The main unit, which weighs 48 pounds, fits snugly in the forward area of the trunk, leaving a good percentage of the usable luggage space intact. Outside air intake is through chrome scoops located in the rear quarter body panels. Twin ducts feed the washed air thru outlets above the rear seat. Upkeep of the unit is at a minimum, with only water added as needed. Installation costs in the neighborhood of $250, depending on make of car. If you're interested in getting rid of the discomforts of polluted air (see page 62 for more information on this subject), one of these units may be for you.

Motor Trend April 1955

The Smiley Air Conditioner was not a refrigeration unit. It was a water evaporative system that sold for $250.

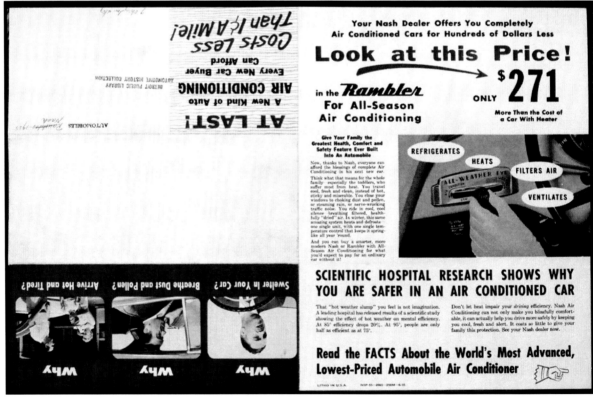

1955 Nash brochure

The 1955 Nash air conditioner system

## ROAD TEST

# NASH
# AMBASSADOR V-8

*A new V-8 engine has been added to the host of features that make Nash a genuinely different car*

THE NASH has always been noted for comfort, reliability and some "different" touches unobtainable on most other cars. The 1955 Ambassador V-8 retains all these features and adds one that will be new to most loyal Nash followers—performance that is far better than that of any Nash in the past.

From the outside the two major changes from 1954 are the grille-mounted headlights and the wraparound windshield, which Nash claims is the largest in the industry. The headlight treatment is derived from the Nash-Healy and looks good.

To go with its fine performance the Nash also offers good handling qualities. With power steering it has a steering ratio of 22.2:1 and goes lock to lock in four turns. Steering is light and easy yet retains a remarkable degree of road feel, more so than in many of its competitors. Turning circle radius is about 44 feet and, while this is about par for the course with its competition, at least one member of the test crew felt that surgery on the front fender cutouts would help a lot.

The coil springs all around on the Nash do permit quite a bit of roll on fast tight turns. Neither driver nor passengers are thrown around uncomfortably at anything approaching normal speeds, however. Even on hard turns during our test-

### SPECIFICATIONS
- Engine type: OHV V-8
- Displacement: 320 cubic inches
- BHP: 208 @ 4200 rpm
- Compression ratio: 7.8-to-1
- Bore: 3 13/16   Stroke: 3 1/2
- Torque: 300 ft.-lbs. @ 2600 rpm
- Transmission: TWIN-ULTRAMATIC
- Rear axle ratio: 3.54
- Wheelbase: 121.25 inches
- Dry weight: 3795 lbs.
- Turning circle: 44 feet
- Steering lock-to-lock: 4 turns

### PRICES
- Car: $2695
- Transmission: $199
- Radio: $93
- Heater: $77
- Power Steering: $107
- Power Brakes: $39
- Air Conditioning: $395

ing the car appeared to be leaning over far more to observers than was apparent inside. And since the Nash was designed to give a soft comfortable ride under normal conditions, not to be driven at speed thru road race courses, it shouldn't be expected to corner like a Ferrari—nor should it be asked to.

The Nash rode uncommonly well over very rough gravel roads—which were in very poor shape due to spring thaw and heaving, incidentally. This will be important to a lot of potential Nash owners because it has traditionally appealed to many sportsmen who use it for hunting and fishing expeditions. They will also be happy to know it retains ample road clearance—eight inches at the rear axle. Although we drove it thru some very soggy territory and over some deeply rutted back roads, it didn't hang up once.

It's in the accessories and unique features department that Nash has always been outstanding. This year is no exception. Although the Michigan weather was a little chilly to make use of the Weather Eye air conditioning system necessary, it has proved its worth well in the past. In fact, many consider it the finest automobile air conditioning system available. The heater still is one of the best, has an excellent defrosting system also, which is a terrific asset in cold climates.

One of the accessories which really impressed the test crew was the radio with its dual speaker setup. The crew includes a couple of hi-fi fans and they listen with a very critical ear. "Great!" was their opinion of this Nash feature.

The Nash twin bed setup is well-known, partly because of the many jokes about this idea. However, jokes or not, it's a good idea. Turning the Nash into a temporary boudoir is quick and easy; takes

Successor to a long line of sixes, the V-8 is pretty hefty, leaves little room to spare in the engine compartment.

Nash heels as driver bends it into a hard corner. Unit frame design makes car exceptionally rattle-free, gives it good safety rating.

*MOTOR Life, July, 1955*

Motor Life   July 1955

The 1955 Nash air conditioner system was a highly rated, low cost unit.
"In fact, many consider it the finest automobile air conditioner available."

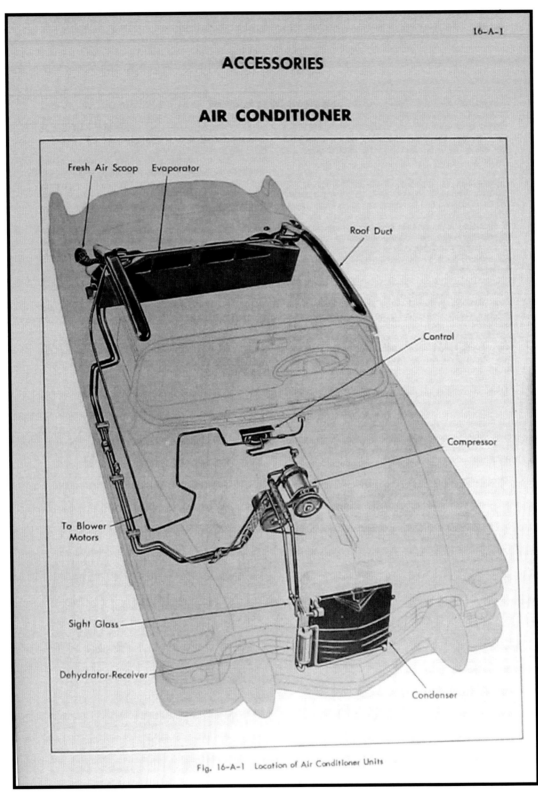

The 1955 Cadillac air conditioner system.

# AIR CONDITIONER

them with glasses or goggles. If Freon-12 liquid should strike the eyeballs:

1. Apply a few drops of sterile mineral oil to the eyes as an irrigator.

2. If irritation continues, wash the eyes with a weak solution of boric acid.

3. See an eye specialist at once.

If liquid Freon-12 comes in contact with the skin, the injury should be treated for frostbite.

## (2) Precautions in Handling Lines

a. Store all lines to avoid crushing or kinking. If a line is kinked, it should not be used.

b. Lines should be kept sealed and dehydrated in stock. Do not remove caps from lines until just before installation.

c. When tightening fittings, use proper size wrenches to avoid over or under tightening. Always use two wrenches, when tightening fittings, to prevent twisting the soft copper tubing. A drop of Frigidaire oil on the pipe flare will allow the flare nut to be tightened without twisting the pipe.

d. Close ends of lines, which have been disconnected for any reason to prevent entrance of moisture or dirt.

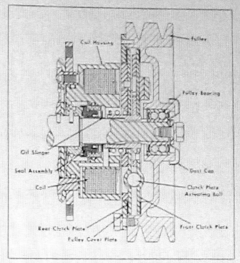

Fig. 16-A-3 Cross-Section of Compressor

e. Gage set and lines should be kept clean and free from moisture.

f. Do not leave Frigidaire oil container open any longer than necessary, as the special oil is moisture-free and will absorb moisture from the air if left uncapped.

g. Use the Vacuum Pump, Tool No. J-5428, to remove any air or moisture which may have entered the system when it was opened to replace a part.

## (3) Maintenance and Inspection

### a. Preliminary Check

1. High and low pressure shut-off valves at compressor must be fully open.

2. Drive belts must be installed properly to prevent slippage.

3. Make certain clutch is engaging and disengaging.

4. Using Leak Detector, Tool No. J-5419, test entire system for leaks, and make necessary repairs.

5. If there is evidence of oil leaks, check oil level.

6. Check operation of blower fans at all control knob positions.

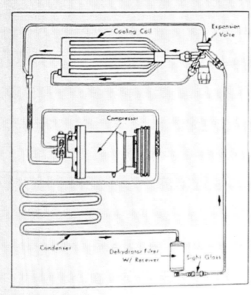

Fig. 16-A-2 Cycle of Operation

1955 Cadillac service manual

The 1955 Cadillac air conditioner electromagnetic clutch

1955 JC Whitney

Two window-mounted evaporative coolers, one was a ram air type, and the other was motorized.

1955 brochure

A French windshield washer system.

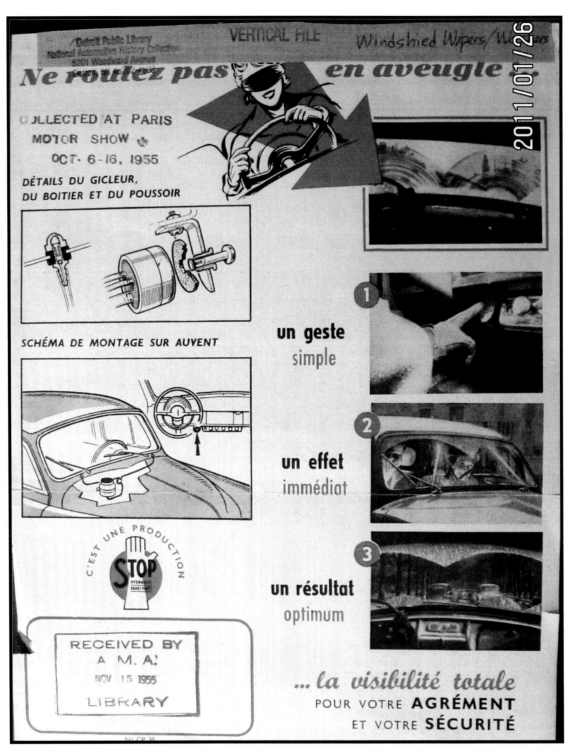

1955 brochure

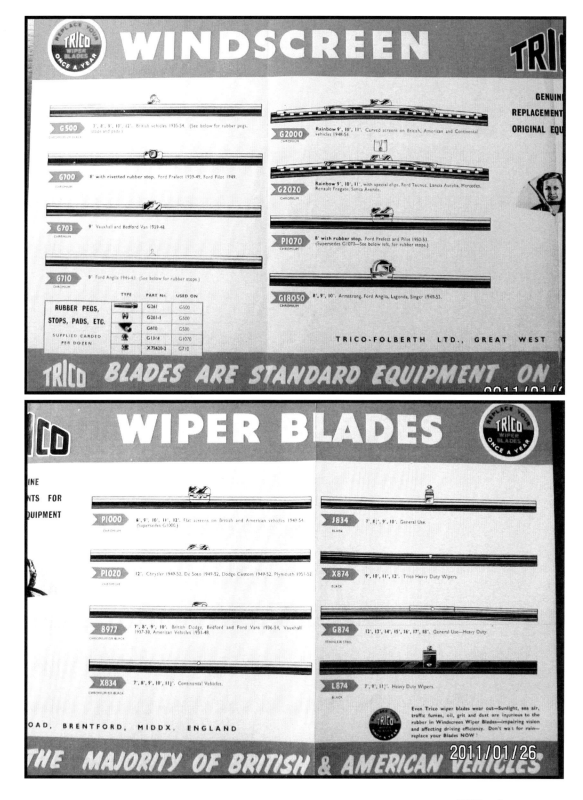

1955 brochure

Trico-Folberth Ltd. offered a variety of windshield wiper blades. Note the rubber pegs, stops, pads, etc. in the lower left corner of the upper image.

1955 brochure

American Bosch Corp. offered the WWB Dual Electric windshield wiper motor along with the WWA single unit.

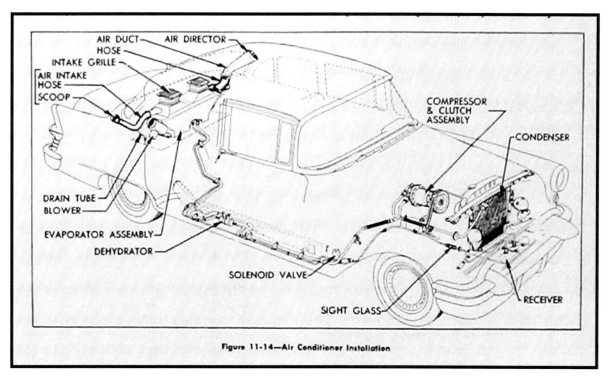

1955 Buick service manual

The 1955 Buick air conditioner system.

1955 JC Whitney

A 1955 aftermarket vent kit. Note that it was called an Air Conditioned VENT KIT. Also note the $4.29 pair "Makes car 'look' air conditioned."

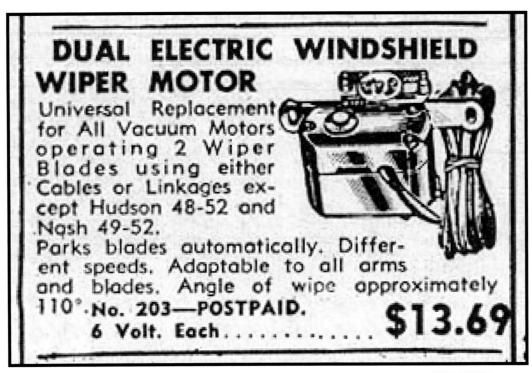

1955 JC Whitney

A 1955 aftermarket electric windshield wiper motor to replace a vacuum unit.

Motor Life   March 1956

Frigiking offered this hang-on air conditioner for $298 plus installation.

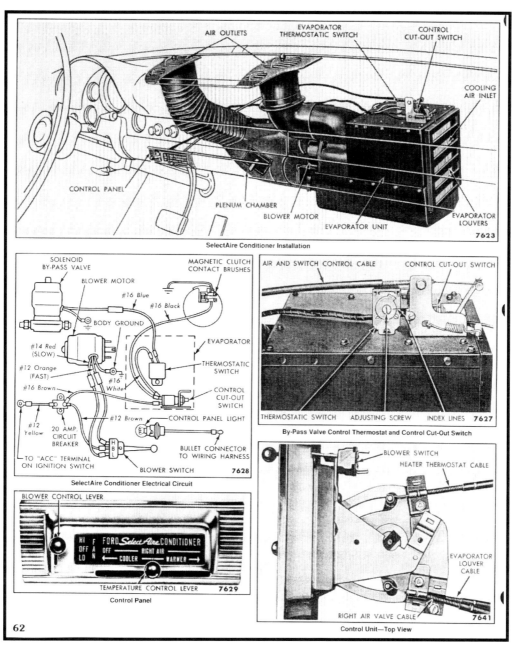

1956 Ford service manual

The 1956 Ford air conditioner system with the electrical schematic.

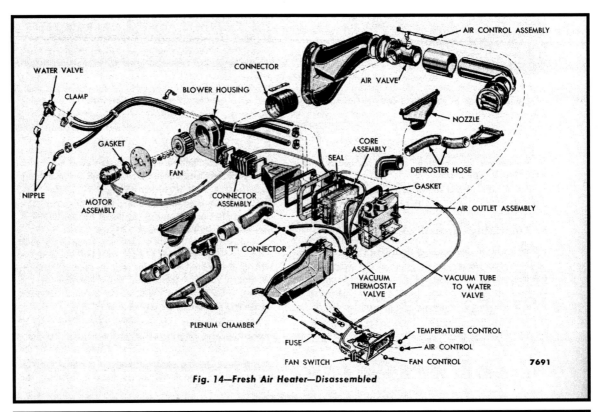

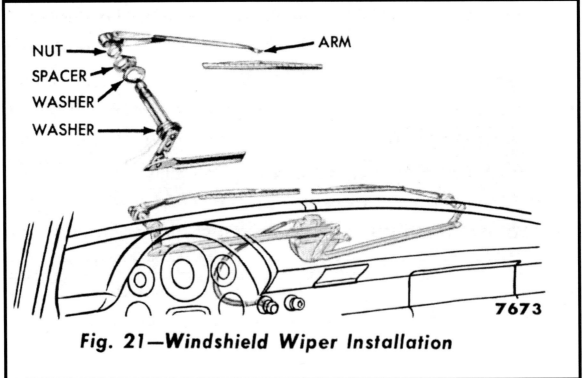

1956 Ford service manual

The 1956 Ford fresh air heater and windshield wiper system.

366

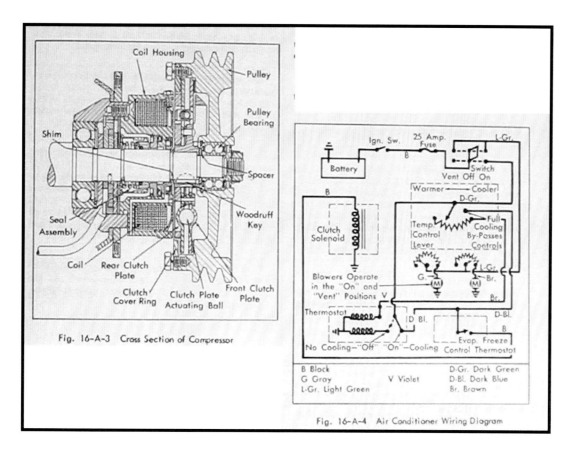

1956 Cadillac service manual

The 1956 Cadillac air conditioner clutch and electrical schematic.

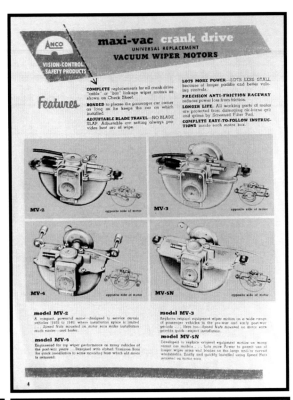

1956 brochure

Anderson Co. offered a wide assortment of vacuum windshield wiper motors in 1956.

# crank drive motors continued

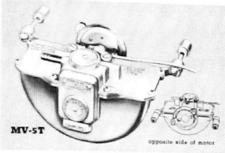
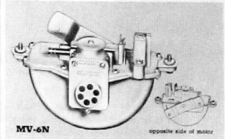
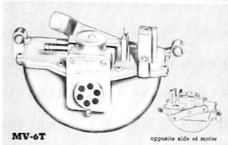
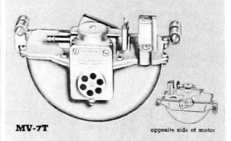

### model MV-5T
Similar to MV-5N in added power—designed to service other recent vehicle models having long wiper arms and blades... Round Trunnion Ears assure Serviceman of rapid, expert installation.

### model MV-6T
Powerful enough to be used on late model cars. Round Trunnion Ears mean fast, sure installation for Serviceman. Designed to be used with washer co-ordinator.

### model MV-6N
One of ANCO's most recently developed motors for replacement on some of the lastest-model cars with wrap-around windshields. Rubber grommets on motor ears ensure cushion mounting. Designed to be used with washer co-ordinator.

### model MV-7T
Like the MV-6N and MV-6T, this motor is designed to carry the heavy load of longer arms and blades and for the large and/or wrap-around windshields. Also equipped with Round Trunnion Ears for ease of installation. Designed to be used with washer co-ordinator.

### allowance for old motors replaced
Each ANCO Motor Box has on it a special Box Flap which is equal to $1.00 in Box Flap value when Dealer sends Box Flap direct to Factory with old vacuum wiper motor. Dealer is authorized—but not required—to make to the car owner a $1.00 allowance, off the retail price of New Motor, for his old Motor.

## prices

| motors—crank drive | retail prices | cartons of | shipping cases | shipping weight |
|---|---|---|---|---|
| motors | | | | |
| Model MV-2 | $7.45 each | One motor | any quantity | 1½ lbs. ea. motor |
| Model MV-3 | 7.95 each | One motor | any quantity | 1¾ lbs. ea. motor |
| Model MV-4 | 8.45 each | One motor | any quantity | 1¾ lbs. ea. motor |
| Model MV-5N | 9.55 each | One motor | any quantity | 1¾ lbs. ea. motor |
| Model MV-5T | 9.55 each | One motor | any quantity | 1¾ lbs. ea. motor |
| Model MV-6N | 9.55 each | One motor | any quantity | 1¾ lbs. ea. motor |
| Model MV-6T | 9.55 each | One motor | any quantity | 1¾ lbs. ea. motor |
| Model MV-7T | 9.55 each | One motor | any quantity | 1¾ lbs. ea. motor |

All Prices shown herein are Retail Prices. See Distributor Salesman's Price List for full schedule of Dealer's Cost Prices. Prices maintained under the Fair Trade Acts of all states in which applicable Acts are in effect. A two per cent (2%) reduction from ANCO Retail Prices is permitted where retail sales are for cash and without installation service.

1956 brochure

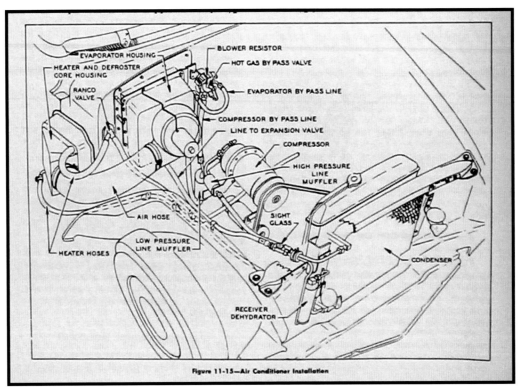

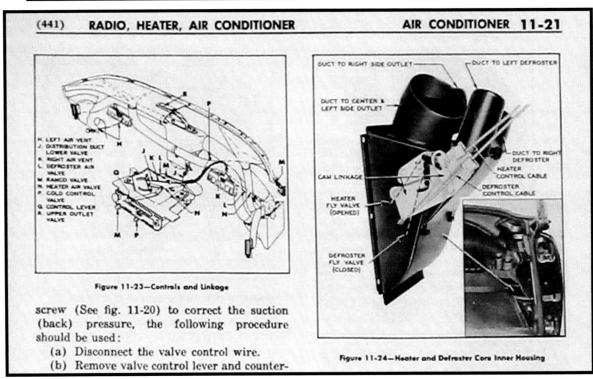

1956 Buick service manual

The 1956 Buick air conditioner system. Note an early application of a cable operated cam door control sub system

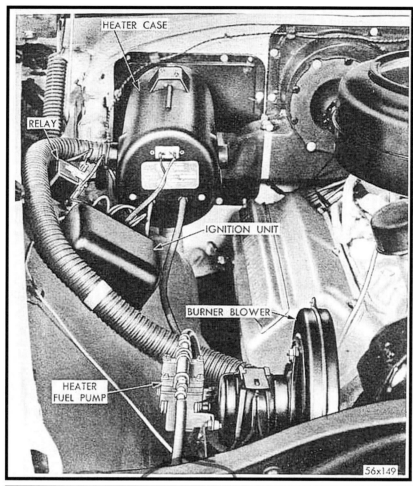
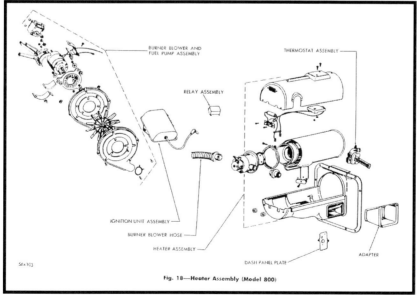

1956 Chrysler service manual

The 1956 Chrysler gasoline heater.

371

1956 JC Whitney

A 1956 electric windshield wiper motor for Ford Model A and others.

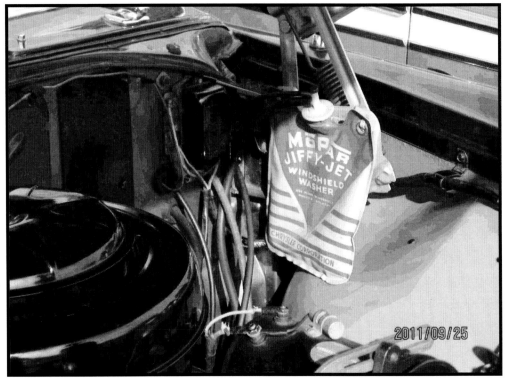

Vehicle Owner Bob Justice

A 1956 DeSoto fitted with MoPar windshield washer bag.

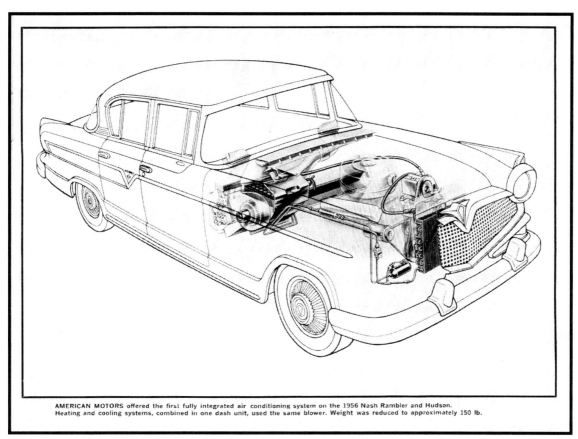

The 1956 Hudson was fitted with a fully integrated air conditioner and heater.

This 1957 MULTI-SPRAY was a forerunner to the 1971 "wet arm" by Trico.

## AIR CONDITIONING

AIR CONDITIONING for automobiles will spread like wildfire in the next few years. It won't be long before air conditioning will be as common on higher priced models as heaters; the next thing to standard equipment. It will be a little longer before this status is reached on low priced cars, but the day is coming. Prices of units are coming down as volume goes up and systems are being simplified. Indeed, some in the industry foresee the day when cars will be designed with integral air conditioning systems. Windows will be sealed and all air entering the passenger compartment will come through the conditioner. This era, they admit, is still a long way off, however. The day when you can get an automotive air conditioning unit for $150 or so is not.

ROLL-DOWN WINDOWS in the rear are likely to be used on more cars in the near future. Mercury's Turnpike Cruiser has this feature now and Buick and Olds reportedly planned to have it for 1957—thus, the ill-fated three-piece rear windows on these cars. Big advantage, aside from the psychological feeling of added spaciousness they give, is the greatly improved ventilation. Only thing that might stop the move to roll-down rear windows is the advent of fully sealed cars with integral air conditioning, but this development seems to be some years away right now (see section on Air Conditioning).

## ROLL-DOWN WINDOWS

Motor Life August 1957

1957 JC Whitney

JC Whitney offered evaporative coolers in three different styles in 1957.
Forecasts for roll-down rear windows and air conditioners in 1957.

374

Vehicle Owner James E. Johnson

A 1957 Chevrolet fitted with factory air conditioning.

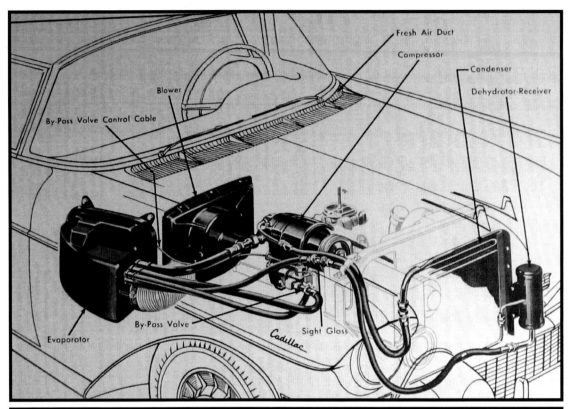
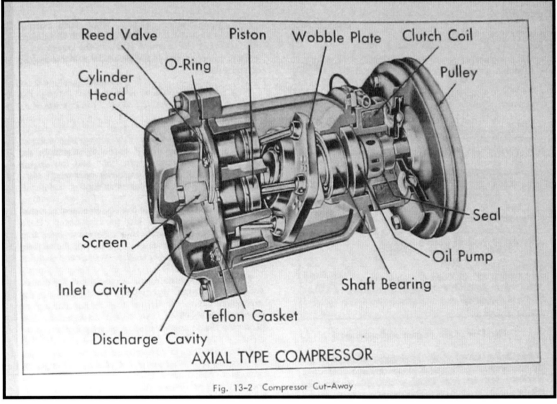

1957 Cadillac service manual

The 1957 Cadillac air conditioner system.

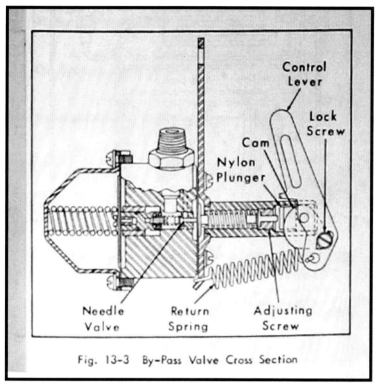

1957 Cadillac service manual

The 1957 Cadillac air conditioner by-pass valve.

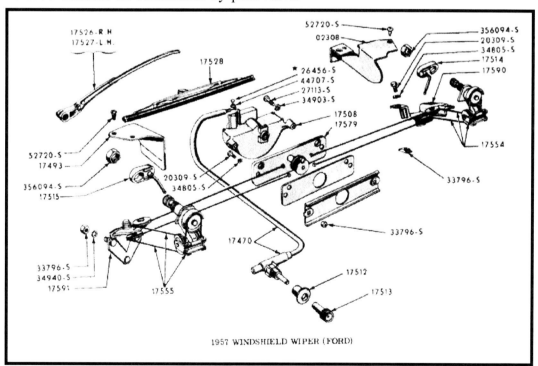

1957 Ford service manual

The 1957 Ford vacuum motor cable wiper system.

## ACCESSORY TYPE AIR-CONDITIONING UNITS

| Name and Manufacturer | Type | Installed List Price | Temperature Control |
|---|---|---|---|
| **AIRTEMP/MOPAR** | Under-Dash | $365 | B |
| Airtemp Division, Chrysler Corp., Dayton, Ohio | | | |
| **ALLSTATE** | Under-Dash | 344 | B |
| Sears, Roebuck & Co., Executive Offices, Chicago, Ill. | | | |
| **A.R.A.** | | | B |
| Direct-Aire | Under-Dash | 335 | |
| Ford Thunderbird | Under-Dash | 399 | |
| President (with grilles) | Trunk | 423 | |
| President (with ducts) | Trunk | 430 | |
| Station Wagon | Overhead | 545 | |
| A.R.A. Manufacturing Co., 1041 Foch St., Fort Worth, Texas | | | |
| **ARCTIC** | | | N.S. |
| Under-Dash Model | Under-Dash | 379 | |
| Trunk Model | Trunk | 489 | |
| Overhead Model | Overhead | 565 up | |
| Arctic Automotive Air Conditioning, 926 S. Sixth St., Tucson, Ariz. | | | |
| **ARTIC-KAR** | | | B |
| Penguin | Under-Dash | 299* | |
| Iceberg | Under-Dash | 339* | |
| Husky | Trunk | 379* | |
| Alaskan Icemaker ① | Trunk | 419* | |
| Polar ③ | Front | 295* | |
| Capitol Refrigeration Manufacturing Co., 3922 Kalloch Dr., Dallas, Texas | | | |
| **CLIMATIC-AIR** | | | B |
| Under-Dash Model | Under-Dash | 336 | |
| Trunk Model | Trunk | 350 | |
| Climatic-Air Manufacturing Co., 804 West Erwin St., Tyler, Texas | | | |
| **COOL QUEEN** | | | B |
| Under-Dash Model | Under-Dash | 379 | |
| Trunk Model | Trunk | N.S. | |
| Klaus-Joyce Inc., 5526 Dyer St., Dallas, Texas | | | |
| **FORSTON** | | | B |
| Under-Dash Model | Under-Dash | 369 | |
| Trunk Model | Trunk | 419 | |
| Forston Corporation, 1400 Conti St., Houston, Texas | | | |
| **FRIGETTE** | Under-Dash | 325 | B |
| Frigiquip Corp., 3724 N. May Ave., Oklahoma City, Okla. | | | |
| **FRIGIKING** | | | B, C |
| Front Mount | Under-Dash | $378 | |
| Rear Unit | Trunk | 419 | |
| Deluxe Rear Unit | Trunk | 439 | |
| Frigikar Corporation, 1602 Cochran St., Dallas, Texas | | | |
| **LO-MERC** | | | B |
| B-300 | Under-Dash | 408 | |
| TB-100 ③ | Under-Dash | 458 | |
| Lo-Merc Corporation, 2402 Houston Ave., Houston, Texas | | | |
| **MARK IV** | | | D |
| Dash Model | Under-Dash | 295* | |
| Trunk Model | Trunk | 345* | |
| John E. Mitchell Co., 3800 Commerce, Dallas, Texas | | | |
| **MOBILETTE** | Under-Dash | 350 | E |
| **WEATHER-MATIC** | Trunk | 450 | E |
| Mobil-Aire Manufacturing Co., Box 122, Denison, Texas | | | |
| **NOVI** | | | H |
| Under-Dash Model | Under-Dash | 310 | |
| Trunk Model | Trunk | 375 | |
| Novi Sales and Service Co., Inc., Novi, Mich. | | | |
| **PARKOMAT** ④ | Under-Dash | 348 | E |
| Parkomat Manufacturing Co., 2000 So. Akard, Dallas, Texas | | | |
| **TOWNE AND COUNTRY** | | | E, F |
| Champion | Under-Dash | 318 | |
| Clipper | Under-Dash | 348 | |
| Statesman | Trunk | 388 | |
| Clardy Automobile Air Conditioning Co., 1728 Layton St., Ft. Worth, Tex. | | | |
| **VORNADO** | Under-Dash | 370 | B |
| O. A. Sutton Corp., 1812 West 2nd St., Wichita, Kansas | | | |
| **WIZARD** | | | A, G |
| Wizard 2 | Under-Dash | 300 | |
| Western Auto Supply Company, 2107 Grand Ave., Kansas City, Mo. | | | |

\* Installation charge not included
N.S. Not stated
A Manual compressor on-off switch
B Manual temperature selector and thermostatic control of magnetic compressor clutch
C Constant cooling position on temperature selector
D Manual temperature selector and modulating valve which unloads compressor as required
E Pre-set temperature control and thermostatic control of magnetic compressor clutch
F Automatic/constant cooling selector switch
G Pre-set temperature control
H Manual temperature selector—magnetic clutch optional
① Has food, beverage, ice cube compartment
② Front unit for Chrysler Corp. cars
③ For Ford Thunderbird
④ Has pushbutton defrost control

### FACTORY-INSTALLED AIR-CONDITIONING PRICES

| | | | | | |
|---|---|---|---|---|---|
| Buick ...$430.00 | Chrysler . 506.00 | Ford .... 412.50 | Lincoln . 475.00 | Olds .... 444.00 | Pontiac . 431.00 |
| Cadillac . 478.00 | DeSoto .. 446.00 | Hudson .. 415.00 | Mercury . 430.00 | Packard . 440.00 | Rambler . 362.00 |
| Chevrolet 430.00 | Dodge ... 380.00 | Imperial . 590.00 | Nash .... 415.00 | Plymouth 446.00 | Stude ... 395.00 |

Motor Trend July 1957

A list of air conditioner manufacturers and factory-installed prices in 1957.

1957 brochure

Trico Marchal windshield washer system.

1958 Mercury service manual

The 1958 Mercury instrument panel controls-one with slide levers and two with rotary dials.

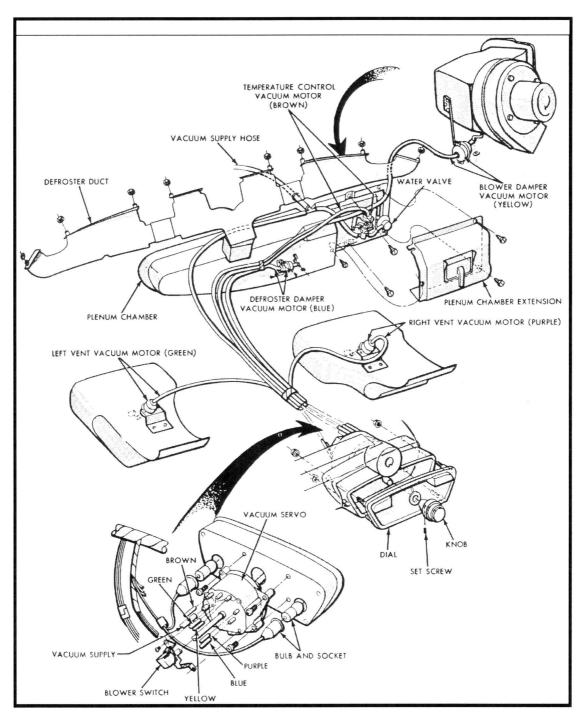

1958 Mercury service manual

An early application of vacuum motors to actuate doors rather than push-pull (Bowden) cables.

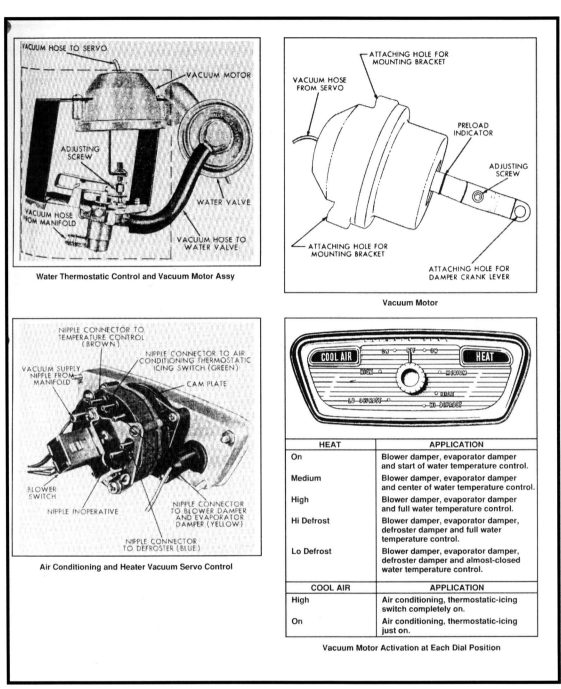

1958 Mercury service manual

A vacuum motor was used to actuate the thermostatically controlled water valve.

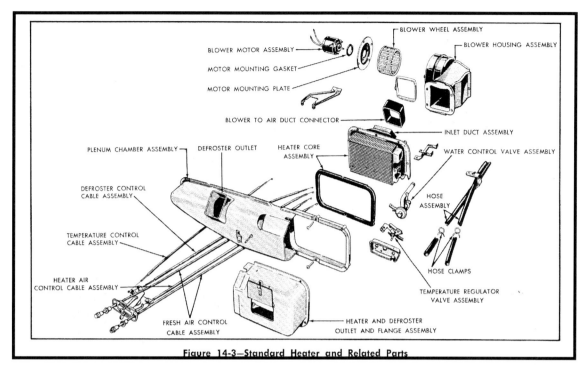

1958 Mercury service manual

The 1958 Mercury fresh air heater.

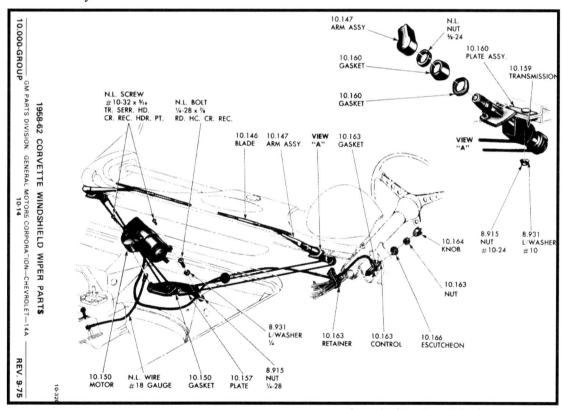

Chevrolet Corvette parts book

The 1958 Corvette wiper cable system.

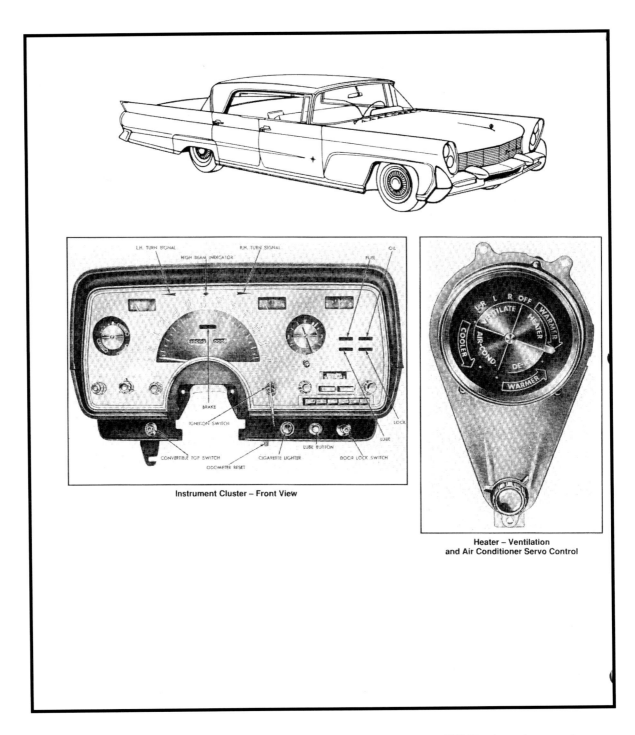

1958 Lincoln service manual

The 1958 Lincoln air conditioner instrument panel control.

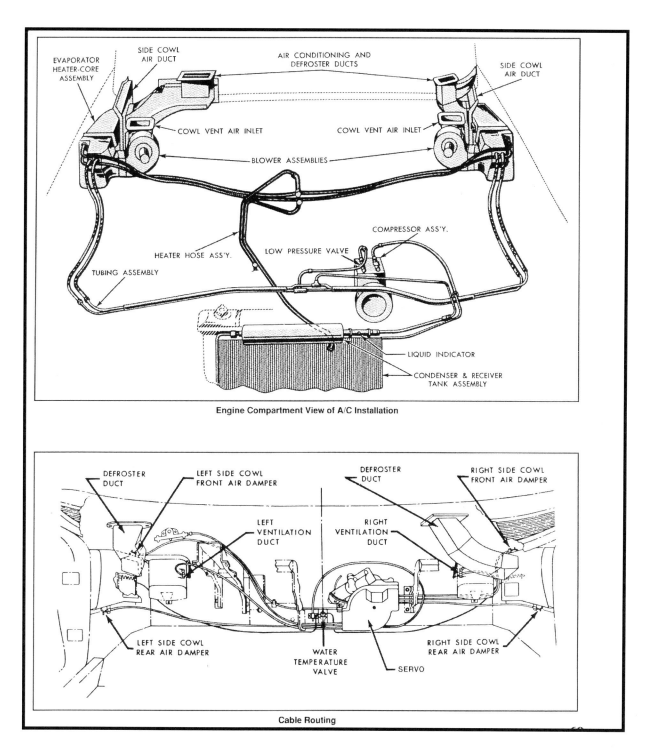

1958 Lincoln service manual

This system of push-pull cables and an electrical servo to actuate the cables along with two evaporator/heater assemblies was surely an engineering challenge for that era. This Lincoln system included dual heaters.

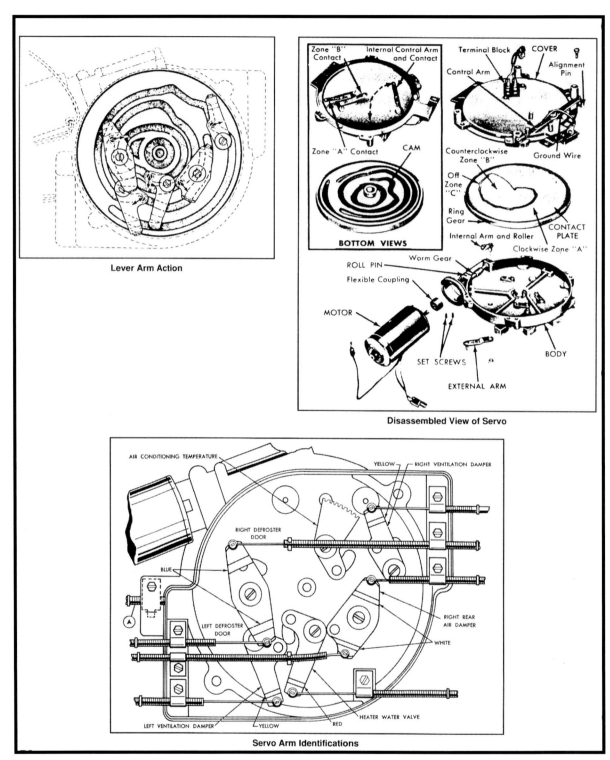

1958 Lincoln service manual

Details of the servo assembly.

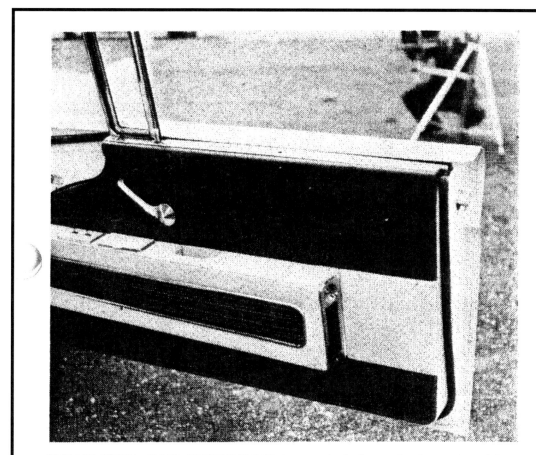

**HEAT FOR CAR INTERIOR** is carried through ducts in either front door and directed to rear seat. This setup was also successful on the Mark II Continentals now carried over to the Mark III cars.

Car Life   March 1958

The 1958 Lincoln door duct, register and control knob.

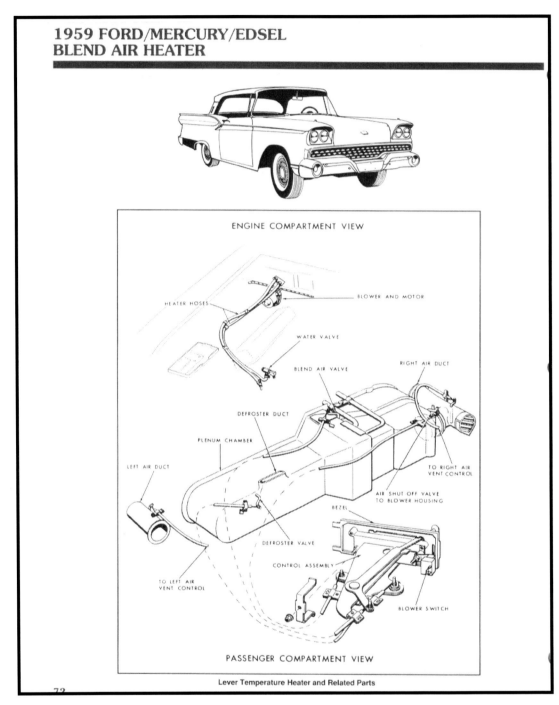

1959 Ford service manual

The 1959 Ford blend air heater system.

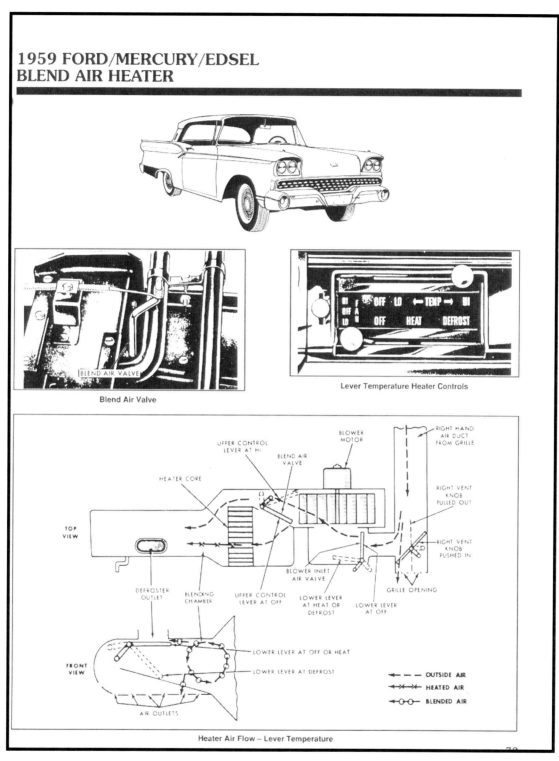

1959 Ford service manual

The 1959 Ford blend air heater system.

# 1959 FORD/MERCURY/EDSEL BLEND AIR HEATER

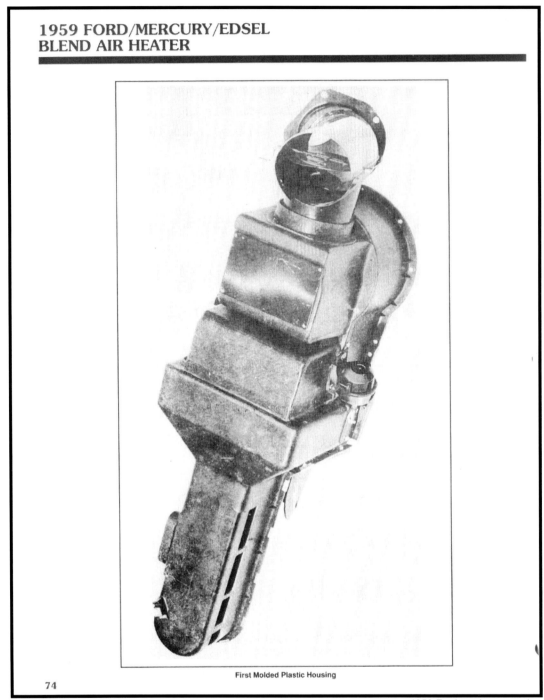

First Molded Plastic Housing

1959 Ford service manual

An early application of molded thermoset plastic housing. This technology would replace all metal housings on future heaters and air conditioners.

## Travel-Aire car cooler helps beat desert heat

WE HAVE FOUND a very satisfactory evaporative-type cooler, Firestone's Travel-Aire for $49.95, that takes the edge off 100° summer heat. Drop legs place the unit over the driveshaft hump, and the only connection is a tap onto a hot electrical lead to power the fan. Fill the tank with approxi-

mately five gallons of water, open a fresh air vent to duct hot air over the cooler, crack a window as an air pressure bleed, and cool air flows through the car.

MOTOR TREND took a Travel-Aire-
*continued*

**continued**

equipped Dodge pickup across the 110° Mojave desert (Dec. '58), and the temperature stayed at a comfortable 80° inside: Awarded MOTOR TREND Seal of Approval

Motor Trend June 1959

A road test report on an evaporative cooler.

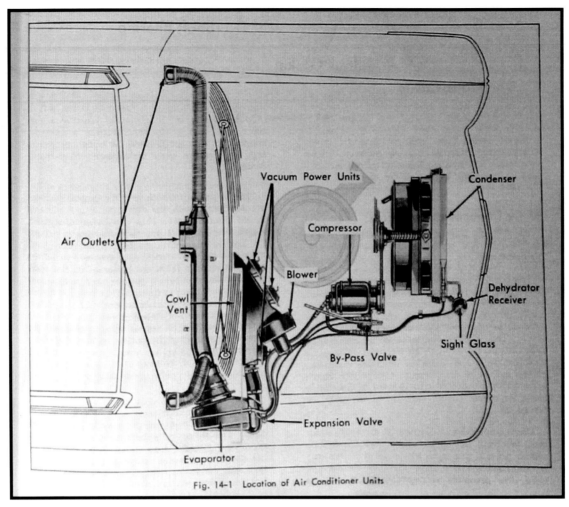

The 1959 Cadillac air conditioner system.

1959 Cadillac service manual

The 1959 Ford air conditioner hang-on type.

Vehicle Owner David Harvey

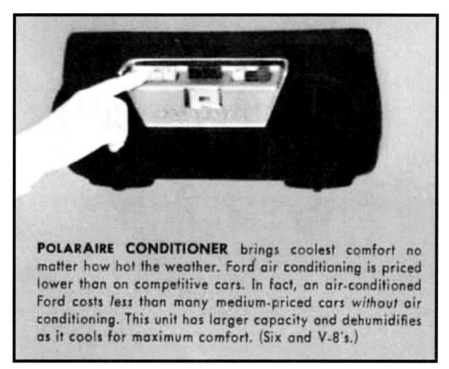

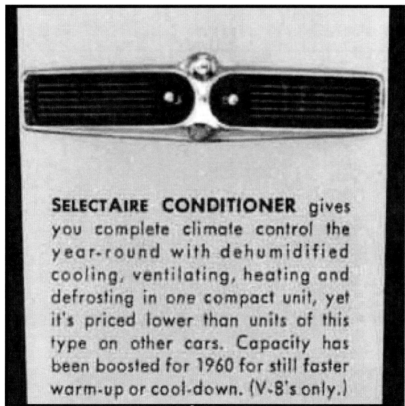

1960 Ford Options and Accessories brochure

The 1960 Ford was available with PolarAire and SelectAire air conditioners.

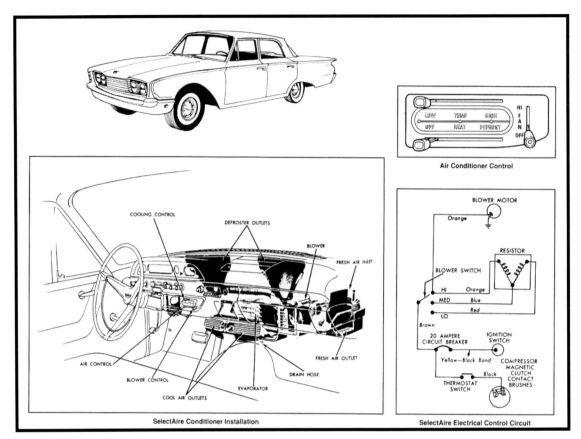

1960 Ford service manual

The 1960 Ford air conditioner system with wiring schematic.

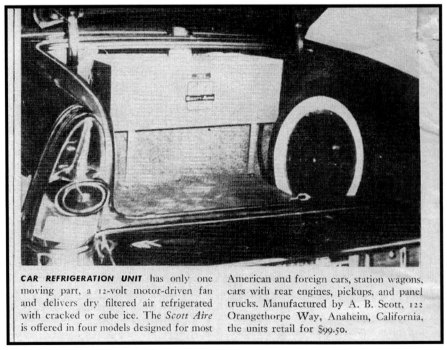

**CAR REFRIGERATION UNIT** has only one moving part, a 12-volt motor-driven fan and delivers dry filtered air refrigerated with cracked or cube ice. The *Scott Aire* is offered in four models designed for most American and foreign cars, station wagons, cars with rear engines, pickups, and panel trucks. Manufactured by A. B. Scott, 122 Orangethorpe Way, Anaheim, California, the units retail for $99.50.

Motor Life February 1960

An A.B.Scott company evaporative cooler called the Scott Aire.

# 12-24 GROUP 12—LIGHTS, INSTRUMENTS, AND ACCESSORIES

**FIG. 22—SelectAire Air Cooling System**

Figure 22 shows the entire Select-Aire cooling system in schematic form. Arrows indicate the direction of refrigerant flow. Figure 23 shows the electrical control circuits.

### RECEIVER UNIT

The air cooling system stores the liquid Refrigerant-12 under pressure in a combination receiver and dehydrator (Fig. 24). The pressure in the receiver normally varies from about 100 to 250 psi (pounds per square inch), depending on the surrounding air temperature and compressor speed. The receiver and condenser comes charged and marked with the total weight, so that any leak can be detected before assembly.

The dehydrator serves the purpose of removing any traces of moisture that may have accumulated in the system. Even small amounts of moisture will cause an air cooling unit to malfunction. A rupture disc is screwed into the top of the receiver (Fig. 24). This disc will rupture and release the refrigerant when the pressure in the system reaches 475 to 575 psi, or when the refrigerant temperature exceeds 212° psi.

### EVAPORATOR UNIT

When the cooling system is in operation, the liquid Refrigerant-12 flows from the combination receiver and dehydrator unit through a flexible hose to the evaporator (cooling unit) (Fig. 25), where it is allowed to evaporate at a reduced pressure. The evaporator is mounted on the engine compartment side of the dash.

### EXPANSION VALVE

The rate of refrigerant evaporation is controlled by an expansion valve (Fig. 25) which allows only enough refrigerant to flow into the evaporator to keep the evaporator operating efficiently, depending on its heat load.

The expansion valve consists of the valve and a temperature sensing capillary tube and bulb (Fig. 25). The bulb is clamped to the outlet pipe of the evaporator. Thus, the operation of the valve is controlled by the temperature of the evaporated liquid at the point where it leaves the evaporator or cooling unit. An equalizer connection at the evaporator outlet applies evaporator outlet pressure to one side of the valve diaphragm. Thus, the

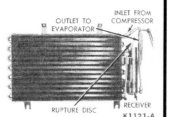

**FIG. 24—Receiver and Condenser**

valve is controlled by both evaporator outlet temperature and outlet pressure.

The restricting effect of the expansion valve at the evaporator causes a low pressure on the low pressure side of the system of 16-55 psi, depending on the surrounding air temperature and compressor speed.

### LIQUID SIGHT GLASS

A liquid sight glass is mounted in the high pressure refrigerant line along the left fender apron (Fig. 26). The sight glass is used to check whether or not there is enough liquid refrigerant in the system. At no time should bubbles be seen in the sight glass.

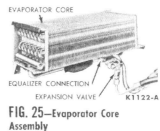

**FIG. 25—Evaporator Core Assembly**

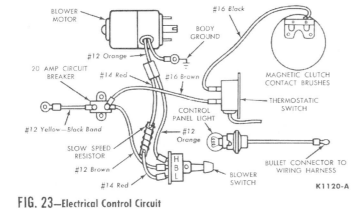

**FIG. 23—Electrical Control Circuit**

1960 Ford Thunderbird service manual

The 1960 Ford Thunderbird air conditioner system with wiring schematic. The system used a tube fin evaporator core.

## PART 12-2 — ACCESSORIES

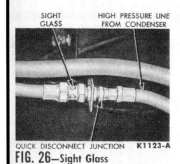

FIG. 26—Sight Glass

### COMPRESSOR UNIT

The evaporated refrigerant leaving the evaporator, now in the form of a gas at a pressure of 16-55 psi, is pumped by the compressor (Fig. 27), located on the engine, (Fig. 35), into the top of the condenser (Fig. 24). The condenser is located in front of the radiator.

The compressor maintains a pressure on its high pressure side of from 100-250 psi, depending on the surrounding air temperature and compressor speed.

As the now heated and compressed refrigerant gas flows down through the condenser, it is cooled by air passing between the sections of the condenser, and the cooled, compressed refrigerant gas condenses to liquid refrigerant which then flows back into the receiver.

### MAGNETIC CLUTCH

It is necessary to control the

FIG. 27—Compressor

amount of cooling that the system produces. To accomplish this, the compressor is cut in and out of operation by the use of a magnetic clutch pulley mounted on the compressor crankshaft (Fig. 28). The magnetic

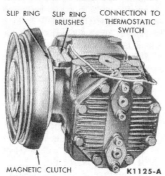

FIG. 28—Magnetic Clutch

clutch is controlled by a thermostatic sensing tube inserted in the fins of the evaporator core.

### THERMOSTATIC SWITCH

The thermostatic switch (Fig. 29), controls the operation of the compressor by controlling the compressor magnetic clutch. The temperature sensing tube of the switch is placed in contact with the evaporator fins. When the temperature of the evaporator becomes too cold, the thermostatic switch opens the magnetic clutch electrical circuit disconnecting the compressor from the engine. Refrigerant continues to flow until the high and low pressures equalize. When the temperature of the evaporator rises to the upper limit at which the thermostatic switch is set, the thermostatic switch closes and energizes the magnetic clutch. This connects the compressor to the engine and cooling action begins again.

When the ignition switch is off, or the cooling control (Fig. 30), is in the off position, the magnetic clutch is not energized, and the cooling system can not operate.

When the ignition switch is ON (engine running), and the cooling control is in the cooling range, the magnetic clutch is energized, the compressor is connected to the engine and the cooling system is in operation.

The thermostatic switch may be adjusted to maintain an average evaporator temperature of from 30°-60°F. The thermostatic switch oper-

ating differential temperature at any one setting is 6°F. The switch is controlled by the cooling control (Fig. 30 middle lever). The further to the left that the control is moved, the cooler the setting of the thermostatic switch.

### SERVICE VALVES

The service valves on the compressor are used to test and service the cooling system (Fig. 22). The high pressure service valve, mounted at the outlet to the compressor, allows access to the high pressure side of the system for attaching a pressure gauge, or a servicing hose.

The low pressure valve, mounted at the inlet to the compressor, allows access to the low pressure side of the system for attaching a pressure gauge, or a servicing hose.

Both service valves may be used

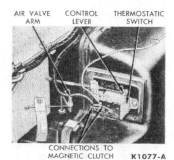

FIG. 29—Thermostatic Switch

to shut off the rest of the system from the compressor during compressor service.

### CONTROL OPERATION

The operating controls for the SelectAire Conditioner consist of the control panel (Fig. 30) and the instru-

FIG. 30—Control Panel

1960 Ford Thunderbird service manual

The 1960 Ford Thunderbird air conditioner compressor and instrument panel control.

## PART 12-2 – ACCESSORIES    12-33

cover the bulb and pipe with the insulation. Check for leaks at the expansion valve connections.

Install the cover and install the evaporator. Check for leaks. Evacuate and charge the system.

### HEATER CORE

The heater used with the Thunderbird SelectAire unit is separate from the evaporator assembly. Follow the procedures for servicing this heater unit (page 12-21).

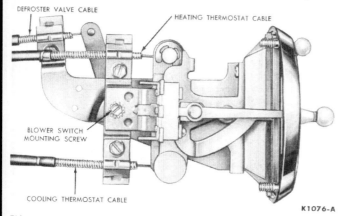

**FIG. 46—Blower Switch Removal**

### EVAPORATOR REMOVAL

Discharge the refrigerant from the system.

1. Remove the thermostatic switch cover from the evaporator case and disconnect the switch control cable and wires from the switch (Fig. 29).

2. Disconnect the evaporator air valve control cable (Fig. 29).

3. Remove the clamp from between the blower and evaporator housing.

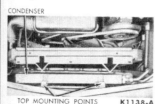

**FIG. 47—Condenser Mounting**

4. Remove the road draft tube from the intake manifold, and disconnect the accelerator linkage.

5. Disconnect the compressor to evaporator and the condenser to evaporator hoses at the evaporator.

6. Remove the two nuts and two screws that hold the evaporator assembly to the dash and remove the evaporator and case assembly.

7. Remove the thirteen clips, two bolts, and two screws that retain the covers and remove the covers (remove the evaporator inlet and outlet pipe cover and gasket first.)

8. The evaporator core may then be removed from the housing by removing the two evaporator core to housing expansion screws.

### EVAPORATOR INSTALLATION

1. Install the old expansion valve on the new evaporator core, leak test the assembly, place the core in the housing and attach the two mounting screws.

2. Attach the housing covers and mount the assembly to the dash.

3. Attach the refrigerant hoses to the evaporator (Fig. 45).

4. Install the road draft tube to the intake manifold and connect the accelerator linkage.

5. Connect the thermostatic switch wires and control cable, and the evaporator air valve control cable and adjust both cables.

6. Install the thermostatic switch cover and then install the blower housing to evaporator housing clamp.

7. Leak test the system, then evacuate and charge the system.

### CONTROL UNIT

When installing a control unit, adjust each Bowden cable for proper operation of the controls. The various cables are attached to the control as shown in Fig. 46. The blower switch is attached to the underside of the control (Fig. 46).

### CONDENSER AND RECEIVER

If the condenser is to be replaced, remove the original receiver along with the condenser as one assembly; because the replacement condenser is supplied only as one assembly with the receiver. The receiver, however, can be replaced as a separate unit.

**Condenser and Receiver Replacement.** Remove the grille to radiator support bracket (this also removes the condenser top mounting bolts (Fig. 47). The condenser lower mounting points, shown by the large arrows in Fig. 47, are held in position by the

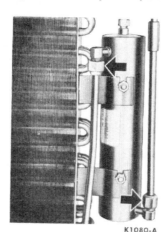

**FIG. 48—Receiver Mounted to Condenser**

---

1960 Ford Thunderbird service manual

The 1960 Ford Thunderbird air conditioner condenser and receiver.

## 8 WINDSHIELD WIPER AND WINDSHIELD WASHER

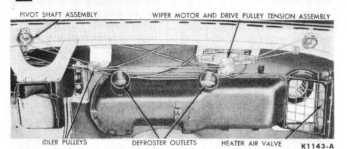

FIG. 54—Windshield Wiper and Heater Plenum

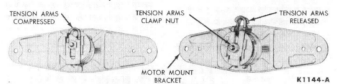

FIG. 55—Wiper Cable Tension Assembly

### WINDSHIELD WIPER

The windshield wiper assembly is shown in Fig. 54.

If service is required on the motor assembly, control assembly, or pivot shaft assemblies they may be removed separately. To remove any of the windshield wiper assemblies it will first be necessary to remove the glove box.

To remove the wiper motor, in addition to the glove box, remove the right defroster air duct. Remove the motor bracket to bracket mount nuts, disconnect the pivot shaft cables, disconnect the vacuum hoses and control head cable and remove the motor. Remove the motor from the shaft assembly.

Before installing a new motor, loosen the motor drive pulley tension clamp nut, compress the tension arms and tighten the nut (Fig. 55). Mount the motor to the drive pulley assembly, install the drive head cable, attach the vacuum hoses, attach the drive cables (Figs. 54 and 56), and mount the assembly. Loosen the tension arms clamp nut to put tension on the cables then tighten it again.

To remove the pivot shaft assemblies, remove the wiper blade assemblies, remove the pivot shaft mounting nuts, slide the pivot assemblies out of the mounting holes, disconnect the cables from the motor drive pulley and remove the assemblies.

To gain access to the left pivot shaft assembly, remove the speedometer assembly. The pivot shaft assembly may then be removed through the speedometer mounting hole.

Before installing new pivot shaft assemblies, loosen the motor drive pulley tension arms clamp nut, compress the tension arms with a "C" clamp and tighten the nut (Fig. 55). Install the cables as shown in Figs. 54 and 56, mount the pivot shaft assemblies, release the tension arms clamp nut, then tighten it.

The control assembly may be removed from the instrument panel by removing the bezel nut after loosening and removing the control knob.

### WINDSHIELD WASHER

The windshield washer unit operates in conjunction with the windshield wiper. The washer control lever is mounted on the left instrument panel extension. Pushing the lever forward opens the vacuum line to a

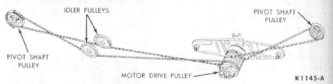

FIG. 56—Wiper Drive Cable Attachment

FIG. 57—Typical Windshield Washer Installation

1960 Ford Thunderbird service manual

The 1960 Ford Thunderbird windshield wiper cable system.

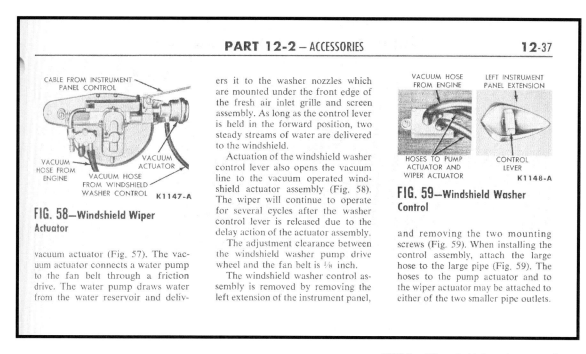

1960 Ford Thunderbird service manual

The 1960 Ford Thunderbird windshield vacuum wiper motor and washer control.

Vehicle Owner Robert Jenkins

A 1960 Chevrolet Corvair fitted with gasoline heater.

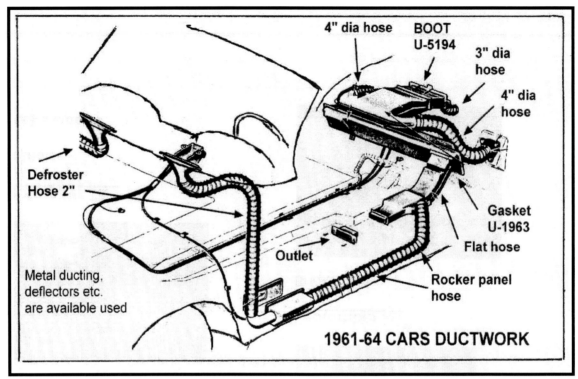

Image courtesy corvairunderground dot com

The 1961 Chevrolet Corvair hot air heater.

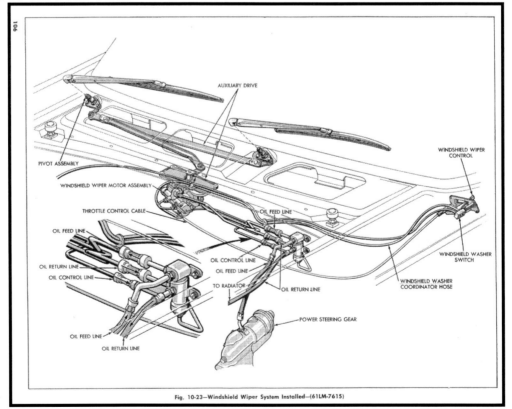

1961 Lincoln service manual

The 1961 Lincoln used a hydraulic system to drive the windshield wipers.

## AIR CONDITIONER

The major components of the 1961 Cadillac Air Conditioner system, Fig. 14-8, are the compressor, condenser, dehydrator-receiver, expansion valve, evaporator, by-pass valve, blower assembly and control panel. The components are dash mounted with the exception of the compressor, condenser and receiver, and the rear evaporator and blower assembly in 75 series cars.

The Air Conditioner permits the selection of either 100% outside air or a combination of 20% outside air mixed with 80% recirculated air. The Air Conditioner may be operated independently, or in conjunction with the heater, to provide maximum comfort control for all conditions. These combinations are made possible by the use of vacuum power units that actuate air valves in the blower housing assembly.

Outside air is supplied through the cowl vent, located directly below the windshield on all 1961 series cars. On 75 series cars, a second cooling

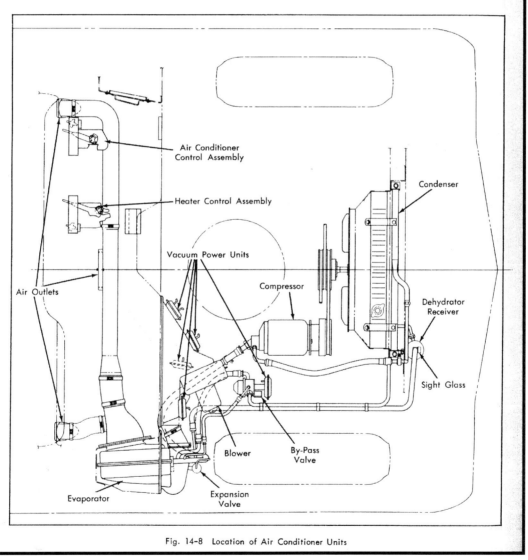

Fig. 14-8 Location of Air Conditioner Units

1961 Cadillac service manual

The 1961 Cadillac air conditioner system.

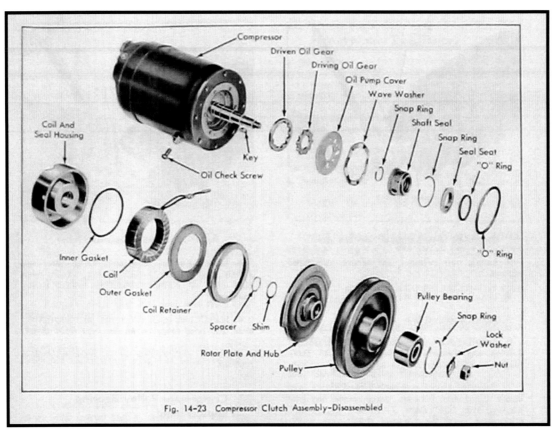
1961 Cadillac service manual

The 1961 Cadillac air conditioner compressor and clutch.

**BEETLE HEATER**

One of the few complaints ever heard from Volkswagen owners concerns the heating system. In cold weather, it seems to take forever to warm the passenger compartment. Central Automotive Sales, Dept. CL-5, 11410 E. Jefferson, Detroit 14, Mich., has an answer for this problem, a controlled heater booster that circulates 100 cu. ft. of air per min. The unit can be installed easily and quickly on any Volkswagen car or truck and comes complete with detailed instructions for $27.50, postpaid anywhere in the United States.

Car Life May 1962

Central Automotive Sales offered this Volkswagen heater booster in 1962.

Image taken at Ypsilanti, MI

The 1963 MG was fitted with 3 windshield wipers.

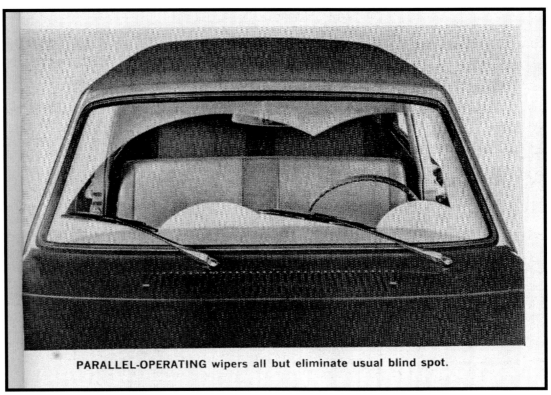

Car Life   March 1963

The 1963 Dodge Dart featured a new windshield wiper pattern.

- 37 -

Airfoil Windshield Wiper

CAUTION

Strict security must be observed for this item. It is imperative that any public or trade announcement of this device be timed in accord with announcement dates negotiated by Corporate Public Relations Staff with the wiper supplier.

A windshield wiper with improved wiping characteristics is provided on 1963 models. Aerodynamic principles are utilized in the design to give improved wiping action under conditions where blades are subject to windlift. For the car owner this means increased safety at expressway speeds as a result of improved vision during wet weather.

The wiper blade employs an airfoil instead of the conventional channel section for the major or central bridge of the blade assembly.

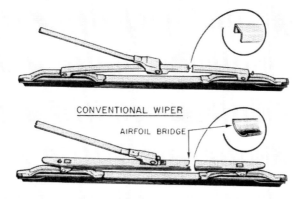

The airfoil bridge acts like an inverted airplane wing. Air flowing over its curved surface produces a force which increases the pressure of the blade on the windshield. The advantage of the airfoil is particularly noticeable at expressway and highway speeds -- where wind and aerodynamic forces tend to reduce the pressure of conventional blades against the glass.

The airfoil windshield wiper is another advanced automotive engineering feature developed by Chrysler Corporation to ensure safer all-weather operation at modern highway driving speeds.

1963 Chrysler Press Release

The 1963 model year Chrysler vehicles were fitted with airfoil windshield wiper blades.

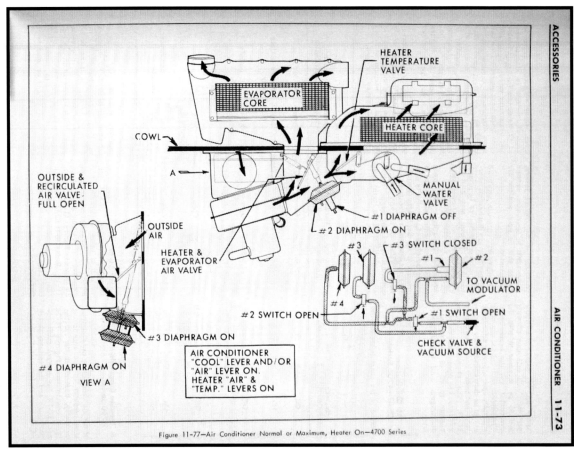

1963 Buick shop manual

The 1963 Buick introduced the reheat system.

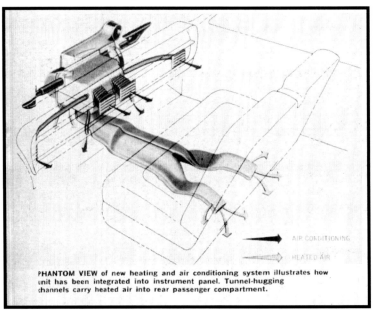

Car Life July 1964

The 1964 Lincoln rear seat heat ducts and integrated air conditioner.

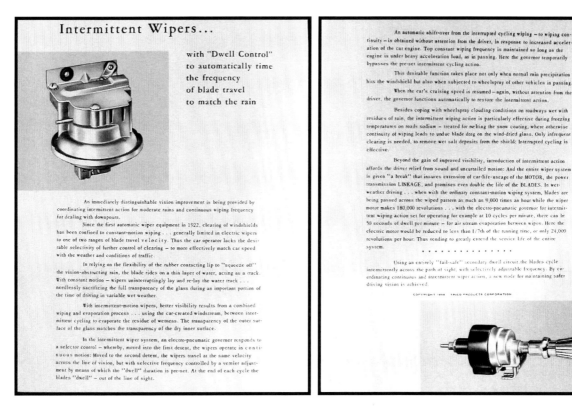

The 1964 Trico mechanical intermittent windshield wiper control.

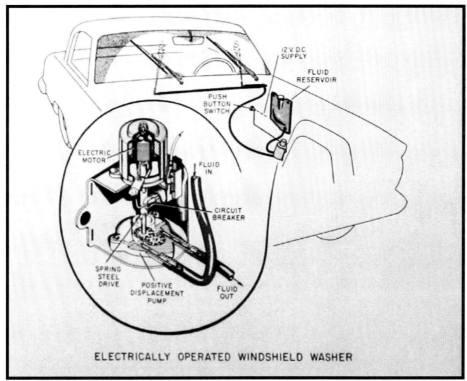

The 1964 Chrysler electrically operated windshield washer.

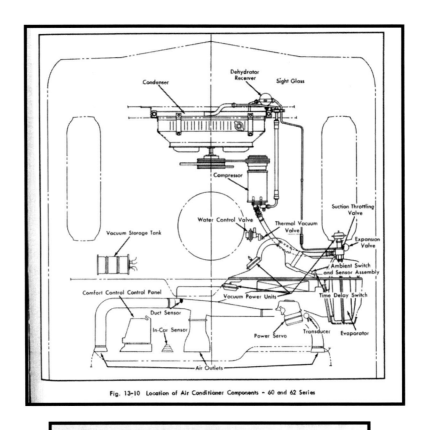

Fig. 13-10 Location of Air Conditioner Components – 60 and 62 Series

The 1964 Cadillac is equipped, on order, with one of two Air Conditioning systems. A new Comfort Control Automatic Air Conditioner system is used only on 60 and 62 Series cars, while 75 Series cars use a manual control Air Conditioner.

Comfort Control is an automatic system that controls the heating and Air Conditioning systems. It permits a constant interior temperature, between 65°F. to 85°F., to be maintained regardless of changes in the ambient air temperature, without any assistance from the driver, once the system has been put into operation.

To accomplish this, the Comfort Control system is composed of four major sections: Three temperature sensors (Thermistors) that sense the in-car temperature, the ambient air temperature and the discharge air temperature; a control panel that contains a transistorized amplifier and temperature dial; a transducer that converts an electrical signal into a modulated vacuum supply; and the power servo unit that controls the vacuum circuitry to operate the system and a circuit board that controls the operation of the blower.

1964 Cadillac service manual

The 1964 Cadillac Comfort Control air conditioner system.

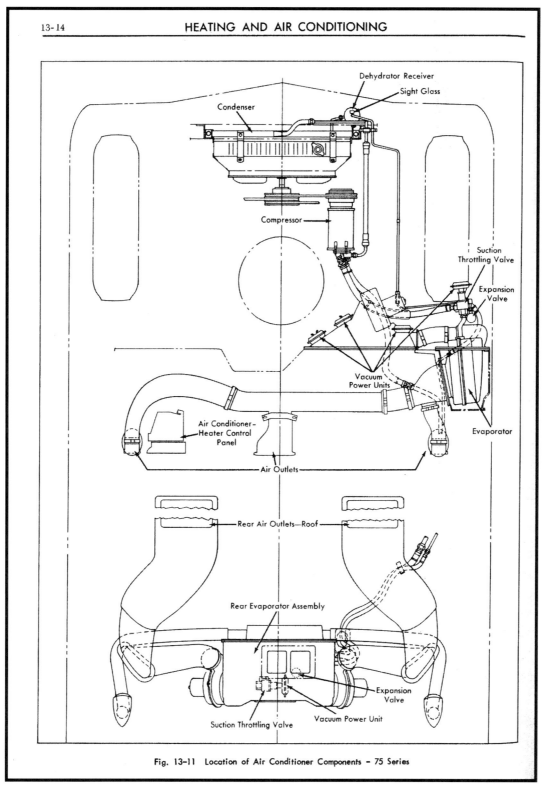

1964 Cadillac service manual

The 1964 Cadillac air conditioner system.

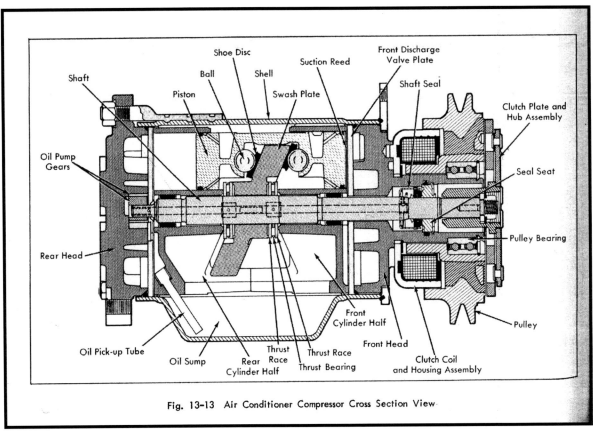

1964 Cadillac service manual

The 1964 Cadillac air conditioner compressor.

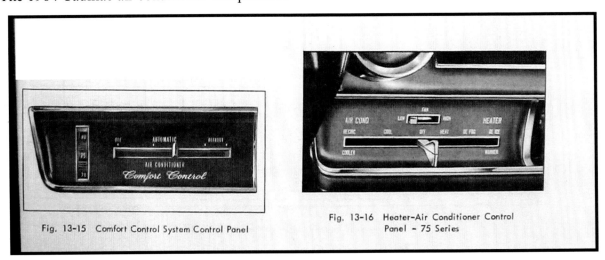

1964 Cadillac service manual

The 1964 Cadillac air conditioner instrument panel controls. The new Comfort Control is on the left.

1965 Ford training manual

This 1965 Ford air conditioner training manual provides an overview of the state-of-the art in that era.

## HOW AIR CONDITIONING WORKS

## OPERATING PRINCIPLES

### THE REFRIGERATION CYCLE

Now that we've seen how heat behaves and have considered the physical laws involved in air conditioning, we're ready to take a closer look at the refrigeration cycle (Fig. 17). By way of a quick review, here are the physical facts involved in mechanical refrigeration, in the order we'll encounter them in the cycle:

1. Whenever a substance changes from liquid to vapor, it must absorb a lot of heat.
2. A liquid can be stored at temperatures higher than the boiling point in a closed container. The pressure of the saturated vapor above the liquid stops the liquid from evaporating.
3. When the pressure is removed from a liquid that is warmer than its boiling temperature, it flashes into vapor and absorbs heat.
4. There is a definite relationship between the temperature and pressure of a saturated vapor.
5. Compressing a vapor that is not in contact with the generating liquid raises the temperature and pressure of the vapor.
6. Heat always flows from hot to cold.
7. When enough heat is removed from a vapor, it condenses back into a liquid.

8. For any appreciable amount of cooling to take place, the refrigerant must change state.

### EXPANSION AND VAPORIZATION

At the start of the refrigeration cycle, liquid refrigerant is about 1/9 the original pressure. So the refrigerant vaporizes readily and absorbs heat from the evaporator coils and fins. This makes the evaporator very cold, so it in turn absorbs heat from the air blown over the fins on the way to the passenger compartment. The air is cooled, and water vapor in the air is condensed out.

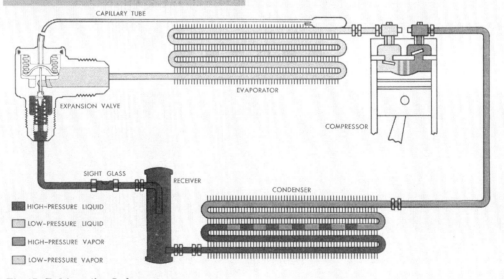

Fig. 17—Refrigeration Cycle

1965 Ford training manual

# OPERATING PRINCIPLES

## HOW AIR CONDITIONING WORKS

For maximum cooling, the flow of refrigerant is controlled so that there is vaporization throughout the evaporator. Just before the refrigerant leaves the evaporator, the last of the liquid changes to vapor.

The vapor in the evaporator is therefore saturated almost throughout. Because of the temperature-pressure relationship of a saturated vapor, the refrigerant, under low pressure, is very cold, despite the heat it has absorbed. Thus, the refrigerant leaving the evaporator picks up some additional heat and enters the compressor slightly superheated.

### COMPRESSION

The compressor operates as a pump to cause circulation of the refrigerant in the system, and it compresses the refrigerant vapor preparatory to removing the heat that was absorbed in the evaporator (Fig. 18).

On the downward or intake stroke of the pistons, the compressor creates a vacuum in the cylinder (Fig. 18). This vacuum causes the refrigerant to be drawn through the intake valve into the compressor, and creates the low-pressure condition in the evaporator.

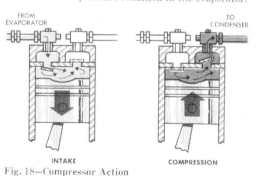

Fig. 18—Compressor Action

On the upward stroke, the piston compresses the vapor. Since the vapor is not saturated, compression increases the temperature and pressure. The vapor, now under high pressure and much hotter than the outside air, is forced past the outlet valve into the condenser.

### CONDENSING THE VAPOR

Outside air is pulled over the condenser fins by the engine fan. This cooler air removes heat from the warmer refrigerant in the condenser. In giving up the heat absorbed in the evaporator and the heat added by compression, the refrigerant condenses back into a liquid.

### DRYING AND STORAGE

Liquid refrigerant from the condenser next goes to the receiver-dehydrator. Moisture in the refrigerant is removed by a drying agent or desiccant. The receiver also acts as a storage chamber for the refrigerant until it's needed at the evaporator. In this way, we always have more refrigerant in the system than is needed for normal service, and small

1965 Ford training manual

# HOW AIR CONDITIONING WORKS

## OPERATING PRINCIPLES

losses do not affect the system operation. Space in this part of the system that is not completely filled with liquid refrigerant contains a saturated vapor—at high pressure, of course. From the receiver-dehydrator, the liquid refrigerant flows to the expansion valve for metering into the evaporator.

### CONTROLLING REFRIGERANT EXPANSION

The expansion valve has two functions. It provides a resistance for the compressor to pump against so pressure can build up, and it controls flow into the evaporator so that the pumping action of the compressor can lower the pressure in the evaporator.

It is easy to see that if the expansion valve were to let too much liquid into the evaporator, there wouldn't be enough room for all the refrigerant to expand and vaporize. So we'd have less than maximum cooling. Also, there would be liquid refrigerant leaving the evaporator and flowing into the compressor inlet. Since a liquid isn't compressible, the compressor would be damaged.

Now if the expansion valve lets too little refrigerant into the evaporator, it all vaporizes part way through. The vapor, being cold, of course still absorbs some heat and leaves the evaporator highly superheated. But there's much less cooling than if the refrigerant were changing state throughout the evaporator.

So, as we've mentioned before, maximum cooling is obtained by metering in just enough refrigerant so that the last of the liquid vaporizes just before the outlet, and the vapor leaves the evaporator slightly superheated. The expansion valve compares the temperature and pressure of the refrigerant and adjusts flow so that the vapor is just slightly superheated. How this is accomplished is described in the next part of this section under "Functional Description."

### CONTROLLING THE CYCLE

There will be times when we want less cooling than the system capacity—for example, when the outside air temperature is comparatively cool, or to prevent condensation from freezing up on the evaporator.

Since the expansion valve is permanently set for maximum cooling, we control the cycle by stopping the compressor when things get too cold. The compressor crankshaft is driven by a **magnetic clutch.** The clutch is energized only when the contacts in the **thermostatic switch** are closed. The thermostatic switch is set by the temperature control knob or lever. If the evaporator gets colder than the selected temperature, the contacts open. The compressor loafs then until the temperature rises and the contacts close.

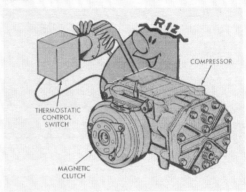

## FUNCTIONAL DESCRIPTION

### EVAPORATOR

The evaporator must serve as a heat exchanger to transfer heat from the air in the passenger compartment to the vaporizing refrigerant. It is constructed of several (3 or 4) rows of copper cooling coils which make several passes through many closely spaced aluminum fins (Fig. 19).

A large number of fins provides a maximum surface for transferring heat. Of course, aluminum and copper are both good heat conductors.

The evaporator housing serves as a duct for the cooled air. On some units, the controls permit fresh air to be cooled on the way to the passenger compartment, or fresh air may be cut off and the air

1965 Ford training manual

# OPERATING PRINCIPLES

## HOW AIR CONDITIONING WORKS

inside the car recirculated through the evaporator for maximum cooling. Fresh air cooling is usually adequate when the outside temperature is 85 degrees or less. The air is forced over the evaporator by the blower, and then to the various cool air outlets.

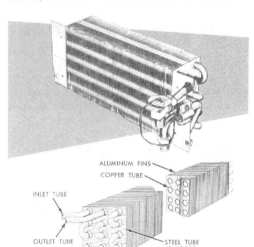

Fig. 19—Typical Evaporator

Since the evaporator also dehumidifies the air, there must be a means of getting rid of condensed moisture. So there is a water-collecting pan with a drain hole and drain tube in the housing under the evaporator.

In some installations, the evaporator is located in the passenger compartment as a separate unit. Often, though, it is an integral part of a combined heating-air conditioning system. In some truck installations, the evaporator is located outside on top of the cab, or in the center of the top inside the cab.

### COMPRESSOR

All compressors (Fig. 20) used in the Ford product line are two-cylinder reciprocating units. The compressor is mounted on an accessory plate at the front of the engine. The compressor crankshaft is driven through the magnetic clutch by a "V" belt or belts from the engine crankshaft.

The compressor bodies and heads are often made of light-weight alloy, and the cylinders are lined with cast-iron sleeves. A valve plate is sandwiched between the head and the top of the cylinder body.

The valve plate contains a spring-loaded intake and outlet valve for each cylinder. Gaskets are installed on both sides of the valve plate.

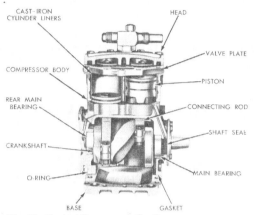

Fig. 20—Typical Compressor Section

The head and valve plate form two chambers on top of the compressor, and a service valve with the port and gauge connections is screwed on the head to connect into each chamber. The chamber on the low-pressure inlet side is connected to the evaporator outlet by a specially designed, flexible nonporous line. This side of the compressor is often referred to as the **suction** side, and contains the inlet valves. As the compressor pistons move down, creating a vacuum or suction, the intake valves open and allow refrigerant into the cylinder. The **discharge** or **high-pressure** side of the compressor contains the outlet valves. These valves open on the upward piston strokes to allow the compressor vapor to flow to the condenser inlet (Fig. 21).

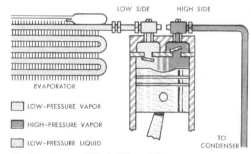

Fig. 21—Low and High Sides of Compressor

Tube fin evaporator and two cylinder compressor.

1965 Ford training manual

# HOW AIR CONDITIONING WORKS

## OPERATING PRINCIPLES

The connection to the condenser inlet is made with the same type of hose as on the suction side. Since the refrigerant now needs less space, this line is smaller in diameter.

Crankshaft and crankcase. The compressor crankshaft is supported by two bearings. A seal is installed on the pulley end to prevent losing compressor oil and refrigerant. A plug in the crankcase permits checking the oil level and adding oil.

Compressor oil is specially refined to be as free from moisture as possible. Refrigerant 12 and refrigerant oil are miscible—that is, they mix together easily. However, they do not react chemically, so one cannot cause the other to break down.

Compressor Mounting. Compressors can be left- or right-hand mounted. A left-hand compressor can be mounted from vertical to 90 degrees left of vertical. Right-hand compressors can be mounted from vertical to 90 degrees right of vertical.

## MAGNETIC CLUTCH

A magnetic clutch (Fig. 22) is used to connect and disconnect the compressor from the engine. The clutch hub is mounted on the tapered end of the compressor crankshaft. A self-locking cap screw is threaded into the end of the crankshaft and retains the hub. The Woodruff key in the crankshaft is indexed with the keyway in the clutch hub. A clutch plate is attached to flat springs on the clutch hub and a double-row ball bearing is pressed on the hub.

The clutch pulley is installed on the double-row ball bearing and is retained by a snap ring. An electromagnet is built into the pulley and is electrically connected to a brass slip ring which is fastened to the pulley with screws. Two carbon brushes are contained in a holder attached to the compressor and these brushes contact the slip ring. A friction facing is incorporated in the front face of the pulley, and is just back of the clutch plate.

Whenever the engine is operating, the pulley is turning since it is connected to the crankshaft by a "V" drive belt, but the compressor does not operate until the clutch is energized. When the air conditioning system is on, and the thermostatic control points are closed, current flows through the carbon brushes, to the brass slip ring, and to the electromagnet. The electromagnet then attracts the clutch plate and pulls the plate against the friction facings on the pulley. The friction between the facings and plate causes the assembly to rotate as a unit, and drive the compressor. When the points open, the magnetic clutch is disengaged, and the compressor stops.

With the pulley rotating at or above engine speeds at all times, and the compressor crankshaft stationary when the clutch is disengaged, a problem is created. If the clutch engages slowly, the life of the friction facings is shortened. If the clutch engages suddenly, it puts a severe load on the drive belt, and could cause the belt to come off the pulley. This problem is solved by engaging the clutch quickly and using a heavy pulley. Enough energy is stored up in the heavy rotating pulley to bring the crankshaft up to pulley speed without placing a sudden heavy load on the drive belt.

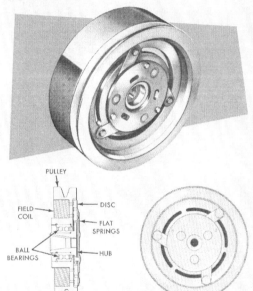

Fig. 22—Typical Magnetic Clutch

## CONDENSER

The condenser also serves as a heat exchanger. You could say that it operates in the reverse of the evaporator. It removes the heat that was absorbed in the evaporator so that the refrigerant condenses back to a liquid.

The construction of the condenser is quite similar to the evaporator—copper tubing coils running back and forth between many aluminum fins for maximum heat transfer (Fig. 23). In the condenser, there are usually two rows of tubing and the tubing diameter is smaller, although some condensers use only one row of flat tubes.

19

1965 Ford training manual

Electromagnetic clutch.

## OPERATING PRINCIPLES

### HOW AIR CONDITIONING WORKS

In most automotive air conditioning installations, the condenser is mounted in front of the radiator. This tends to reduce air flow through the radiator, and of course, the air is much warmer when it flows through the radiator. In most cases, though, the radiator capacity is high enough to handle the extra heat load. In some installations, a higher capacity radiator and cooling fan are part of the air conditioner package. Also, the engine cooling system operates at higher pressure with a high-pressure radiator cap. This prevents the coolant from boiling away up to about 250 degrees.

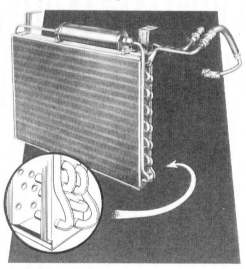

Fig. 23—Typical Condenser

**Truck Installations.** In air-conditioned trucks, the condenser is usually in a housing mounted on top of the cab. Here the condenser is exposed to cooler air flow. When the truck is stopped or moving slowly, an electric blower moves the air through the condenser.

**Sub-cooling.** Some installations use a slightly modified condenser. In the standard-type condenser, the refrigerant enters at the top and flows through the coils to the bottom; then to the receiver-dehydrator. In the **sub-cooling** design, the liquid refrigerant leaves the condenser at an intermediate point and goes through the receiver-dehydrator. Then the refrigerant goes back through the bottom coils of the condenser for additional cooling. This system reduces the tendency of bubbles to form in the liquid.

### RECEIVER-DEHYDRATOR

The receiver-dehydrator is used as a storehouse for liquid refrigerant and to remove moisture which could freeze at the expansion valve. The receiver is made up of a section of heavy steel tubing with a cover brazed in each end. Inlet and outlet fittings are provided and their location is determined by the receiver mounting position. The receiver may be horizontal or vertical, depending on the space available for the installation. In any case, the outlet tube or **quill** is positioned so that the end of the tube is as far below the level of the liquid refrigerant as possible. The receiver in the illustration is designed for a horizontal mounting position, and the quill tube is curved to place it well below the liquid refrigerant level (Fig. 24).

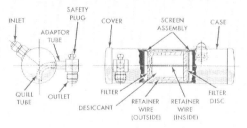

Fig. 24—Receiver-Dehydrator

The dehydrator or dryer portion is made up of two screens, filter discs and a dryer. This dryer is referred to as a desiccant and is used to remove any traces of moisture from the refrigerant. The material used in most of the system is calcium sulphate. The calcium sulphate is enclosed in a filter or **molecular sieve**. The molecular sieve is similar in appearance to a sintered bronze fuel filter. The filter and desiccant assembly is held together with a wire retainer as shown in Fig. 24.

A fusible plug is incorporated in all receiver assemblies. The purpose of this plug is to discharge the system when temperatures go above a safe limit. This fusible plug will open the system at a temperature of 232 degrees.

Condenser and receiver-dehydrator.

1965 Ford training manual

## HOW AIR CONDITIONING WORKS — OPERATING PRINCIPLES

### SIGHT GLASS

A sight glass is installed in the high-pressure line between the receiver and expansion valve, or in the expansion valve itself. The sight glass permits inspection of the refrigerant on its way to the expansion valve. Bubbles or foaming in the sight glass indicates that the refrigerant is low. Some bubbling is normal when the system is first started, but a solid stream of refrigerant should show in the glass after a minute or so. Otherwise, the system must be charged (filled) to the correct amount.

### EXPANSION VALVE

The expansion valve (Fig. 25) controls expansion of the refrigerant so that vaporization takes place throughout the evaporator and the refrigerant gas leaves the evaporator slightly superheated.

Operation of the expansion valve depends on the temperature of the refrigerant leaving the evaporator, and the pressure inside the evaporator. Refrigerant is metered by the ball valve, which is positioned by the spring and diaphragm. The amount of valve opening depends on the relative strength of the pressure and temperature signals.

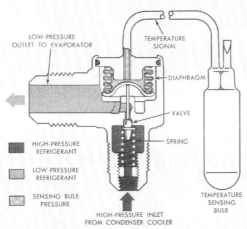

Fig. 25—Expansion Valve Construction

Evaporator pressure, tending to close the valve, acts directly on the diaphragm, and is assisted by the valve spring. The **temperature** signal comes from a temperature-sensing bulb clipped to the evaporator outlet. The bulb and the capillary tube to the opposite side of the diaphragm are filled with refrigerant or carbon dioxide. An increase in temperature increases pressure against the diaphragm, to open the valve. Decreasing the temperature decreases the pressure on top of the diaphragm, letting the valve close.

Of course, the expansion valve doesn't control the evaporator temperature or pressure. Rather, it compares the two and maintains a relationship which results in maximum cooling at all times. If there is too much liquid in the evaporator, we have a saturated vapor at the outlet. The temperature then is lower with respect to pressure and we get a larger pressure signal which closes the valve. If too little liquid enters the evaporator, we get a highly superheated vapor at the outlet. Temperature is too high with respect to pressure then, and the higher temperature signal opens the valve to let in more liquid.

The valve spring calibrates the valve for the design superheat. Changing the valve spring force would change the temperature-pressure relationship of the refrigerant leaving the evaporator.

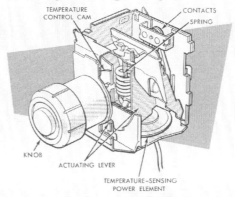

Fig. 26—Typical Thermostatic Control

### THERMOSTATIC CONTROL

The thermostatic switch (Fig. 26) controls the evaporator temperature by controlling the compressor magnetic clutch. When the points in the switch are closed, the compressor clutch is energized and the compressor operates. With the points open, the compressor stops.

The position of the cooling control (knob or slide-lever cable) determines the switch setting. If the control is set at OFF, the contact points are open. In the cooling range, the switch can be set to regulate the evaporator temperature between 30 and 60 degrees.

1965 Ford training manual

Expansion valve and thermostatic control.

| Ford Motor Company Windshield Wiper Technology 1965-1970 | | | |
|---|---|---|---|
| Car Line | Model Years | Speeds | Motor Type |
| Ford/Mercury | 1965/66 | 1 standard | electric |
| | | 2 optional | electric |
| Fairlane | 1965 | 1 standard | electric |
| | | 2 optional | electric |
| Mustang | 1965/68 | 1 standard | electric |
| | | 2 optional | electric |
| Falcon | 1965 | 1 standard | electric |
| | | 2 optional | electric |
| Bronco | 1966/69 | | vacuum |
| Thunderbird | 1965/69 | | hydraulic |
| Thunderbird | 1970 | | hydraulic |
| | | | electric |

Compiled from 1965 Ford parts book

This chart outlines how the types of Ford Motor Co. windshield wiper motors were evolving in 1965.

Note the service ports on the compressor.

Ford PDC car show both images

A 1965 Ford Mustang fitted with a Ford hang-on air conditioner.

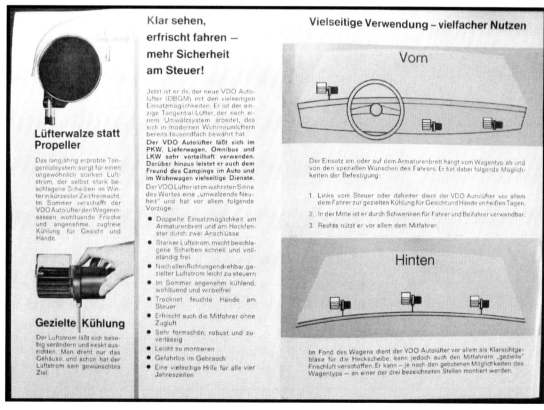

1965 brochure

Small fans were popular worldwide, including Germany.

Northville car show

This 1965 Pontiac GTO shows how the windshield washer nozzle was fitted thru the cowl grill.

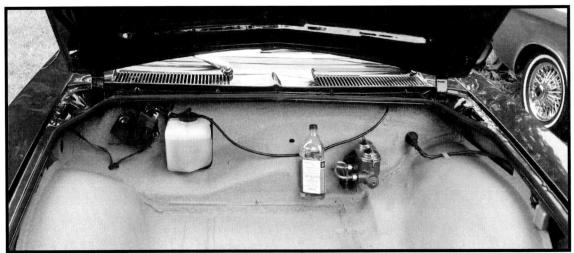

Vehicle Owners Phil & Audrey Raker

The 1965 Chevrolet Corvair windshield washer system.

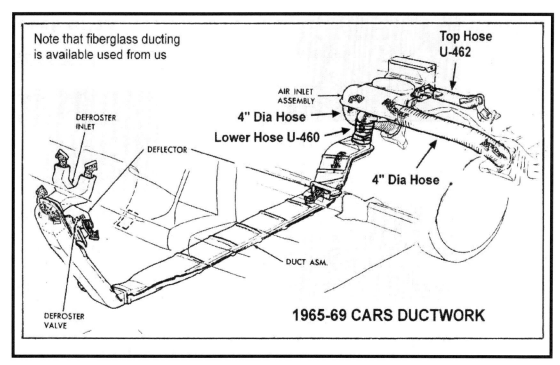

Image courtesy corvairunderground dot com

The 1965 Chevrolet Corvair hot air heater.

Vehicle Owners Phil & Audrey Raker

A 1965 Chevrolet Corvair fitted with a hot air heater duct to the passenger compartment.

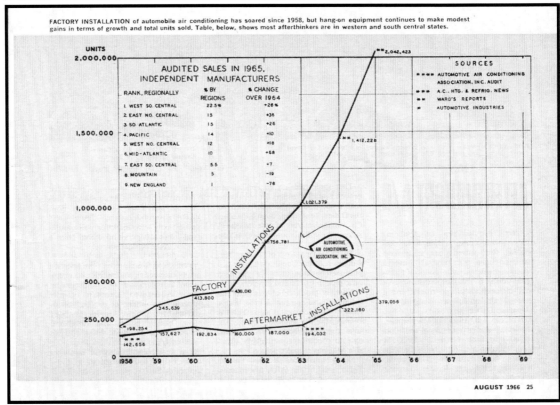

Car Life  August 1966 both images

Note the dramatic increase in factory-installed air conditioners from 1961.

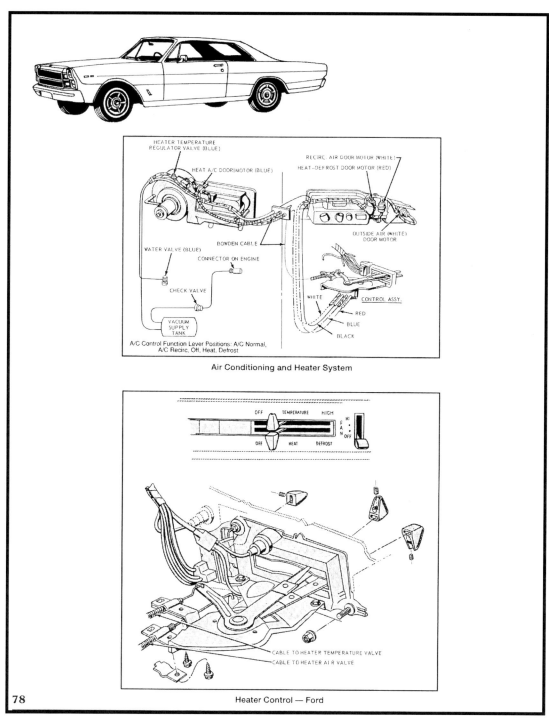

1966 Ford service manual

The 1966 Ford integrated air conditioning system.

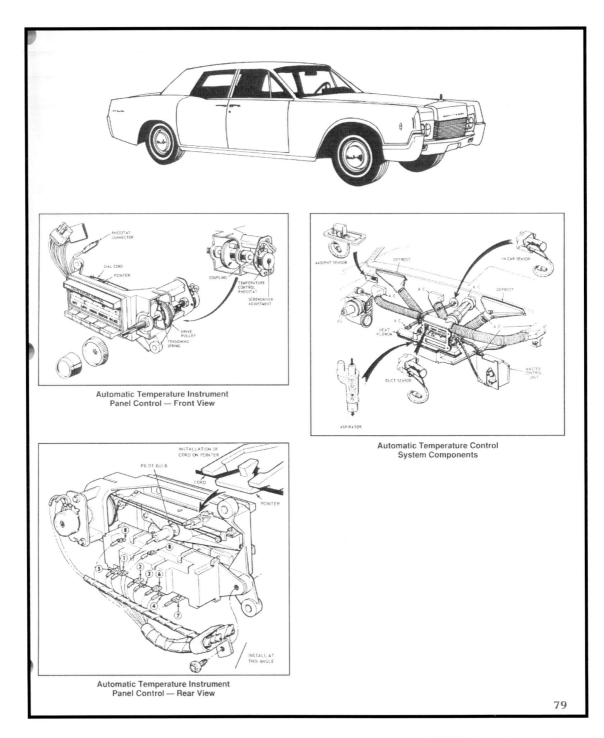

1966 Lincoln service manual

The 1966 Lincoln automatic temperature control air conditioner system.

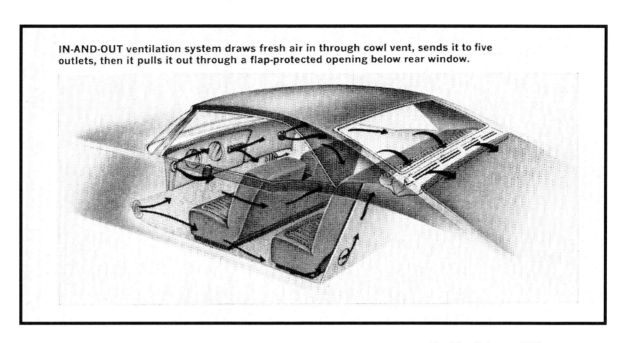

Car Life February 1966

1966 Buick Riviera featured flow thru ventilation. Note this was the beginning of the end for ventipanes.

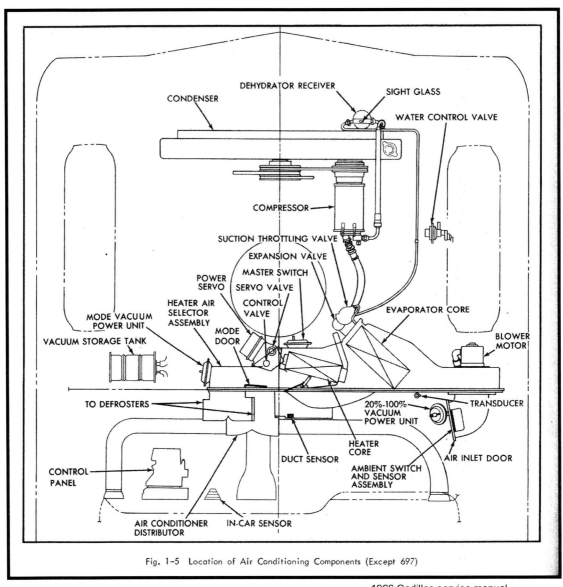

1966 Cadillac service manual

The 1966 Cadillac air conditioner system.

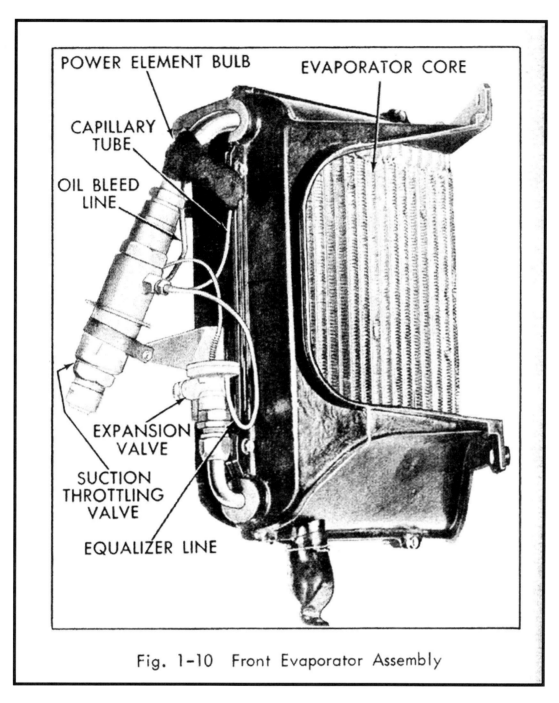

Fig. 1-10 Front Evaporator Assembly

1966 Cadillac service manual

An early example of a plate fin evaporator core. This technology replaced the tube fin type that had been used since 1940.

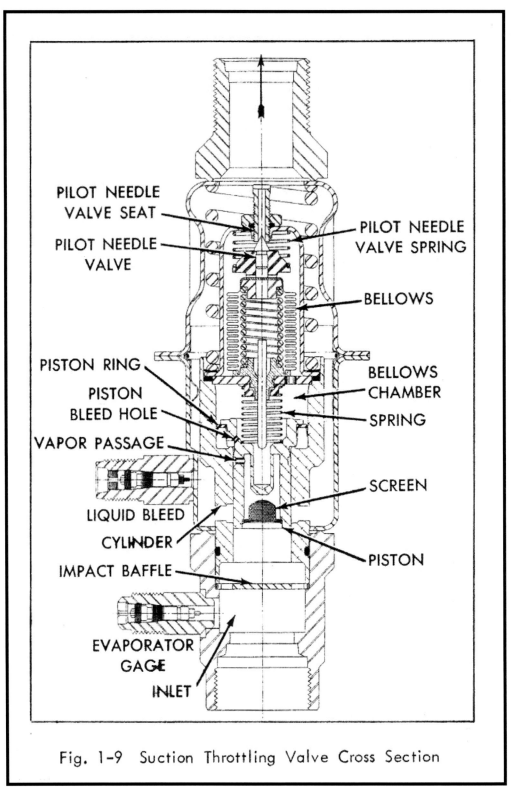

Fig. 1-9 Suction Throttling Valve Cross Section

1966 Cadillac service manual

The air conditioner suction throttling valve used on the 1966 Cadillac.

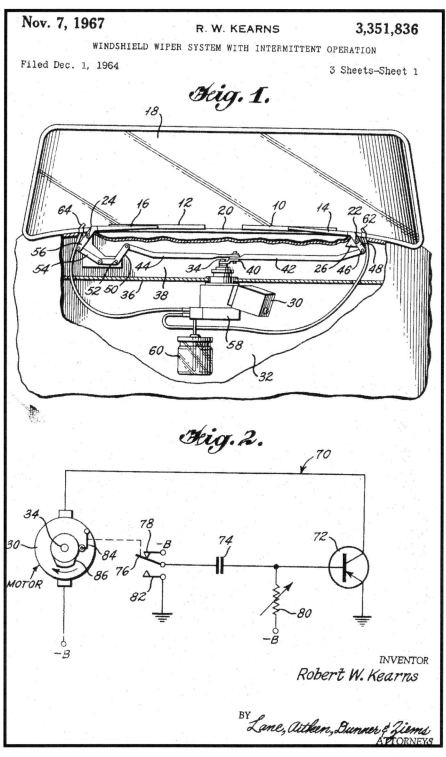

This patent for an electronic intermittent wiper control led to several lawsuits and many millions of dollars in settlements for the inventor. The story is documented in the movie "Flash of Genius" released in October 2008.

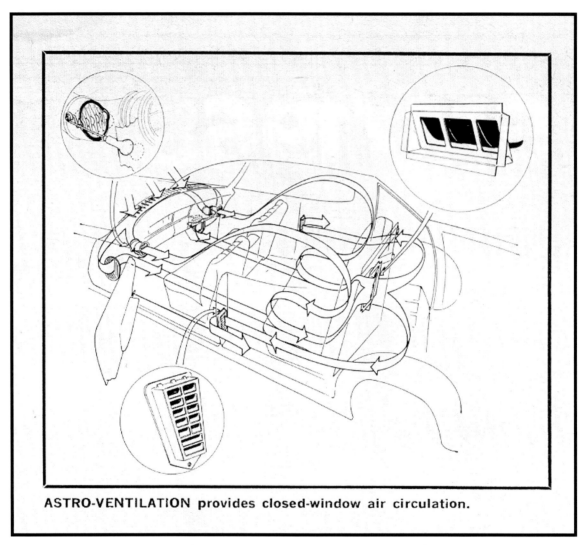

Car Life October 1967

Astro ventilation was released on 1968 model year General Motors cars that no longer utilized venti panes.

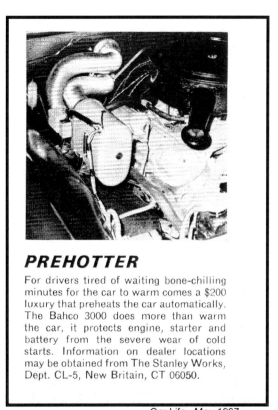

Car Life May 1967

The Stanley Works offered the Bahco 3000 pre heater in 1967.

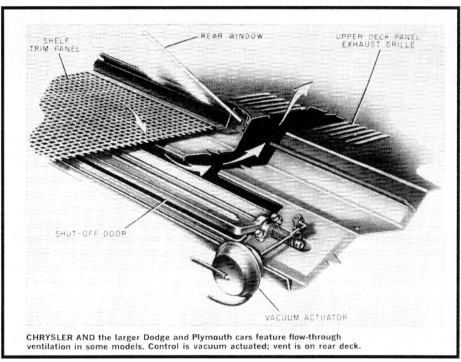

Car Life August 1966

The 1967 Chrysler Flow-Through ventilation system.

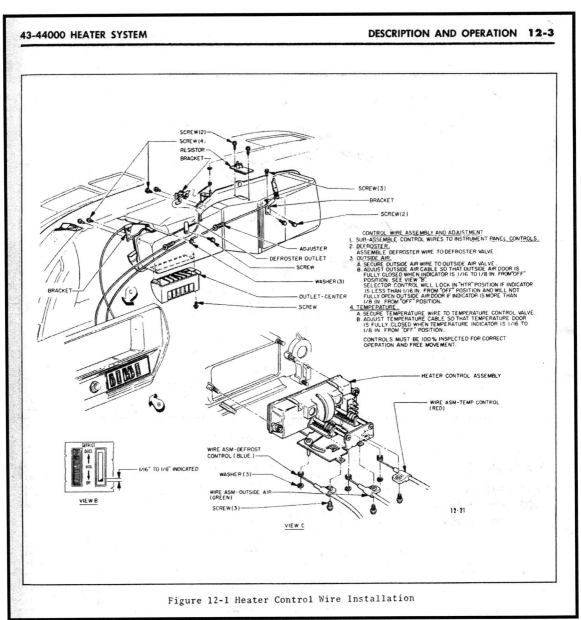

Figure 12-1 Heater Control Wire Installation

1968 Buick service manual

Turnbuckle adjusters were used on the heater control wires (Bowden cables) on the 1968 Buick.

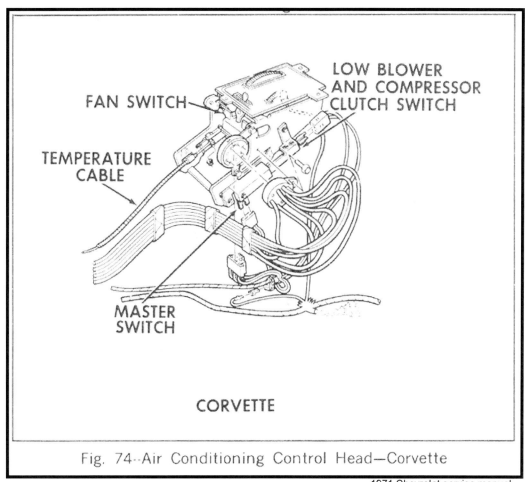

Fig. 74 -- Air Conditioning Control Head—Corvette

1971 Chevrolet service manual

Image taken 2011

The 1968 Corvette featured an instrument panel heater and air conditioner control with two thumb wheels.

Car Life November 1967

Image taken 2011

The 1968 Chevrolet Corvette windshield wiper cover system.

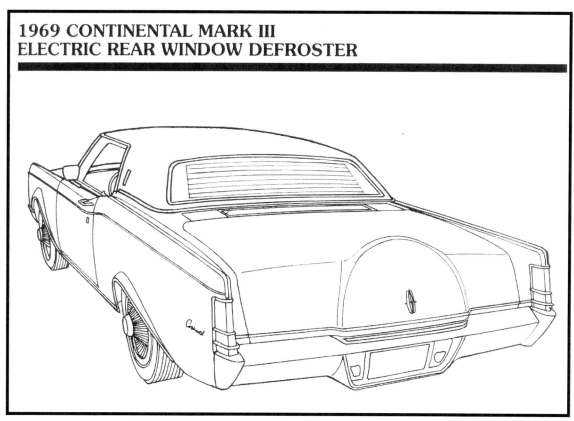

Ford Motor Company image

The 1969 Continental Mark III offered an electric grid rear window defroster.

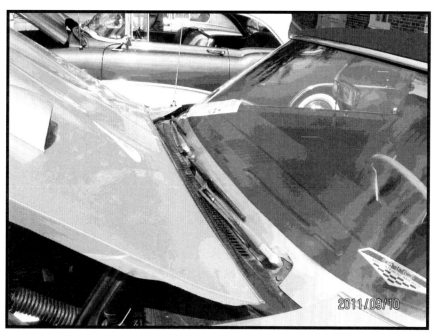

Vehicle Owner Philip Fischer

1969 Pontiac GTO featured wipers parked on the windshield trim molding.

1970 Saab brochure

Saab invented headlamp wipers and released them on their 1970 model year vehicles.

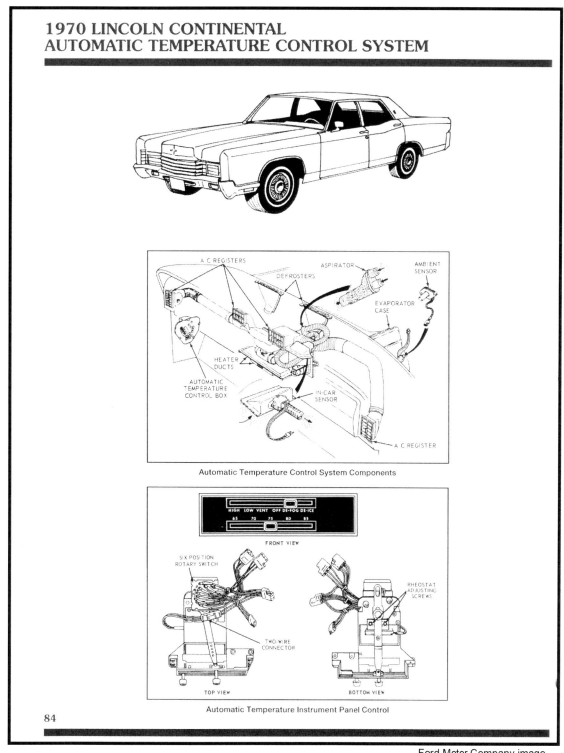

The 1970 Lincoln Continental automatic temperature control system.

1971 Ford with Ford hang-on air conditioner that spans instrument panel.

Saab introduced electrically heated seats on its 1972 models.

440

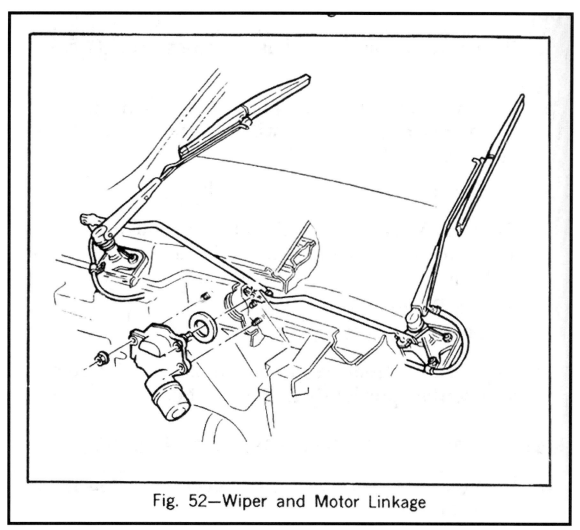

Fig. 52—Wiper and Motor Linkage

Chevrolet shop manual

The 1971 Chevrolet Corvette featured wiper arms with integral washer nozzles. This system was referred to as "wet arms".

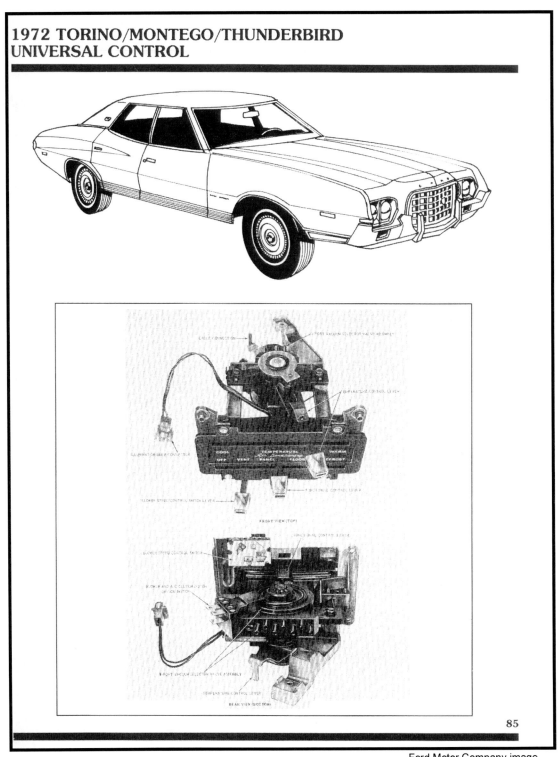

Ford Motor Company image

The Ford Universal Control instrument panel heater and air condition control was introduced on the 1972 Torino. It was subsequently used for several model years on numerous Ford vehicle lines.

# 1974 THUNDERBIRD/MARK IV
# QUICK DEFROST WINDSHIELD AND REAR WINDOW

## DESCRIPTION AND OPERATION

### DESCRIPTION

The Quick Defrost Windshield and Rear Window system is available on the Thunderbird and Continental Mark IV. The system (Fig. 1) consists of a special front windshield and rear window glass, a separate three phase high voltage supplemental alternator, a relay (same as used in heated rear window system), an electrical timer, a control switch, an indicator light, and special wiring and connectors to carry high voltage current between the high output alternator and the glass.

Heater elements in the windshield and rear window are composed of metallic gold, deposited in a thin film. There are no fine wires to distract vision. Almost the entire area of the window is uniformly heated.

Since high voltage is required for the heater elements, new red plastic shielded wiring, with high voltage warning tags at each junction, is used from the alternator output to the heat elements. This completely isolates the system from the car normal electrical system to prevent any accidental feed of hazardous high voltage electricity.

### OPERATION

A relay, low-current timer and switch provide automatic control of the system. Switch positions are: OFF, NORMAL, ON. The ON and OFF positions are springloaded and returns to NORMAL when released. When the switch is turned to ON by pushing the control switch to the right, an indicator light will glow, indicating that the system is operating. The indicator light will remain on at all times during operation of the system. The system is turned off by pushing the control switch to the left.

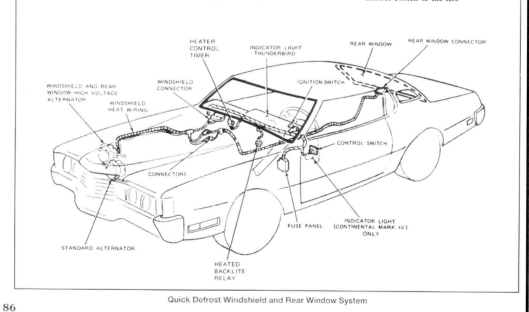

Quick Defrost Windshield and Rear Window System

1974 Ford service manual

The Quick Defrost Windshield and Rear Window featured gold film on the glass and a second alternator to defrost rapidly. It was used on the Thunderbird and Mark IV.

443

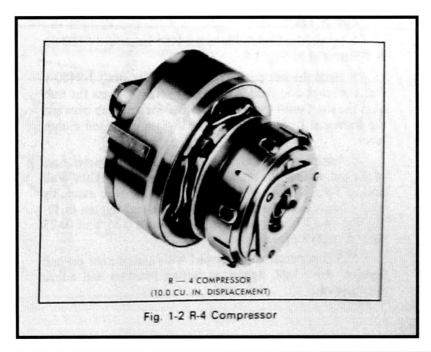

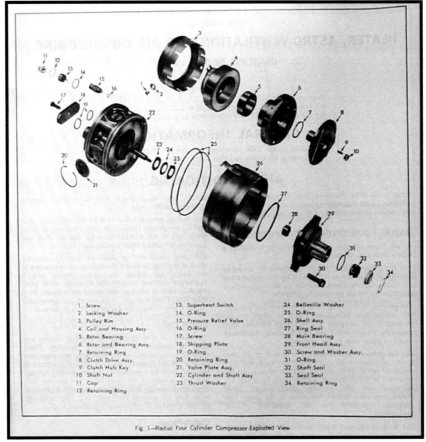

1975 Chevrolet service manual

The General Motors R-4 air conditioner compressor.

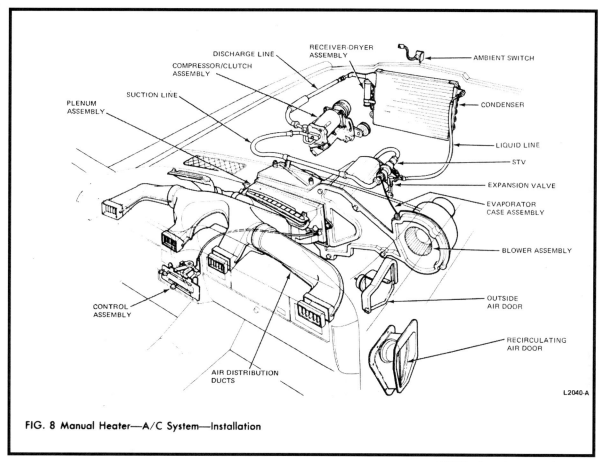

1975 Ford service manual

The 1975 Ford heater and air conditioner system.

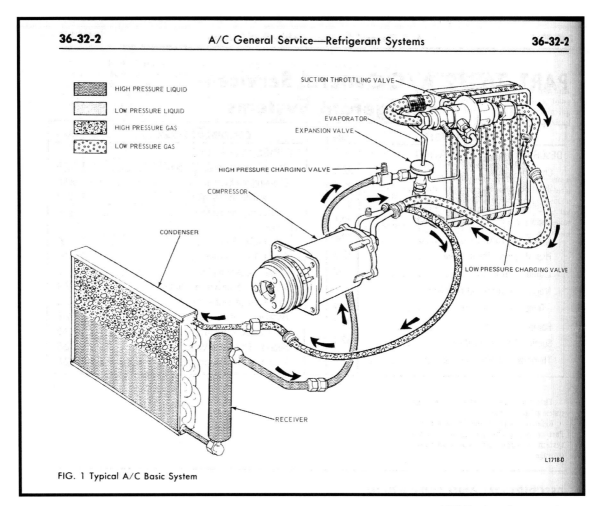

1975 Ford service manual

The 1975 Ford air conditioner compressor, condenser and evaporator.

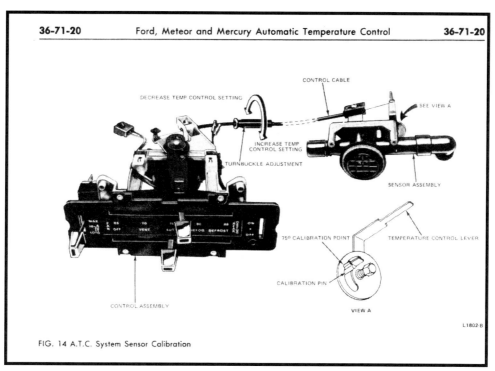

1975 Ford service manual

A control cable with a turnbuckle was used to precisely adjust the sensor.

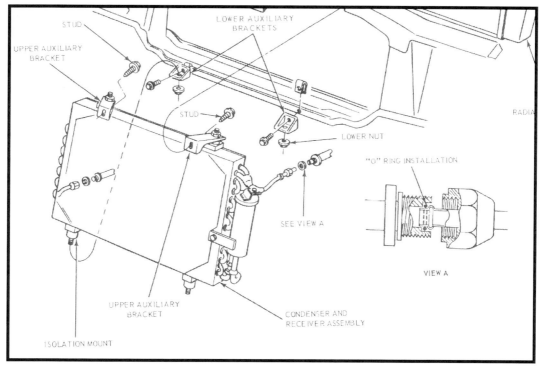

1975 Ford service manual

VIEW A shows a tube-o refrigerant fitting.

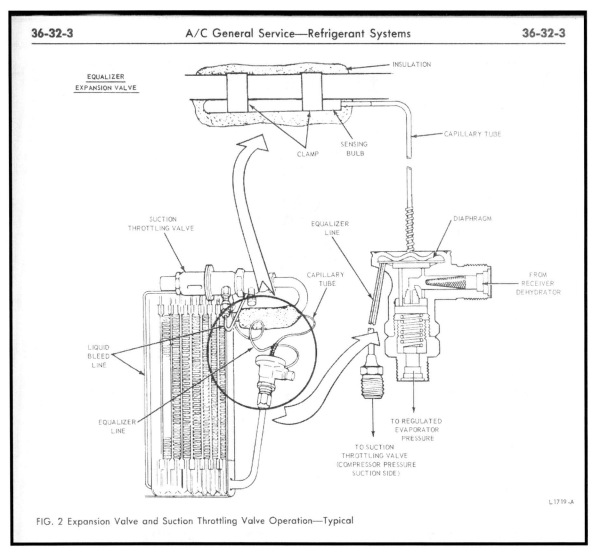

The 1975 Ford air conditioner expansion valve and suction throttling valve.

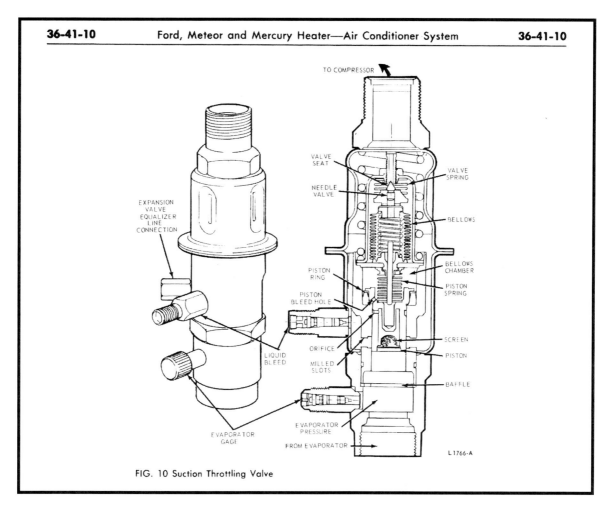

FIG. 10 Suction Throttling Valve

1975 Ford service manual

The 1975 Ford air conditioner suction throttling valve.

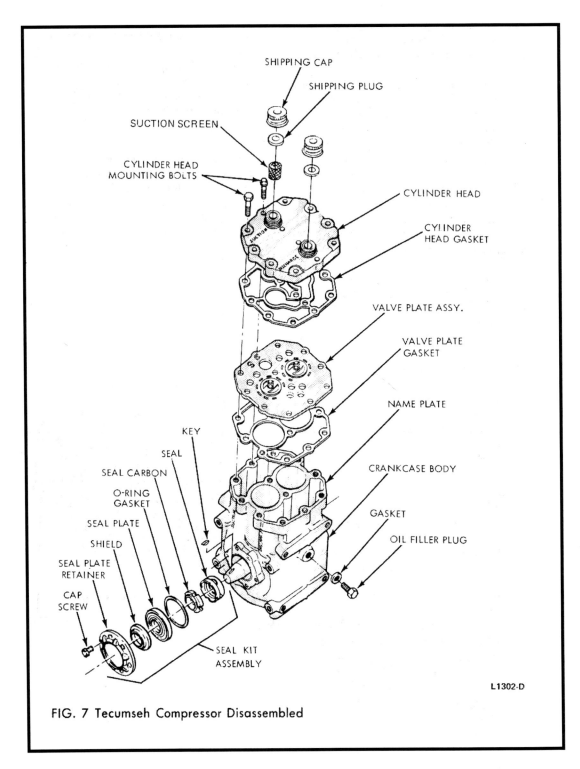

FIG. 7 Tecumseh Compressor Disassembled

1975 Ford service manual

The 1975 Ford Tecumseh two-cylinder air conditioner compressor.

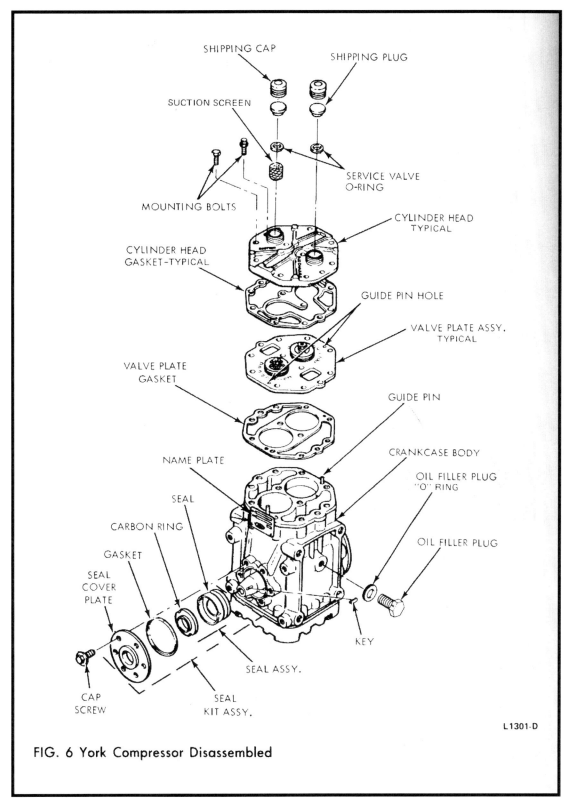

1975 Ford service manual

The 1975 Ford York two-cylinder air conditioner compressor.

# MAJOR REPAIR OPERATIONS

## COMPRESSOR

*Disassembly*

1. Remove the compressor from the car and the clutch assembly from the compressor as described under Removal and Installation.
2. Remove the shaft seal as described under Removal and Installation.
3. Invert the compressor and holding fixture so that the front end of the compressor shaft is down. **Additional oil may leak from the compressor at this time. All oil must be drained into a container so that total amount can be measured. A liquid measuring cup may be used for this purpose. Drained oil should then be discarded.**
4. Remove four lock nuts from the threaded studs on the compressor shell and remove the rear head (Fig. 22). Tap uniformly around the rear head if the head is binding.
5. Wipe excess oil from all sealing surfaces on the rear head casting webs, and examine the sealing surfaces (Fig. 23). If any damage is observed, head should be replaced.
6. Remove suction screen and examine for any damage or contamination. Clean or replace if necessary.
7. Paint an identifying mark (Prussian blue or other suitable marking material may be used) on exposed face of the oil pump inner and outer gears and then remove the gears. Identifying marks are to assure that the gears, if reused, will be installed in identical position.
8. Remove and discard the rear head-to-shell O-ring.
9. Carefully remove the rear discharge valve plate assembly. Use two small screwdrivers under the reed retainers and pry up on the assembly (Fig. 24). Do not position the screwdrivers between the reeds and reed seats.

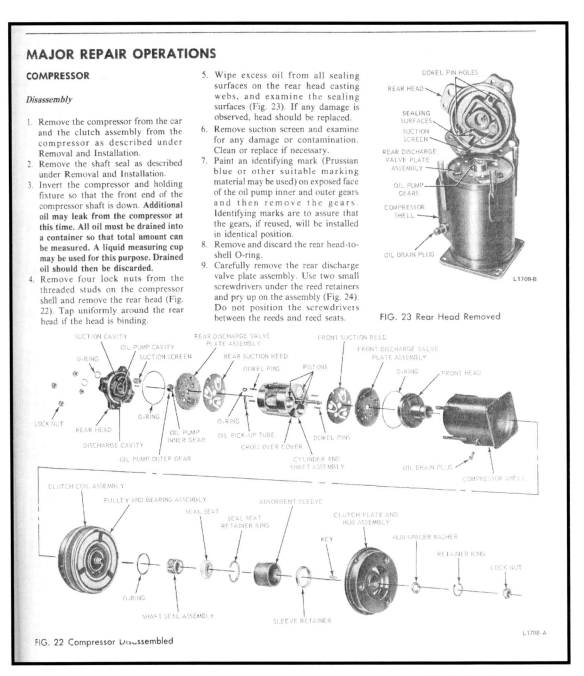

FIG. 23 Rear Head Removed

FIG. 22 Compressor Disassembled

1975 Ford service manual

The 1975 Ford 6 cylinder air conditioner compressor.

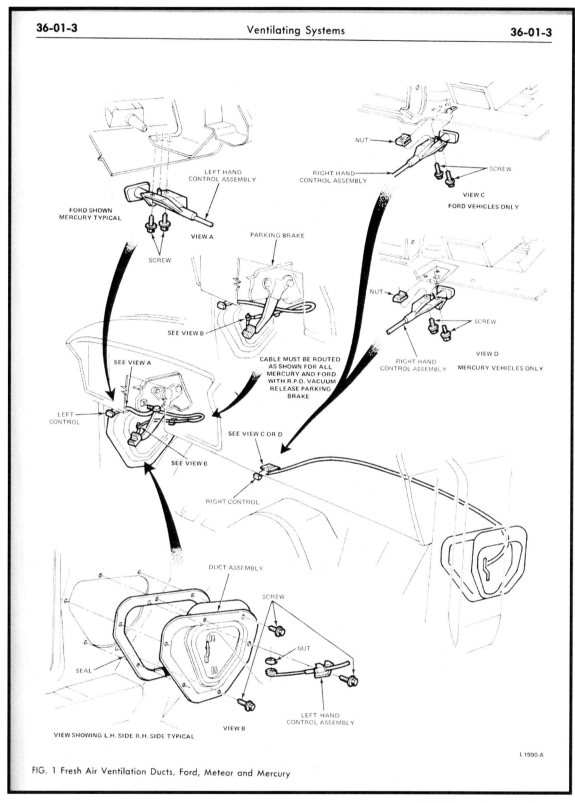

FIG. 1 Fresh Air Ventilation Ducts, Ford, Meteor and Mercury

1975 Ford service manual

The 1975 Ford ventilation system.

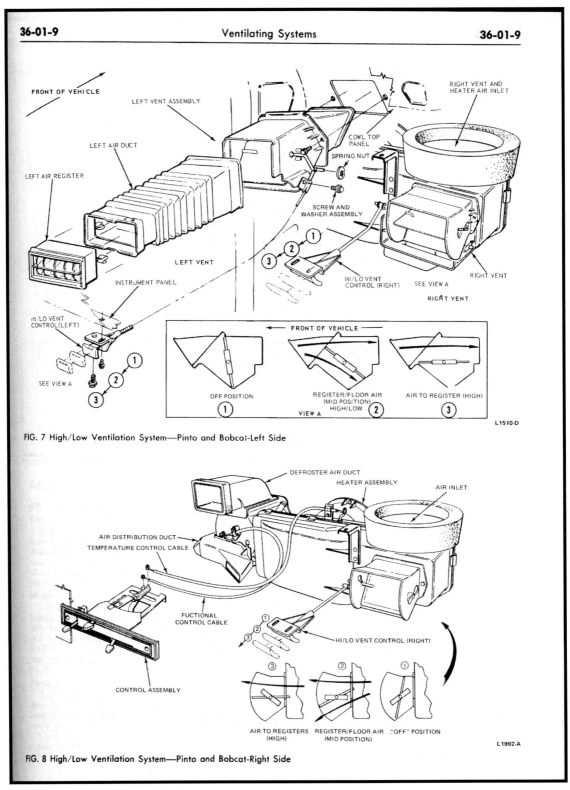

The 1975 Pinto and Bobcat heater and ventilation system.

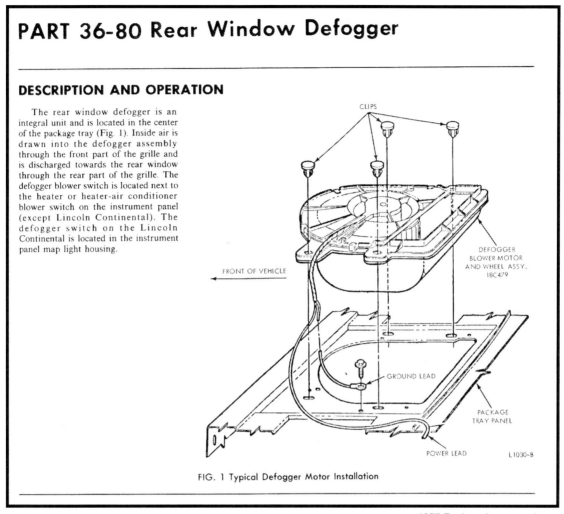

1975 Ford service manual

The 1975 Ford rear window defogger.

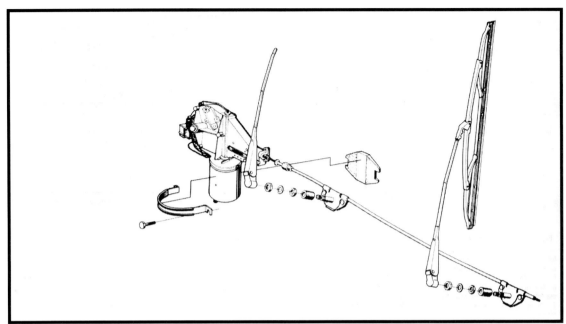

Saab image

The 1975 Saab wiper system used a rotary cable.

Image taken Ypsilanti, MI 2011

1975 Ferrari 512 BB wiper system has two blades on one articulated arm. This could be referred to as a one and a half blade system.

DuPont brochure Circa 1977

This was an early application of plastic replacing the traditional metal instrument panel heater and air conditioner control bracket.

## Climate Control (Cont'd)

bracket. "Minlon" qualified on all counts and design engineer Gene Dickirson tells why:

"There are more than 225 separate dimensional specifications on this one component," says Dickirson. "Meeting those specs and holding to tight tolerances over long thin sections meant that we had to have a virtually warp-free part. That, plus stiffness, high strength, low mold shrinkage and resistance to high temperatures. All the materials tried had most of the attributes we were seeking; only 'Minlon' had them all."

One potential problem, the danger of light "leaking" through the lamp housing, was eliminated by "beefing" up the wall thickness in that area to .070 inches (1.78 millimeters) and adding three percent titanium dioxide to "Minlon" 10B-40 WT 113 resin to increase its opacity. In other sensitive areas, design enhanced the natural capabilities of the resin.

"Undercuts around the locating studs and staking posts allowed for the recommended fillet radii while permitting a flush fit," Dickirson explains. "Cutout sections around the pivot posts enabled us to avoid sinks and reduce material requirements while providing the desired bearing surface.

"It pays off. Once the face plate and lens have been secured to the bracket via ultrasonic heat staking, the assembly can withstand vibration tests of 50 Hz at an .05 amplitude. It's also capable of withstanding external forces as great as 40 pounds (18 kilograms).

"The use of self-tapping screws at regular assembly line speeds and torques was a real time saver," Dickirson goes on. "Self-tapping screws hold electrical components in place with the help of locating pins. To get top holding power, we specify dichromate and zinc finish screws and no lubricant. Power wrenches apply a torque of 12 to 18 inch-pounds (1.3 to 2.0 newton-meters) at speeds of 600 to 2000 rpm."

"Minlon" proves itself in other ways in the Ford bracket. Wear resistance and low friction were affirmed by a test in which control and temperature levers move the length of molded-in tracks through more than 10,000 cycles. In addition, the bracket must withstand temperatures from −40° to 140°F (−40° to 60°C), plus the heat generated by the assembly's own lamp.

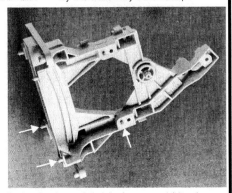

Radiused undercuts (arrows) enhance capabilities of "Minlon". Cutout sections around pivot posts reduce material requirements, leave adequate bearing surface.

---

Jack McAllister

Whether you need actual design assistance or just a rundown on property values for a specific resin, there's a lot of professional know-how available to you through the members of our Automotive Development Group—a total of nearly one and a half centuries experience with Du Pont's First Family of Engineering Plastics, more than 77 years of that in direct involvement with auto design.

Senior Marketing Representative Jack McAllister is typical. A chemical engineer who got his degree from the University of Texas *after* being mustered out of the World War II Army as a first lieutenant in the Chemical Corps, he joined Du Pont in 1949. He arrived in Detroit as a plastics industry technologist seven years later and has had a "piece of the action" in at least every model year from 1957 on. Like a lot of Michiganeans, Jack is an outdoor type—golfing, fishing or trailering about the continent with his family. At home, he's an amateur musician and—according to his friends— a professional wine maker.

But even in his off hours, Jack continues to be a very real part of the "autoplastic" scene. He's active in the Society of Automotive Engineers and the Detroit Chapter of the Society of Plastics Engineers and has been a board member, program chairman and session organizer for the latter group as well as president of its Plastics in Autos Division.

If Jack or one of his colleagues in the Du Pont Automotive Development Group can help you, they're just a telephone call away at 29201 Telegraph Road, Southfield, Michigan 48037—(313) 559-6000. Or write:

Du Pont Company
Plastic Products and Resins Dept.
Wilmington, DE 19898

The Du Pont Company assumes no obligation or liability for any advice furnished by it, or results obtained with respect to these products. All such advice is given and accepted at the buyer's risk. Du Pont warrants that the material itself does not infringe the claims of any United States patent; but no license is implied nor is any further patent warranty made.

**ZYTEL** Nylon Resins   **DELRIN** Acetal Resins   **MINLON** Engineering Thermoplastic Resins   **RYNITE** Thermoplastic Polyester Resins

DuPont brochure Circa 1977

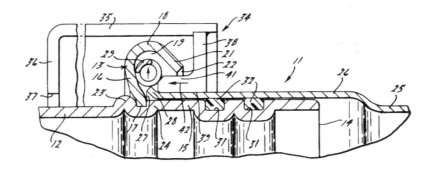

## United States Patent [19]
### McWethy

[11] 4,055,359
[45] Oct. 25, 1977

[54] QUICK-CONNECT TUBULAR COUPLINGS
[75] Inventor: Irvin E. McWethy, Gravette, Ark.
[73] Assignee: Ford Motor Company, Dearborn, Mich.
[21] Appl. No.: 632,874
[22] Filed: Nov. 17, 1975
[51] Int. Cl.² ........................... F16L 35/00
[52] U.S. Cl. ........................... 285/39; 285/318; 285/347; 285/351; 285/DIG. 25
[58] Field of Search ............... 285/318, 317, 316, 315, 285/308, 39, 321, DIG. 25, 13, 14, DIG. 7, 382.5, 347, 35

[56] References Cited
U.S. PATENT DOCUMENTS

| | | | |
|---|---|---|---|
| 2,914,344 | 11/1959 | Anthes | 285/321 X |
| 2,939,728 | 6/1960 | Bitel | 285/315 |
| 3,304,104 | 2/1967 | Wiltse | 285/321 X |
| 3,453,005 | 7/1969 | Foults | 285/DIG. 25 |
| 3,532,101 | 10/1970 | Snyder | 285/318 X |
| 3,569,903 | 3/1971 | Brishka | 285/315 X |
| 3,924,883 | 12/1975 | Frank | 285/382.5 |
| 3,929,357 | 12/1975 | Devincent et al. | 285/DIG. 25 |

FOREIGN PATENT DOCUMENTS

1,277,984  10/1961  France ........................ 285/318

Primary Examiner—Dave W. Arola
Attorney, Agent, or Firm—John J. Roethel; Keith L. Zerschling

[57] ABSTRACT

A quick-connect tubular coupling for a pair of tubes one of which has an end portion telescopically disposed within an end portion of the other. An annular cage is externally mounted on the one or inner tube in axially spaced relation to the free end of its end portion. The cage is held on the one tube against axial displacement along the latter. The end portion of the other or outer tube is flared outwardly at its free end and extends in the cage. A circular spring means is interposed between the flared end portion of the other tube and the cage and when so disposed prevents axial movement of the one tube relative to the other in telescopic disengagement direction. Between the telescoped end portions are a plurality of "O" rings that are compressed in a sealed mode to prevent leakage through the coupling.

7 Claims, 7 Drawing Figures

1977

The patent for the spring lock coupling that would be used on millions of applications.

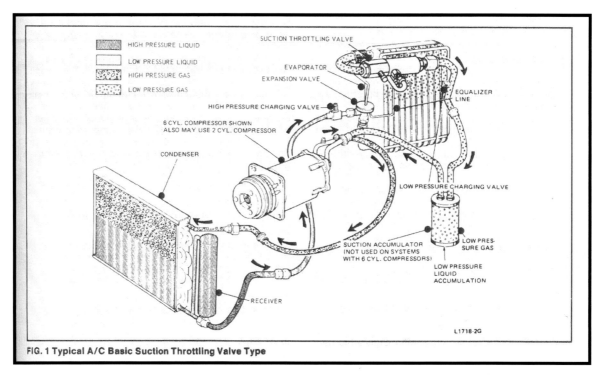

1979 Ford service manual

The 1979 Ford air conditioner system.

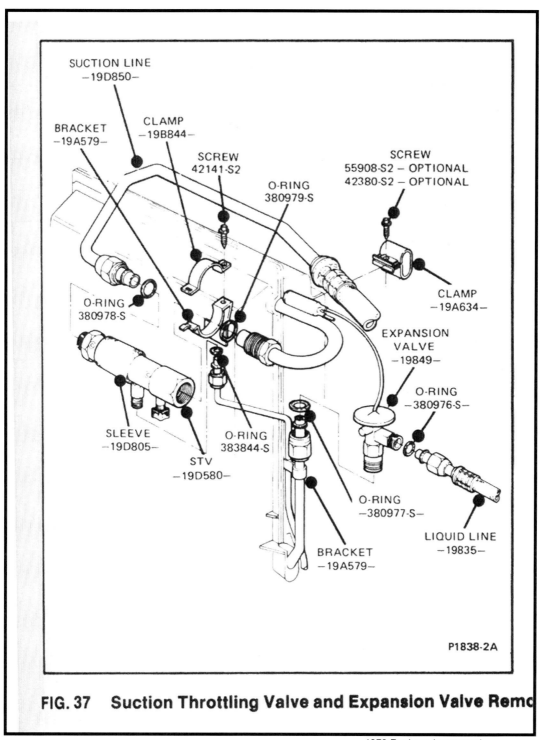

**FIG. 37 Suction Throttling Valve and Expansion Valve Remo[val]**

1979 Ford service manual

The tube-o refrigerant fittings have been popular for many years because of their excellent sealing capability.

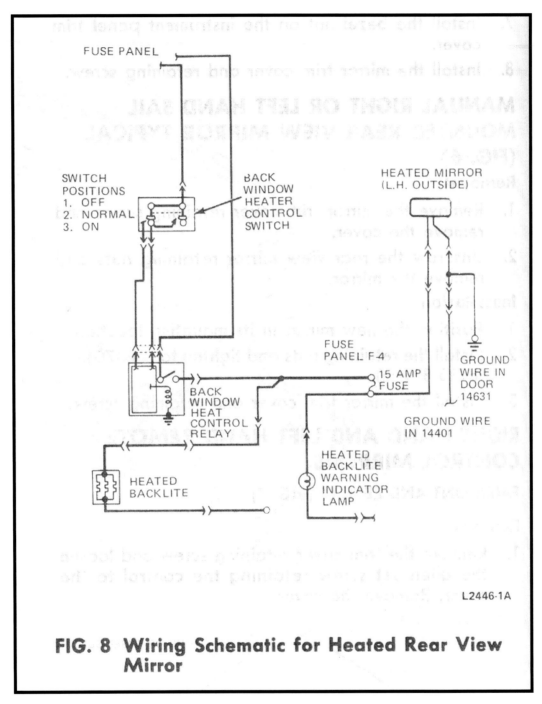

1979 Lincoln Mark V service manual

The 1979 Lincoln Mark V offered an early application of a heated rear view mirror. The electrical circuitry is shown in this wiring schematic.

This 2-cylinder design saw some car and truck applications during model years 1980 and 1981. During these same years, the fixed orifice tube system replaced the expansion valve in Ford, Mercury, Lincoln, Mark VI, Escort and Lynx passenger cars. Only the Econoline continues to use an expansion valve as a component of its auxiliary A/C system.

Other compressors used during the 1980-1990 period are listed by model number in the text which follows:

- The FS-6 – a Ford (CCD), swash plate, 6-cylinder, 3 double acting piston, axial compressor.

- The FX-15 – a Ford (CCD), swash plate, 10-cylinder, 5 double acting piston, axial compressor.

- The 10P15 – a Nippondenso, swash plate, 10-cylinder, 5 double acting piston, axial compressor.

- The HR-980 – a Tecumseh, scotch yoke, 4-cylinder, 4 piston, radial compressor.

- The 6P148 – a Nippondenso, swash plate, 6-cylinder, 3 double acting piston, axial compressor.

- The 10PA17 – a Nippondenso, swash plate, 10-cylinder, 5 double acting piston, axial compressor.

- The 6E171 – a Nippondenso, swash plate, 6-cylinder, 3 double acting piston, axial compressor.

With exception of the 2-cylinder compressor, each of the other units listed has one or more current applications. Brief descriptions of these compressors follow.

1990 Ford Motor Company A/C REFRIGERANT SYSTEMS REFERENCE MANUAL 1980-1990

Ford air conditioner compressor usage chart.

**1979 REFRIGERANT SYSTEM COMPONENTS AND CAPACITIES—CAR**

| Vehicle | Compressor | | Combination Valve | | Expansion Valve | | STV | | Ambient Switch | Low Pressure Switch | Suction Accumulator | Discharge Muffler | Refrigerant Capacity* | | |
|---|---|---|---|---|---|---|---|---|---|---|---|---|---|---|---|
| | 6 Cyl. | 2 Cyl. | Std. | BPO | Std. | BPO | Std. | BPO | | | | | (lbs.) | (oz.) | (kg.) |
| Ford | X | | | | X | | X | | X | | | | | | |
| Mercury | X | | | | X | | X | | X | | | | | | |
| Lincoln | X | | | | | X | | X | X | | | | | | |
| Mark V | X | | | | | X | | X | X | | | | | | |
| LTD II | | X | | | X ① | X ② | X ① | X ② | X | | X ① | | 4¼ | 68 | 1.928 |
| Cougar | | X | | | X ① | X ② | X ① | X ② | X | | X ① | | | | |
| Thunderbird | | X | | | X ① | X ② | X ① | X ② | X | | X ① | | | | |
| Ranchero | | X | | | X ① | X ② | X ① | X ② | X | | X ① | | | | |
| Granada | X | | X | | | | | | X | | | † | | | |
| Monarch | X | | X | | | | | | X | | | † | 4 | 64 | 1.814 |
| Versailles | X | | X | | | | | | X | | | X | | | |
| Fairmont | | X | X | | | | | | X | | | | | | |
| Zephyr | | X | X | | | | | | X | | | | | | |
| Mustang | | X | X | | | | | | X | | | | 3½ | 56 | 1.588 |
| Capri | | X | X | | | | | | X | | | | | | |
| Pinto | | X | X | X | | | | | X | | X ① | | 2¼ | 36 | 1.021 |
| Bobcat | | X | X | X | | | | | X | | X ① | | | | |

*Plus ¼ lb. (4 oz.) (.113 kg.) minus zero.  †8 cyl. engine only  ①Before 3-1-79  ②3-1-79 and after

Fig. 36—Components and Capacities—1979 Car

3631-001-036

**1980 REFRIGERANT SYSTEM COMPONENTS AND CAPACITIES—CAR**

| Vehicle | Compressor | | Combination Valve | | Fixed Orifice Tube | Clutch Cycling Pressure Switch | Ambient Switch | Suction Accumulator Drier | Discharge Muffler | Refrigerant Capacity* | | |
|---|---|---|---|---|---|---|---|---|---|---|---|---|
| | 6 Cyl. | 2 Cyl. | Std. | BPO | | | | | | (lbs.) | (oz.) | (kg.) |
| Ford | X | | | | X | X | | X | X | | | |
| Mercury | X | | | | X | X | | X | X | 3¼ | 52 | 1.474 |
| Lincoln | X | | | | X | X | | X | X | | | |
| Mark VI | X | | | | X | X | | X | X | | | |
| Versailles | X | | X | | | | X | | X | | | |
| Granada | X ① | X ② | X | | | | X | | | 4 | 64 | 1.814 |
| Monarch | X ① | X ② | X | | | | X | | | | | |
| XR-7 | | X | | X ③ | | | X | | | | | |
| Thunderbird | | X | | X ③ | | | X | | | | | |
| Fairmont | | X | | X ③ | | | X | | | | | |
| Zephyr | | X | | X ③ | | | X | | | 3½ | 56 | 1.588 |
| Mustang | | X | | X ③ | | | X | | | | | |
| Capri | | X | | X ③ | | | X | | | | | |
| Pinto | | X | X | | | | X | | | | | |
| Bobcat | | X | X | | | | X | | | 2¼ | 36 | 1.021 |

*Plus ¼ lb. (4 oz.) (.113 kg.) Minus Zero  ①6 Cyl. Engines  ②8 Cyl. Engines  ③Mini Comb. Valve

**1980-81 REFRIGERANT SYSTEM COMPONENTS AND CAPACITIES—LIGHT TRUCK**

| Vehicle | Compressor 2-Cylinder | Std. Expansion Valve | Fixed Orifice Tube | Clutch Cycling Pressure Switch | De-Icing Switch | Suction Accumulator Drier | Refrigerant Capacity (lbs.) |
|---|---|---|---|---|---|---|---|
| F-100 through F-350 | X | | X | X | | X | 3¾ |
| Bronco | X | | X | X | | X | 3¾ |
| Econoline Main System | X | X | | | | X | 3½ |
| Auxiliary System | | X | | | | | 4¼ ① |

①Total refrigerant charge for main and auxiliary systems.

Fig. 37—Components and Capacities—1980 Car and 1980-81 Light Truck

3631-001-037

1981 Ford Motor Company Climate Control REFRIGERATION SYSTEMS

Ford refrigerant control systems for 1980 model year cars.

### 1981 REFRIGERANT SYSTEM COMPONENTS AND CAPACITIES—CAR

| Vehicle | Compressor 6 Cyl. | Compressor 2 Cyl. | Combination Valve Std. | Combination Valve BPO | Fixed Orifice Tube | Clutch Cycling Pressure Switch | Ambient Switch | Suction Accumulator Drier | Discharge Muffler | Refrigerant Capacity* (lbs.) | (oz.) | (kg.) |
|---|---|---|---|---|---|---|---|---|---|---|---|---|
| Ford | X | | | | X | X | | X | | 3¼ | 52 | 1.474 |
| Mercury | X | | | | X | X | | X | | 3¼ | 52 | 1.474 |
| Lincoln | X | | | | X | X | | X | X | 3 | 48 | 1.361 |
| Mark VI | X | | | | X | X | | X | X | 3 | 48 | 1.361 |
| Granada | | X | | X † | | | X | | | 3½ | 56 | 1.588 |
| Cougar | | X | | X † | | | X | | | 3½ | 56 | 1.588 |
| XR-7 | | X | | X † | | | X | | | 3½ | 56 | 1.588 |
| Thunderbird | | X | | X † | | | X | | | 3½ | 56 | 1.588 |
| Fairmont | | X | | X † | | | X | | | 3½ | 56 | 1.588 |
| Zephyr | | X | | X † | | | X | | | 3½ | 56 | 1.588 |
| Mustang | | X | | X † | | | X | | | 3½ | 56 | 1.588 |
| Capri | | X | | X † | | | X | | | 3½ | 56 | 1.588 |
| Escort | X | | | | X | X | | X | | 2½ | 40 | 1.134 |
| Lynx | X | | | | X | X | | X | | 2½ | 40 | 1.134 |

*Plus ¼ lb. (4 oz.) (.113 kg.) Minus Zero   †Mini Comb. Valve

### 1980-81 REFRIGERANT SYSTEM COMPONENTS AND CAPACITIES—LIGHT TRUCK

| Vehicle | Compressor 2-Cylinder | Std. Expansion Valve | Fixed Orifice Tube | Clutch Cycling Pressure Switch | De-Icing Switch | Suction Accumulator Drier | Refrigerant Capacity (lbs.) |
|---|---|---|---|---|---|---|---|
| F-100 through F-350 | X | | X | X | | X | 3¼ |
| Bronco | X | | X | X | | X | 3¼ |
| Econoline Main System | X | X | | | X | | 3½ |
| Auxiliary System | | X | | | | | 4¼ ① |

① Total refrigerant charge for main and auxiliary systems.

Fig. 73—Refrigerant Charge Capacities

1981 Ford Motor Company Climate Control REFRIGERATION SYSTEMS

These charts show the transition of Ford air conditioner systems from suction throttling valve systems to cycling clutch fixed orifice tube systems.

## VEHICLE APPLICATION

Ford/Mercury, Lincoln Continental/Continental Mark VI.

## DESCRIPTION AND OPERATION

A swash plate design six cylinder compressor will be used on some models. The compressor is designated as the FS-6 and is shown in Fig. 1.

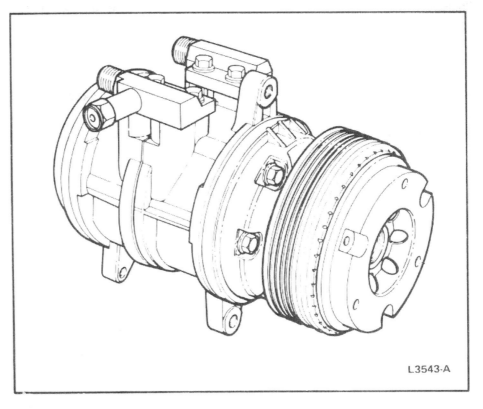

FIG. 1 FS-6 Compressor

1981 Ford service manual

The Ford FS6 (Ford Swash plate 6 cylinder) was used on millions of vehicles.

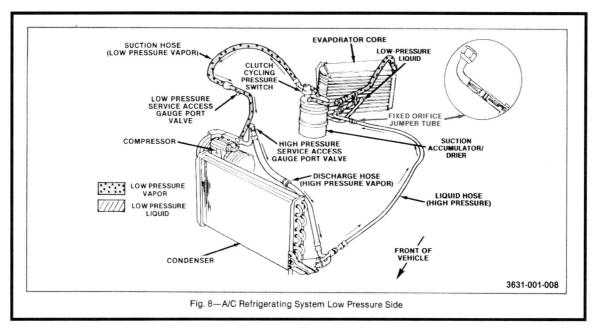

Fig. 8—A/C Refrigerating System Low Pressure Side

1981 Ford Refrigeration Systems

The Ford cycling clutch orifice tube air conditioner system introduced on some the 1980 models.

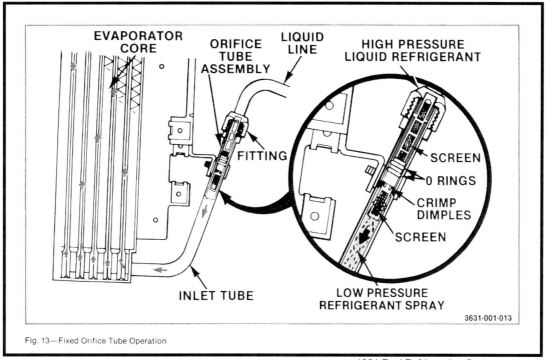

Fig. 13—Fixed Orifice Tube Operation

1981 Ford Refrigeration Systems

The Ford air conditioner orfice tube.

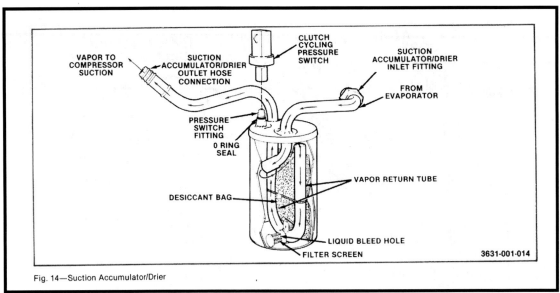

The Ford air conditioner accumulator.

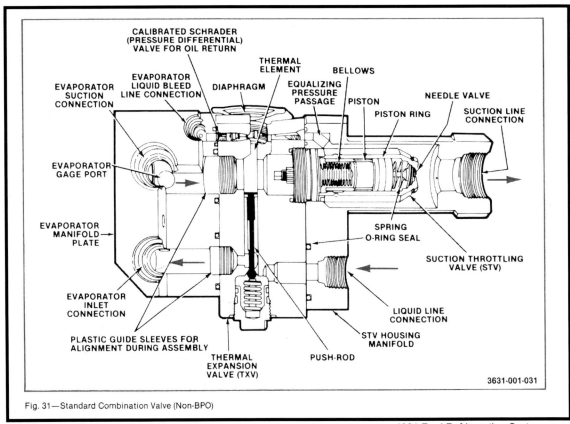

The Ford air conditioner standard Non-BPO combination valve.

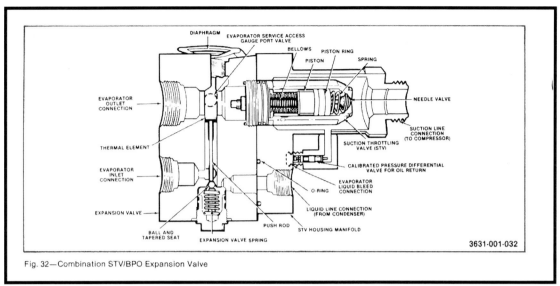

Fig. 32—Combination STV/BPO Expansion Valve

1981 Ford Refrigeration Systems

The Ford air conditioner combination STV/BPO expansion valve.

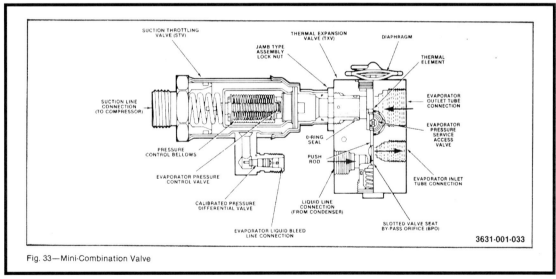

Fig. 33—Mini-Combination Valve

1981 Ford Refrigeration Systems

The Ford air conditioner mini combination valve.

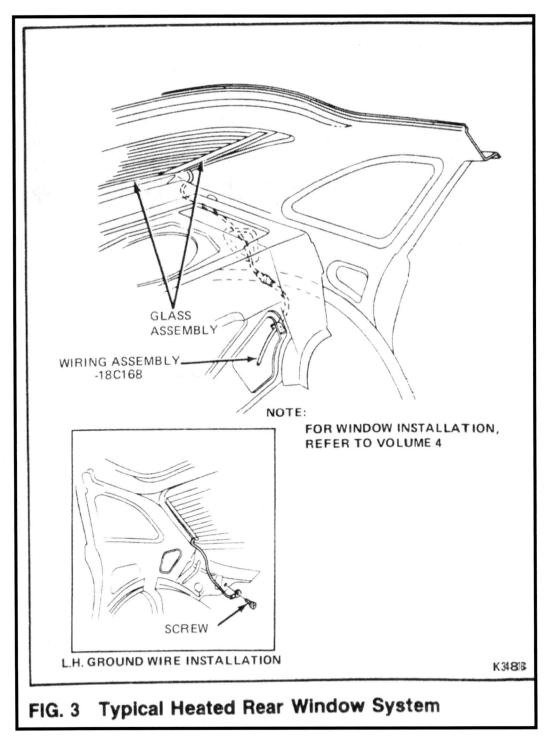

The 1981 Ford heated rear window defroster system.

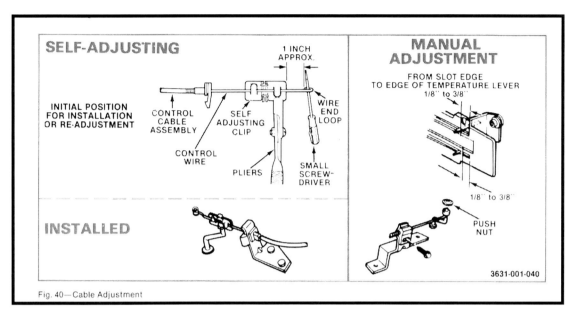

Fig. 40—Cable Adjustment

1981 Ford Motor Company Climate Control REFRIGERATION SYSTEMS

Ford vehicles used two types of cable adjustment in the 1981 era. The self adjust type featured a clip that slid along the push-pull wire. The other manual adjust type required the screw to be loosened on the cable end flag.

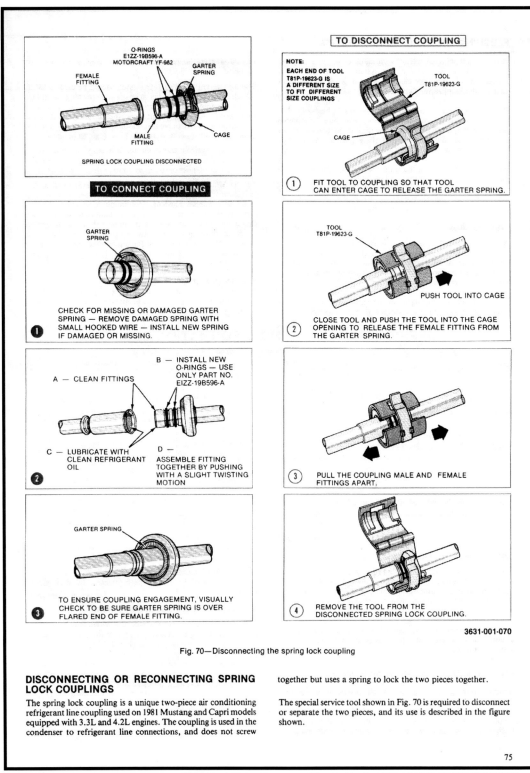

Fig. 70—Disconnecting the spring lock coupling

## DISCONNECTING OR RECONNECTING SPRING LOCK COUPLINGS

The spring lock coupling is a unique two-piece air conditioning refrigerant line coupling used on 1981 Mustang and Capri models equipped with 3.3L and 4.2L engines. The coupling is used in the condenser to refrigerant line connections, and does not screw together but uses a spring to lock the two pieces together.

The special service tool shown in Fig. 70 is required to disconnect or separate the two pieces, and its use is described in the figure shown.

1981 Ford Motor Company Climate Control REFRIGERATION SYSTEMS

The Ford Spring Lock Coupling was introduced on the 1981 Mustang/Capri.

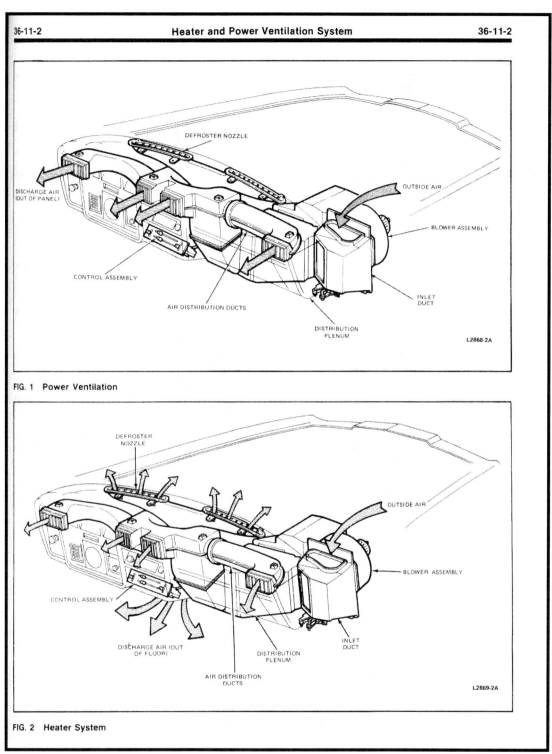

FIG. 1 Power Ventilation

FIG. 2 Heater System

1981 Ford service manual

Power ventilation forces outside air (non-refrigerated) thru the instrument panel registers using the blower motor.

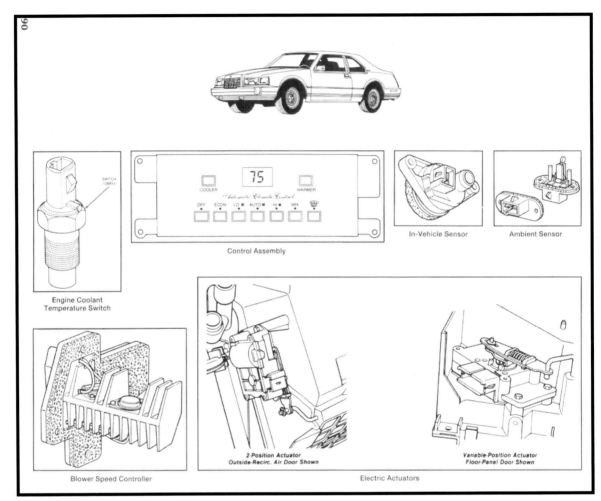

1984 Lincoln Electronic Automatic Temperature Control system.

# REFRIGERANTS

### Refrigerant 12 (R-12)

**In past years, Refrigerant 12 (R-12) has been the type of refrigerant used most with automotive air conditioning systems** because it has most of the previously listed properties for a good refrigerant. The boiling point for R-12 is -22 degrees Fahrenheit (-30 degrees Celsius); it has a latent heat value of 66.75 BTU's/pound at 18 degrees Fahrenheit (-8 degrees Celsius); it liquifies at 74 degrees Fahrenheit (23 degrees Celsius) when at 75 psi, and it boils at 32 degrees Fahrenheit (0 degrees Celsius) when at 30 psi.

However, **one of the major downfalls of R-12 is that it is a chlorofluorocarbon-based (CFC) refrigerant**. Scientists' research has proven that chlorofluorocarbons like R-12 have harmful effects on the environment. One of the biggest concerns about chlorofluorocarbons is that they contribute to the depletion of the ozone layer of the atmosphere. The ozone layer is the outer layer of the atmosphere that protects us from harmful ultraviolet rays from the sun that could damage our bodies. Continued use of chlorofluorocarbons may cause holes to form in the ozone layer, subjecting us to these harmful rays.

### Refrigerant 134a (R-134a)

Because of the concern over the harmful effects of using the chlorofluorocarbon-based (CFC) R-12 in air conditioning systems, **a non-chlorofluorocarbon-based (Non-CFC) refrigerant known as R-134a was developed**. R-134a is a hydrofluorocarbon-based (HFC) refrigerant. **Because of the absence of chlorine in its molecular structure, R-134a doesn't contribute to the depletion of the ozone layer**. The remaining properties of R-134a such as its boiling point, vapor pressure, latent heat value, etc. are very similar to those of R-12. Over the next few years, automotive manufacturers will be phasing in air conditioning systems that use R-134a and other types of Non-CFC-based refrigerants while phasing out systems that use the CFC based R-12.

1985 Circa Ford booklet "The ABC's of Air Conditioning"

A brief discussion of why R-12 refrigerant would be replaced with R-134a

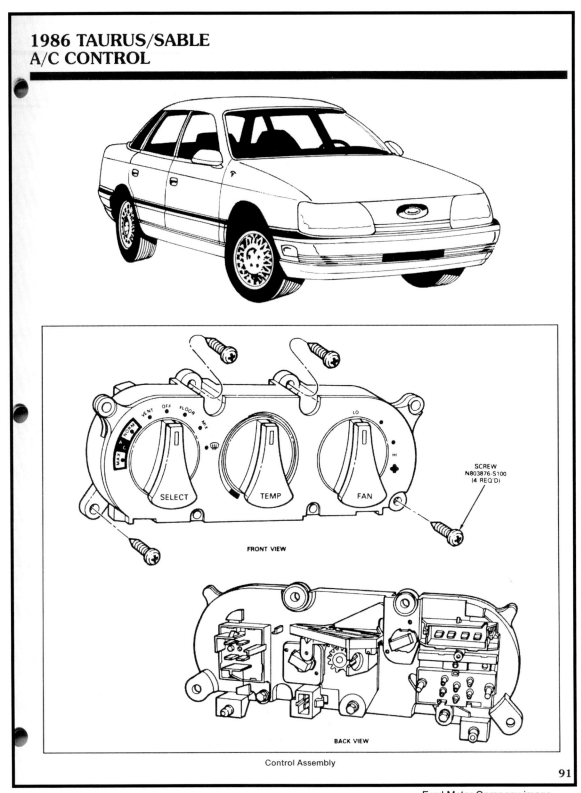

The earliest application of a 3 knob rotary control on a North American Ford vehicle was on the 1986 Ford Taurus and Mercury Sable.

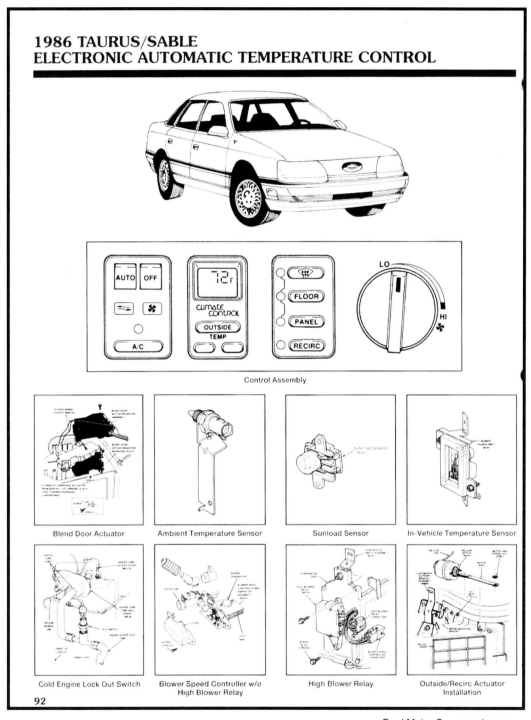

The 1986 Taurus/Sable automatic temperature control system.

Thomas Doerfer image wikipedia dot org Daimler_XJ40.jpg

Numerous vehicles have used single wipers including this 1986 Daimler. Some other applications include: 1996 Lotus Elise, 1985 Subaru, 1988 Isuzu Impulse, 1977 VW Scirocco, 1984 Mercedes-Benz W124, 1985 Subaru XT, 1967 Ferrari Competizione, 2011 Lotus.

Ford Motor Company image

A Ford Motor Co. drafting room circa 1987. The people are designing heater and air conditioner components. L to R: Wayne Schnaidt, Gene Dickirson, Dan La Palm, Ken Sisk.

Images taken 2011

The 1987 U.S. Postal Service Long Life Vehicles were fitted with fans.

# 1989 RANGER, 1990½ EXPLORER
## A/C CONTROL

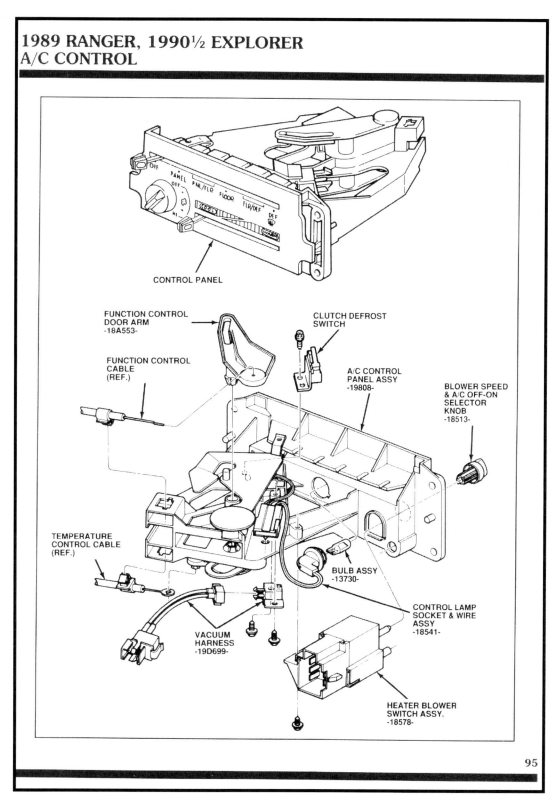

Ford Motor Company image

This instrument panel control with slide levers would be one of the last of the type used by Ford Motor Co. in North America.

General Motors image

### Heating and Cooling

The Impact includes an electrically driven heat pump system to provide heated or cooled air to the occupants. The small rotary compressor is driven by a 1-hp Magnequench motor designed by the GM Research Laboratories and built by Delco Remy. The system is capable of operating when the vehicle is unattended, allowing the car to be cooled or heated at a predetermined time.

General Motors press release January 3, 1990

The 1990 General Motors Impact electric vehicle used an electric motor to drive the air conditioner compressor. It also used a heat pump.

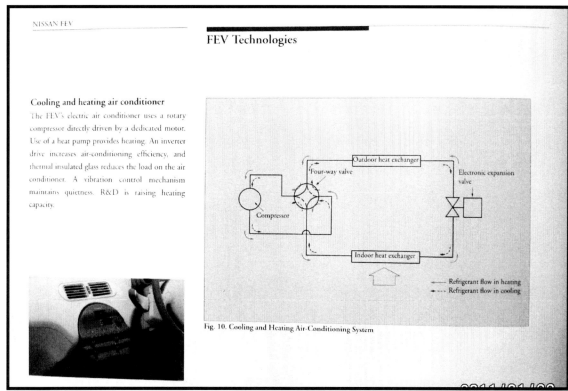

1991 Nissan FEV brochure

The 1991 Nissan FEV electric vehicle used a heat pump.

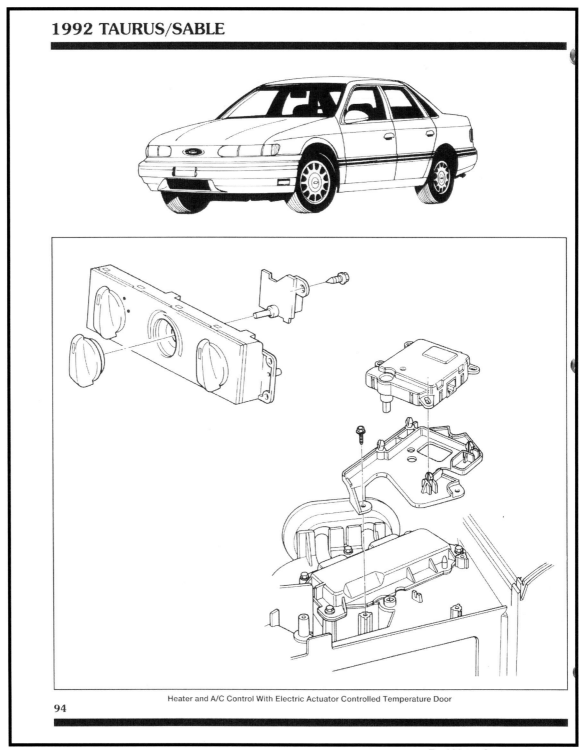

Ford Motor Company image

An electric actuator and potentiometer replaced a mechanical push-pull control cable to control the air conditioner temperature door as a running change on the 1992 Taurus and Sable vehicle lines. This was done as a product upgrade.

# FORECAST FROM 1992

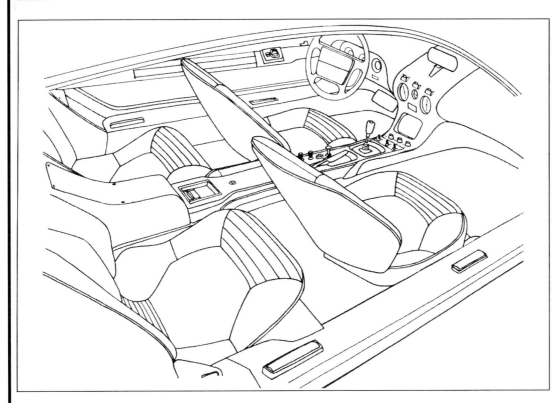

## FUTURE CONTROLS

- Integrated A/C and Radio Controls — "Integrated Control Panels"
- Push Button — Electrical/Electronic
- Voice Recognition/Voice Controlled
- Individual Zone Control
- Fuzzy Logic
- Electroluminescent Lighting

## FUTURE CLIMATE CONTROL SYSTEM FEATURES

- Instant Comfort — Instant Heating/Cooling
- Total Quiet Operation
- Infrared Sensors for Occupants' Skin Temperature and Moisture Level
- Sensors for Fog, Frost, Ice, Hydrocarbons, Rain

CCD/PEO G.D. DICKIRSON 1992 MARCH

Ford Motor Company image

This forecast was made by the author in 1992.

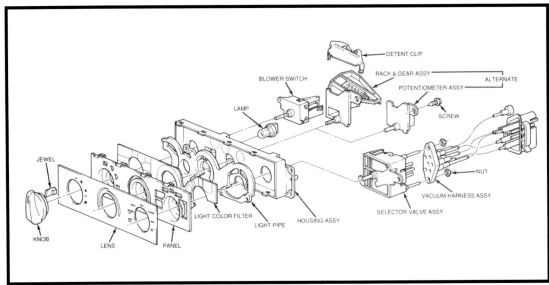

Ford Motor Company image

The 1992 model year three knob rotary control was released on many Ford vehicle lines over the years and one other vehicle manufacturer used it. It is still in production on the 2011 Ford Ranger pickup.

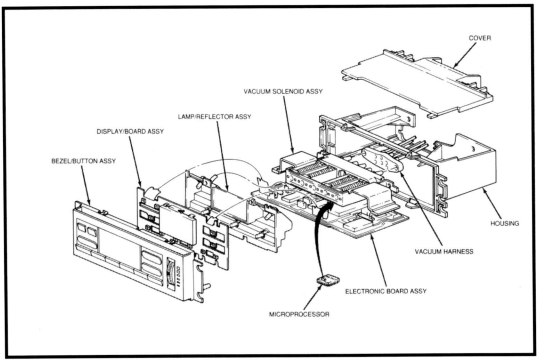

Ford Motor Company image

The 1993 Ford Electronic Automatic Temperature Control assembly was used for several years on various Ford vehicles.

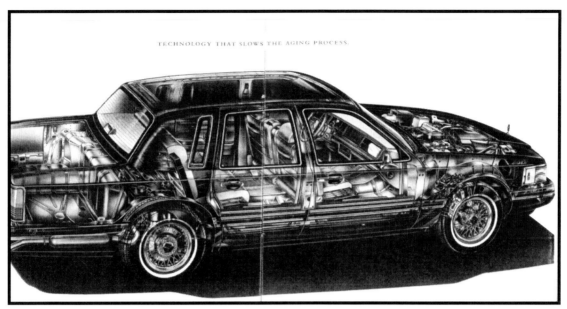

1994 Lincoln brochure

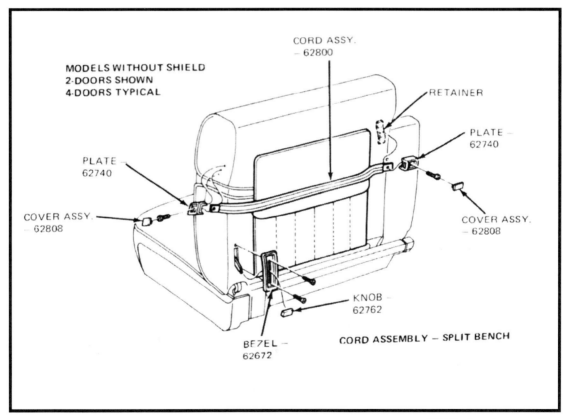

1981 Ford service manual

Robe cords were still being fitted to luxury vehicles in the 1990's.

The 1995 Lexus LS 400 featured one of the earliest dual zone controls. (2007 model shown)

A critically important part of the comfort equation is maintaining pleasant temperature and humidity levels in the interior. This is a particularly difficult challenge because of the amount of energy needed to heat and cool the confines of any automobile.

Conventional cars use waste heat from the internal combustion engine's cooling system to heat their interiors. Since little energy is wasted in an electric car and minimal heat is produced by the propulsion system, a completely different approach had to be developed for the EV1. Engineers selected a heat pump approach, common in Southern U.S. homes but never before used in an automobile.

A heat pump is an air conditioner capable of both heating and cooling. To do so, the ozone-friendly R-134a refrigerant sealed within the system must be capable of flowing both ways. Major components include a compressor located under the hood and two heat exchangers — one under the instrument panel and one under the hood. The compressor is directly driven by a computer-controlled three-phase AC motor. A reversing valve, also computer-controlled, switches refrigerant flow as needed.

In the cooling mode, the refrigerant expands through the interior heat exchanger to remove unwanted heat, which is then ejected into the atmosphere through the under-hood heat exchanger. In the warming mode, the exterior heat exchanger picks up heat from the atmosphere (even on cold days) for delivery to the interior. A climate-control panel permits EV1 occupants to select from five blower speeds and three power settings.

Maximum energy consumption is purposely limited so the car's range is not seriously impacted by use of the climate control system. As long as the ambient temperature remains between 30 and 100 degrees F, this system does an excellent job of maintaining comfort. Special window glass helps filter the sun to reduce the solar load on the interior. Instead of a conventional hot-air defroster, the EV1 uses an electrically heated metallic layer to dissipate windshield frost and a resistor grid to clear the rear window.

The ability to precondition the interior before any trip is an interesting climate-control feature. The energy needed

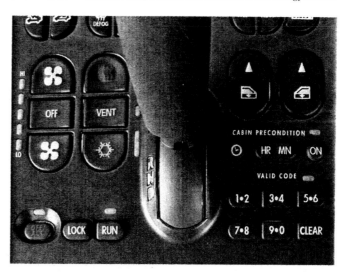

for preconditioning comes from the car's charging hookup instead of from its battery pack. All the driver has to do is to enter the anticipated time of departure and the EV1's climate-control system automatically goes to work 15 minutes early to heat or cool the interior as needed. When the car is disconnected from its charger for the trip, the driver can buckle up and depart with a fully charged battery and a comfortable interior climate.

More than any past attempts to interest consumers in alternative propulsion, General Motors' EV1 is a thoroughly thought-out, well-equipped, very comfortable automobile. Extra added benefits are a fun-to-drive personality, ultralow maintenance requirements and pollution-free operation.

But the greatest joy of all in the EV1 is driving nonchalantly past each and every gas station.

GM Evolution magazine

The 1996 EV1 (electric vehicle) by General Motors used a sophisticated heater and air conditioning system. An electric motor drove the air conditioner compressor.

General Motors 1996 EV1 brochure

Saab image http://www.saabhistory.com/2007/11/17/ventilated-seats-saab-innovation

1997 Ventilated Seats – Saab 9-5 is the first car with ventilated seats. As a compliment to air conditioning, this provides an outstanding level of comfort and helps the driver to stay fit and alert.

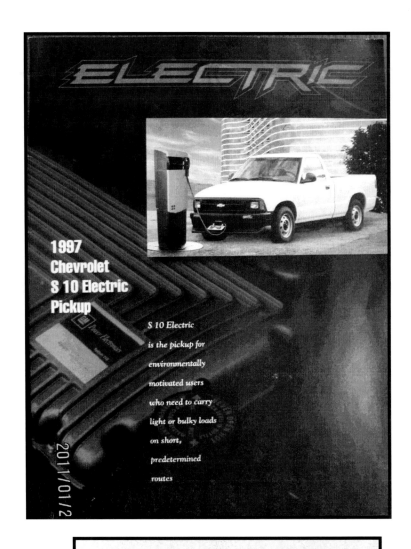

1997 Chevrolet S10 Electric brochure

The 1977 Chevrolet S10 electric vehicle used a heat pump and a diesel heater.

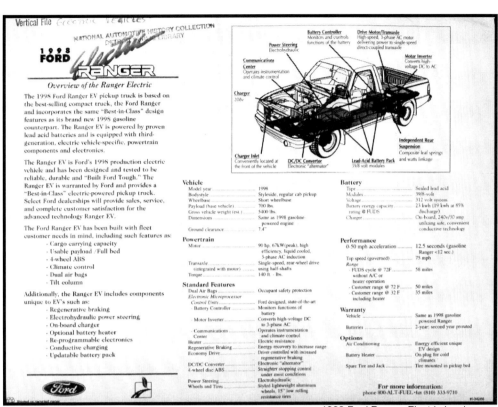

1998 Ford Ranger Electric brochure

**Climate Control Systems - Heating System**

Since the Ranger EV's powertrain is much more efficient than that of a conventional gasoline engine, waste heat cannot be captured in coolant and used to heat the passenger compartment. As a result, the electric Ranger uses an electric device called a Positive Temperature Coefficient (PTC) heater. Much like a high voltage toaster oven, the PTC heater uses direct current from the traction battery to heat resistive elements to defrost the windshield and warm the passenger compartment. In order to conserve battery pack energy, the passenger compartment is only allowed to heat up to 80 degrees Fahrenheit.

**Climate Control Systems - Air Conditioning System**

The absence of an internal combustion engine means there is no drive belt to power a pulley-driven compressor or air-conditioner. Instead, the Ranger EV uses a unique high-voltage, variable speed compressor powered by alternating current from its own Inverter Motor Controller. A 3.5 kW electric motor is used to circulate the refrigerant through the cooling system. The vehicle's air conditioning system can bring the passenger compartment down to just 65 degrees Fahrenheit in order to conserve energy.

Compiling each of these state-of-the-art technologies in the 1998 Ford Ranger EV results in a vehicle that will meet or exceed EV customer expectations.

\#\#\#

12/20/96

Ford Motor Company press release

The 1998 Ford Ranger electric vehicle used a Positive Temperature Coefficient heater. An electric motor was used to drive the compressor.

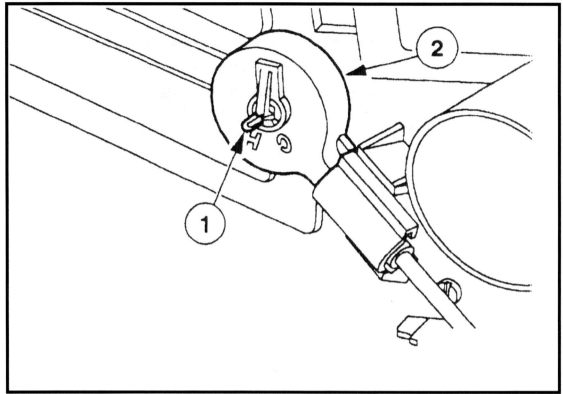

1998 Ford Escort service manual

This 1998 Ford Escort pull-pull cable replaced the traditional push-pull (Bowden) cable because it had lower hysteresis and was immune to kinking.

Image taken at Ypsilanti, MI 2011

A Circa 1998 Porsche fitted with parallel wipers.

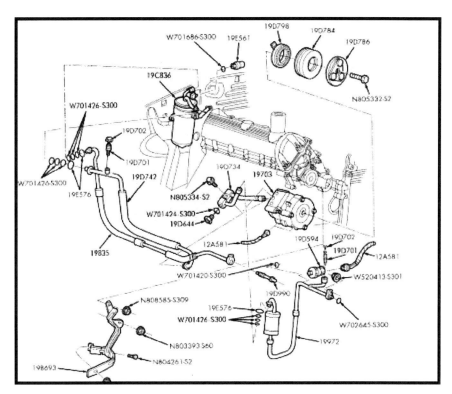

1998 Ford Mustang parts manual

The peanut refrigerant fitting saw an early application on the 1998 Ford Mustang.

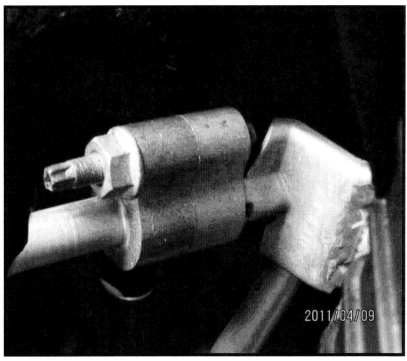

Image taken 2011

The peanut refrigerant fitting is named for the shape of its cross section.

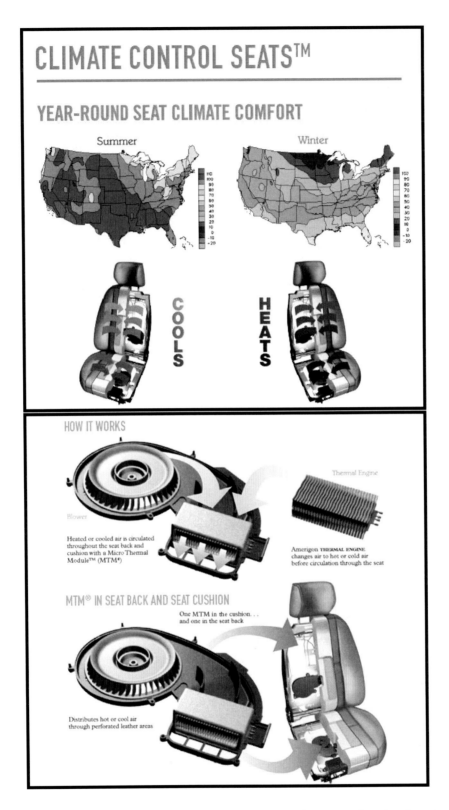

The 1999 Lincoln Navigator offered Amerigon heated and cooled seats.

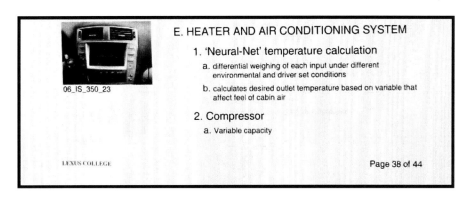
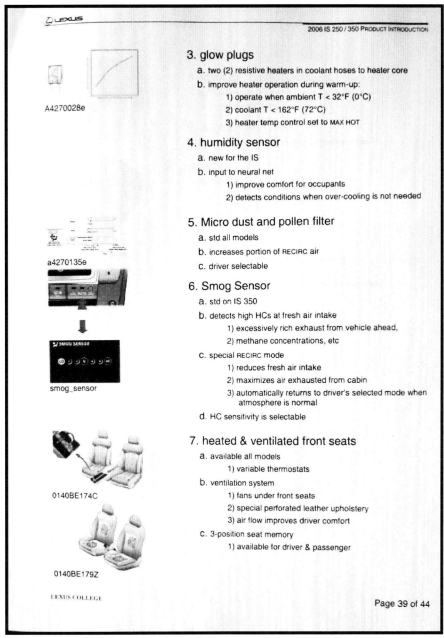

2006 Lexus IS 350/250 Product Introduction brochure

The 2006 Lexus featured a list of stat-of-the-art features.

Both images taken 2011

The 2007 Subaru featured fully styled windshield wiper arms and blades.

The 2009 Nissan G37 X also featured fully styled windshield wiper arms.

DENSO Corporation announced today that it has developed a new air conditioning unit, which is approximately 20 percent smaller in volume than the conventional product. The new system will be installed on Toyota's new compact car, the "iQ," to be launched in Japan in November and in Europe in early 2009. With improvements including new resin molding technologies, DENSO reduced the size of the blower fan for the new air conditioning unit. This allows the blower unit containing the blower fan to be reduced by almost half the size of the conventional unit, contributing to the size reduction of the air conditioning unit.

Denso press release dated October 22, 2008.

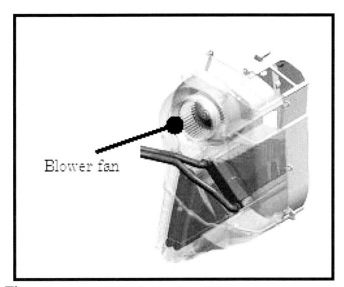

The new compact system.

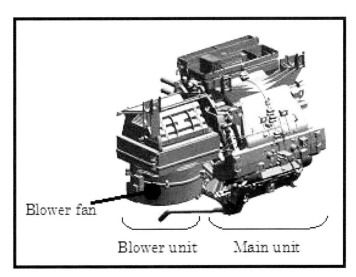

The older system.
The 2008 Denso compact air conditioner system.

Small automotive fans are still available in 2011 in several different styles.

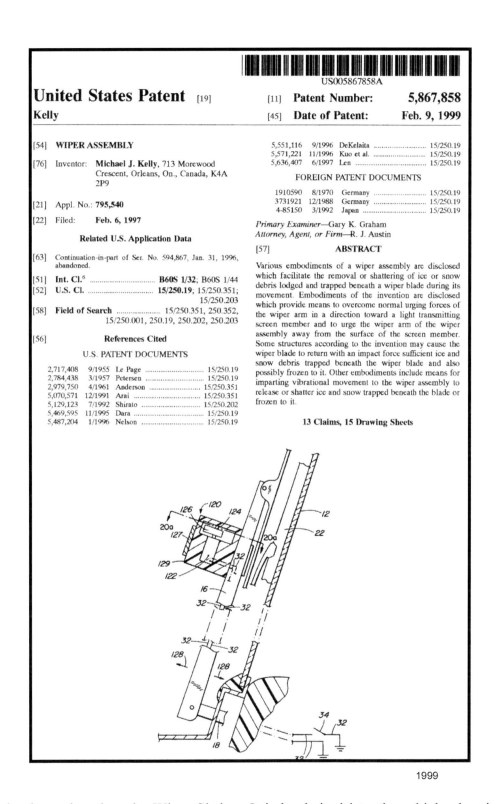

This device is marketed as the Wiper Shaker. It is hardwired into the vehicle electrical system. It is listed as an option in several of the 2011 Ford and Lincoln brochures.

## 2011 Lexus LS Climate Control features

Climate-comfort front seats with heat/cool knob
Climate-comfort rear seats with heat/cool knob
Enhanced interior heater
Four-zone automatic climate control with rear-seat infrared temperature sensor, interior air filter, smog sensor and automatic recirculation mode
Headlamp washers
Heated rear seats
Heated steering wheel
Power rear sunshade
Power rear-door sunshades
Rain-sensing intermittent windshield wipers with mist
Rear-seat Air Purifier
Rear-seat cool box
Rear-window defogger with auto-off timer
Water-repellent front-door glass
Windshield-wiper deicer
Wood- and leather-trimmed heated steering wheel

http://www.lexus.com/models/LS/ 2011 Apr 06

The list of climate control features offered on the 2011 Lexus is extensive and impressive.

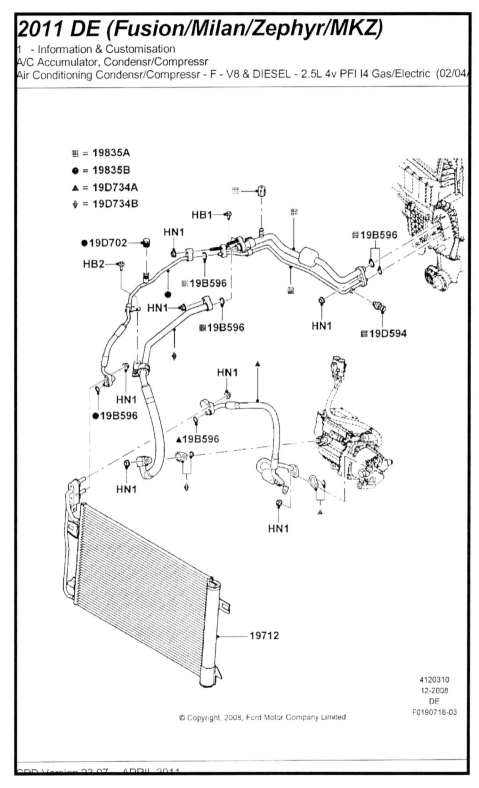

Ford/Lincoln Dealer service parts catalog

The 2011 Ford/Lincoln MKZ Fusion Hybrid refrigeration system.

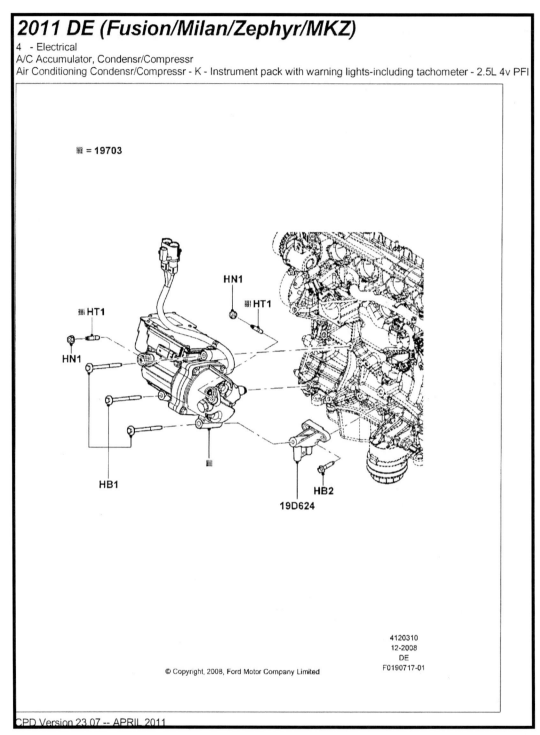

The 2011 Ford Fusion/ Lincoln MKZ Hybrid electric motor driven compressor.

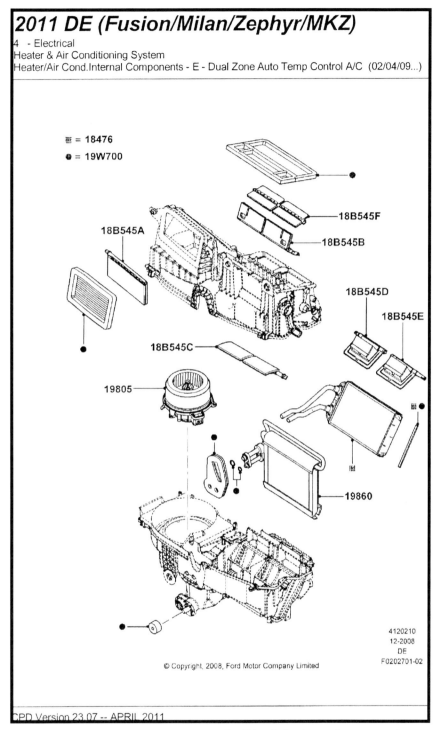

The 2011 Ford Fusion/ Lincoln MKZ Hybrid air handling system.

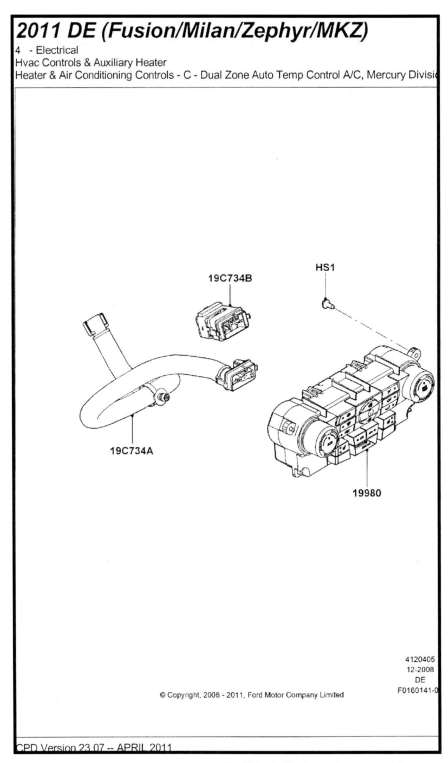

The 2011 Ford Fusion/ Lincoln MKZ Hybrid instrument panel control assembly

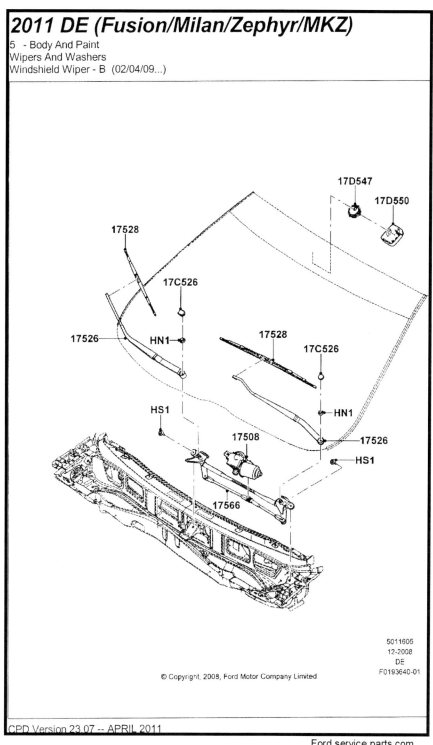

The 2011 Ford Fusion/ Lincoln MKZ Hybrid windshield wiper system

Denso media image

The 2011 Denso Ejector evaporator.

# BIBLIOGRAPHY

## BOOKS
*Standard Catalog of American Cars* 1946-1975 John Gunnell, 1992
*Dykes Automobile and Gasoline Encyclopedia* 1915, 1922, 1940
*100 Years of the American Auto*, James M. Flamming, 2002
*Packard A History of the Motor Car and the Company*, Beverly Rae Kimes, June, 1978
*Original Ford Model A*, Jim Schild, 2003

## BROCHURES
American Bosch 1948, 1952, 1955
Anderson Company 1948, 1956
Buick 1923, 1934, 1940, 1954
Chevrolet 1926, 1930, 1997
Chrysler 1940, 1963, 1964
EV1 1996
Fisher Body 1933
Ford Motor Co. Accessories 1947, 1960
Ford Motor Co., 1927, 1928, 1934, 1935 (2), 1940, 1949, 1952, 1998
Lava Pare-Brise 1955
Lincoln 1994
MoPar 1950, 1954
Motorola 1946
Nash 1938, 1955
Nissan FEV 1991
Oldsmobile 1917, 1918, 1934, 1935, 1939, 1942
Oldsmobile Price Lists 1938, 1939
Packard 1931, 1933, 1934, 1940, 1941, 1954
Plymouth 1963
Pontiac 1942, 1954
Robert Bosh, A.G. 1929, 1952
Saab 1970, 1999
Smith 1929
Smith Machine 1935
Stewart Warner South Wind 1940, 1941, 1948
Studebaker 1939, 1941, 1946
Trico 1946, 1954 (2), 1955
Tropic-Aire 1946
VDO Autolufter 1965

## ENGINEERING DOCUMENTS
Automotive Climate Control the first 100 Years, Gene D. Dickirson, P.E., 1993
Buick Engineering Information for 1940, 1940
1953 Buick New Engineering Features, 1953

Chevrolet letter from R.W. Connor, 1950
Chrysler Heating and Ventilating System 1940 Models, 1940
DuPont Plastic News, Circa 1977
Modern Developments in Vehicle Heating, F.A. Ryder, 1949
Oldsmobile Engineering Features- 1938
Progress of Automobile Heating Through the Years, Harold V. Joyce, 1960
Wiping Wrap-Around Windshields Was a Problem, John W. Anderson, 1954

## MAGAZINES
American Automobile- December 1901
Automobile Magazine- March 1902; August, September, November 1903
Automobile Quarterly, Forth Quarter 1985, Volume 51, Number 1
Automobile Review- December 1904
Automobile Trade Journal- November 1916
Automotive Industries- October 1918
Car Life- July, October 1964; February, August 1966; July, October 1967
Ford Dealer and Owner- November, December 1925
Ford Dealer and Service Field- September, November 1925; October, December 1927; May 1928; January, June, August, September, October, November, December 1929; January, July, August, September, October, November, December 1930
Ford Field- September 1939
GM Evolution- 1996
Hemming Classic Car- January 2010
Horseless Age- 1903
JC Whitney- catalog number-124, 1957
Motor- January 1953
Motor Age- October 1916; January 1917; October 1920; January 1924
Motor Life- July, August, November 1955; March 1956; August 1957; February 1960
Motor- November, December 1905; October, November 1906; February 1907; September, November 1941
Motor Service- October 1941
Motor Trend- August 1952; March, May, July, August, November 1954; April, July 1955; July 1957; June 1959
Motor World- October 1917
Popular Mechanics- February 1934
Popular Mechanics- November 1936
Sears Catalog- 1908
Specialty Interest Autos- May, June 1971
The Autocar- September 15, 1900
The Automobile & Motor Review- December 1902
The Automobile and Horseless Vehicle Journal- May 1897
The Automobile- March 1902; October, December 1903; January, March 1908; September, October, December 1909; December 1911; January, December 1912; December 1914; November, December 1916; January 1917

## NEWSPAPERS
Dallas Morning News- March 4, 1992
Detroit Free Press- November 15, 1990; June 12, 1992
Detroit News- March 11, 1992
New York Journal and American- July 24, 1940

## ORGANIZATIONS
Benson Ford Research Center located at the Henry Ford, Greenfield Village, Dearborn, MI
Buick Gallery & Research Center, Flint, MI
Cadillac LaSalle Museum & Research Center, Farmington Hills, MI
Ford Motor Company Engineering Library, Dearborn, MI
Fountainhead Antique Auto Museum, Fairbanks, AK
Hines Park Lincoln, Inc. in Plymouth, MI
National Automotive History Collection, Detroit Public Library, Detroit, MI
Stahls Automotive Foundation, Chesterfield, MI

## OWNERS MANUAL
Ford- 1939

## PARTS MANUALS
Ford- 1965
Ford Mustang- 1998
Ford Service Parts dot com
Ford Truck- 1948-1950, 1951
Lincoln- 1953, 1954, 2011

## PRESS RELEASES
Denso- October 22, 2008
General Motors- January 3, 1990; July 23, 2010
Lexus- 2006
Toyota- September 1, 2000

## SERVICE MANUALS
A/C Refrigerant Systems Reference Manual 1980-1990, Ford Motor Co., Climate Control Div.-1990
Buick- 1954, 1955, 1956, 1968
Cadillac- 1953, 1954, 1955, 1956, 1957, 1959,1961,1964,1966
Chevrolet-1971, 1973, 1974
Chrysler- 1953
DeSoto- 1953, 1954
Dodge- 1953
Fisher Body Service News- 1954 Buick
Ford- 1956, 1959, 1960, 1966, 1974, 1975, 1979, 1981
Ford Escort- 1998
Ford heater installation instructions- 1937

Ford Refrigeration Systems- 1981
Ford Service Bulletin- 1939
Ford Taurus-1996
Ford Thunderbird- 1960
Lincoln- 1958, 1961, 1966
Lincoln Mark V- 1979
Mercury- 1958
Nash- 1940
Oldsmobile- 1953
Studebaker- 1941
The ABC's of Air Conditioning, Ford Motor Co., Climate Control Div., Circa 1985
Tropic-Aire- 1932

## WEBSITES
autopartsway dot com
chevychevy dot com
classicaccessories dot org
corvairunderground dot com
fordparts dot com
forwhatyouneed dot com
glasslinks dot com
glassonline dot com
google dot com
jcwhitney dot com
lexus dot com/models/LS/2011apr06
oldcarmanualproject dot com
saabhistory dot com/2007/11/16/saabhistory
sae.org/mags/aei/6741
tricoproducts dot co
wikipedia dot org

# INDEX

## AIR CONDITIONER
3 knob rotary control, 476
accumulator, 468
air handling system Ford Fusion 2011, 505
automatic temperature control, 14, 22, 28, 29, 31, 32, 35, 227, 265, 266, 351, 426, 439, 477
Buick system 1953, 312
Buick system 1955, 363
Buick system 1956, 370
Cadillac system 1941, 261, 262
Cadillac system 1953, 315
Cadillac system 1954, 22, 343
Cadillac system 1955, 356
Cadillac system 1957, 376
Cadillac system 1959, 392
Cadillac system 1961, 401
Cadillac system 1964, 409
Cadillac system 1966, 428
Chevrolet system 1957, 375
Chrysler system 1953, 320, 321
clutch, 15, 22, 251, 322, 343, 344, 346, 357, 367, 402, 416, 465, 467
combination STV/BPO expansion valve, 469
Comfort Control Cadillac 1964, 408
Comfort Control instrument panel control 1964 Cadillac, 410
compact system Denso 2008, 499
compressor, 15, 20, 21, 22, 27, 28, 31, 32, 33, 34, 35, 251, 262, 263, 322, 343, 396, 402, 410, 415, 420, 444, 446, 450, 451, 452, 463, 482, 489, 493, 504
condenser, 21, 342, 397, 446
condenser and receiver-dehydrator, 417
DeSoto system 1954, 345
Dodge system 1954, 323
dual zone Lexus 1995, 33, 488
ejector evaporator, 508
electric actuator and potentiometer, 33, 484
electric motor driven, 33, 489, 493, 504
electrical schematic, 340, 365, 367
electromagnetic clutch, 322, 357, 416
expansion valve, 325, 418, 448, 469
flare fitting, 31
Ford 6 cylinder compressor, 452
Ford air conditioner compressor usage chart, 463
Ford FS6 compressor, 466
Ford fusion Hybrid system 2011, 503
Ford System 1954, 335
Ford System 1956, 365
Ford system 1960, 394
Ford system 1966, 425
Ford system 1975, 445
Ford system 1979, 460
Ford Thunderbird system 1960, 395
Ford Universal Control, 442
four zone, 502
Freon, 24, 180, 326
Freon handling instructions, 326
hang-on, 25, 27, 28, 30, 311, 364, 392, 420, 440
heat pump, 32, 33, 482, 483, 492
HFO-1234yf refrigerant, 35
ice bunker test, 11
in-plant charge, 28
installation rate 1958-1965, 26, 424
instrument panel control, 4, 25, 26, 27, 28, 29, 30, 31, 32, 264, 277, 380, 384, 396, 410, 481, 506
instrument panel control Ford Fusion 2011, 506
Lexus features 2011, 502

Lincoln system 1953, 317, 318, 319
Lincoln system 1954-55, 331, 332, 333
Lincoln system 1958, 27, 384
Lincoln system 1966, 426
Lincoln system 1970, 439
manufacturers 1957, 378
mini combination valve, 469
Nash system 1954, 329
Nash system 1955, 354, 355
Neutral Net, 34
Non-BPO combination valve, 468
Oldsmobile system 1953, 313
orfice tube, 467
package size reduced by 20%. Denso, 34
Packard system 1940, 250, 251
Packard system 1940 road test, 252
Packard system 1941, 261
peanut refrigerant fitting, 34, 495
plate fin evaporator core, 29, 429
Pontiac system 1954, 328
pre charge, 28
R-12 refrigerant, 10, 20, 22, 32, 180, 346, 475
R-134a refrigerant, 32, 33, 35, 475
R134a replaces R-12, 475
R-4 compressor, 444
refrigerant, 10, 20, 21, 22, 28, 31, 32, 33, 34, 35, 180, 346, 447, 461, 464, 475, 495
refrigerant control, 31, 464
reheat system, 28
reheat system 1963 Buick, 406
sensors, 35
service ports, 420
Specifications 1954 Dodge, 324
spring lock coupling, 31, 459, 472
suction throttling valve, 430
Tecumseh two-cylinder compressor, 450
temperature control valve, 305, 325
thermostatic control, 418
transition of refrigerant control systems Ford 1979-1981, 31
tube fin evaporator, 415
tube-o fittings, 31, 447, 461
two cylinder compressor, 415
two thumb wheels instrument panel control, 30, 435
variable capacity compressor, 34
York two-cylinder compressor, 451

## ANTIFREEZE
alcohol, 2, 54, 347
calcium chloride, 2, 53
Concentrated Non-Freezing Fluid, 9
denatured alcohol, 2, 54
glycol, 347
oil, 2, 8, 53
Prestone, 141, 241
Thermo Royal, 278
windshield washer, 348
wood alcohol, 2, 54

## COMPONENTS
air filters, 20
Bowden cable, 18, 19, 25, 26, 27, 30, 32, 33, 34, 305, 381, 434, 494
cam, 26, 32, 370
control bracket, 31, 457
control cable, 30, 33, 447, 484
humidity sensor, 34
pull-pull cable, 33, 494
thermoplastic, 27
transistorized amplifier and thermistors, 28
vacuum motor, 5, 11, 18, 19, 20, 25, 26, 27, 28, 30, 101, 144, 183, 304, 341, 377, 381, 382

## DEFROSTERS
double, 15
ducts, 9, 14, 150, 151, 277, 280
electric sleet frost shield, 15
Ford rear window defogger 1975, 455
integrated defroster nozzles, 13
molded plastic defroster nozzles, 341
nozzles, 9, 13, 224, 225, 229, 233, 341
Quick Defrost Windshield and Rear Window, 31, 443

## HEATERS
air filter, 13, 14, 16, 17, 34, 216, 227, 236, 271, 277, 502

air-to-air heat exchanger, 3, 70, 78, 109
automatic temperature control, 227, 265, 266
Bahco 3000 pre heater, 433
blend air type, 27
blower, 7, 10, 12, 16, 17, 18, 21, 27, 31, 162, 165, 170, 175, 243, 255, 277, 328, 473, 499
booster, 403
Cadillac automatic temperature control 1941, 265, 266
charcoal, 1, 37, 38, 44, 46, 49, 57, 58, 59
Climatizer, 14, 16, 17, 236, 269, 270, 271, 276, 277, 278
coal briquettes, 1
Conditioned Air, 14
diesel, 33, 492
distribution chamber, 9
door duct, 334
dual, 327, 385
electric, 84
exhaust, 8, 48, 51, 55, 60, 63, 65, 66, 68, 69, 80, 81, 82, 85, 86, 93, 95, 99, 105, 106, 108, 109, 112, 115, 121, 125, 127, 128, 129, 130, 132, 148, 149, 154, 163, 170, 172, 177, 182, 200, 222
exhaust with blower, 161, 162, 165, 175
exhaust with fresh air, 167, 173
exhaust-to-air heat exchanger, 5, 7
factory-installed, 3, 21, 26, 335, 378, 424
fan belt driven blower, 176
Ford system 1956, 366
Ford system 1959, 388, 389
fresh air, 3, 9, 14, 16, 17, 18, 19, 22, 27, 154, 167, 172, 173, 245, 255, 273, 274, 278, 282, 290, 292, 306, 351, 366, 383
gasoline, 16, 17, 18, 26, 27, 257, 258, 279, 283, 284, 285, 371, 399
glow plug, 34
glow plugs in heater hoses, 34
heat pump, 32, 33, 482, 483, 492
hot air, 126, 157, 175
hot water, 7, 8, 83, 84, 117, 123, 124, 148, 157, 212
Hudson system 1951, 305
hybrid, 169, 210
integrated, 13, 14, 16, 17, 190, 227, 233, 373, 406, 425
Magic Air, 18, 292, 294, 295
metal tubes, 15
Nash Weather Eye system 1938, 227
performance curve, 7
positive pressure, 14, 16
Positive Temperature Coefficient, 493
PTC, 33
rear seat, 11, 13, 14, 15, 26, 28, 35, 216, 232, 334, 406, 502
rear seat duct, 406
rear seat hot water, 232
recirculation, 18, 22, 27, 295, 351, 502
rubber hose, 5, 15
steam, 12, 13, 212, 248
thermostatically controlled water valve, 282, 382
water- to- air heat exchanger, 6
water valve, 17, 18, 19, 26, 227, 282, 306, 382

## MANUFACTURERS & BRANDS

A.R.A., 26
Air Temp, 26
All-Season, 24
Allstate, 26
American, 5, 11, 18, 22, 23, 24, 25, 37, 55, 101, 181, 198, 252, 288, 301, 310, 362, 476, 509, 510, 511
American Austin, 11, 181, 198
American Automatic Wiper Co., 101
American Bosch Corp., 288, 310, 362
American Motors, 23, 24
Amerigon, 496
Anderson Co. (The), 286, 287, 368
Apco, 142
Artic, 26
Artic-Kar, 26

Arvin, 6, 9, 13, 14, 157, 160, 212, 240
Ashco Corp., 132, 170
Atlas Brass Foundry Co. (The), 115
Autocraft-Bovey, 3, 70
Auto-Lite, 8, 149, 239
Automobile Devices Co., 4, 90
Aveo, 29
Baby Grand, 3, 64
Bacho 3000, 30
Bailey, 69
Ball-Fintze Co. (The), 45
Blackmore, 69
Breeze Filter, 8, 143
Brickey Auto Heater, 3
Buick, 3, 5, 12, 15, 16, 20, 21, 22, 23, 26, 29, 30, 35, 64, 98, 205, 234, 255, 256, 268, 307, 308, 312, 334, 339, 340, 341, 363, 370, 406, 427, 434, 509, 511
Burke Clearsight Centrifugal, 4
Cadillac, 8, 11, 12, 15, 16, 18, 19, 20, 22, 23, 28, 29, 35, 143, 144, 145, 146, 147, 148, 182, 193, 253, 261, 262, 263, 264, 265, 266, 267, 298, 315, 316, 327, 334, 342, 343, 344, 356, 357, 367, 376, 377, 392, 401, 402, 408, 409, 410, 428, 429, 430, 511
CAM-O-MATIC, 22
Carron & Co., 61
Chas. E. Miller, 50
Chevrolet, 3, 5, 6, 10, 14, 19, 23, 26, 27, 29, 30, 31, 33, 35, 64, 109, 110, 111, 112, 118, 120, 177, 189, 195, 301, 302, 375, 383, 399, 400, 422, 423, 435, 436, 441, 444, 492, 509, 510, 511
Chicago Flexible Shaft Co., 57
Clear Advantage, 33
Clear Vision Cleaner Co., 72
Climatic-Air Manufacturing Co., 26
Climatizer, 14, 16, 17, 236, 269, 270, 271, 276, 277, 278
Coleman Manufacturing Co., 173
Columbus Varnish Co., 3
Continental Flyer, 11, 195

Continental Mark III, 30, 437
Cool Queen, 26
Cooper Manufacturing Co., 128
Corvair, 27, 29, 399, 400, 422, 423
Corvette, 26, 30, 383, 435, 436, 441
Cox Brass Mfg. Co., 82
Crew Levick Co., 78
Custom Conditioner Air Heater, 26
Daimler, 32, 478
Davis Robe Co. (The), 47
Dayton Top Improvement Co., 69
Dayton Welding Co., 68
Delco, 12, 13, 248
Demmerle & Co., 43
Denso, 34, 35, 499, 508, 511
DeSoto, 26, 322, 345, 346, 372, 511
Detroit Automobile Heater Co., 60
Detroit Electric, 3, 64
Detroit Leather Specialty Co., 39
Detroit Weatherproof Body Co., 74
Duesenberg, 12, 183
Dunn Manufacturing Co., 129
DuPont, 12, 31, 457, 458, 510
E.A. Laboratories, Inc., 272
Eaton Manufacturing Co., 15, 242
Edsel, 26, 27
Electric Ranger, 33
Emil Grossman Co., 58
Escort, 33, 494, 511
EV1, 33, 489, 490, 509
Eveready Prestone, 141
Ever-Tite Bolt Co., 155
F Series, 32
F. C. Purcell & Co., 175
Ferrari, 31, 32, 456, 478
FEV, 32, 483, 509
Fisher Body, 8, 11, 12, 15, 145, 190, 193, 195, 209, 341, 509, 511
Folberth Auto Specialty Co. (The), 5, 103
Ford Motor Co., 419, 479, 481, 509, 511, 512
Forston Corp., 26
Forsyth Metal Goods Co., 121
Francisco Auto Heater Co., 161
Freon, 24, 180, 326

Frezalene Mfg. Co., 3
Frigette, 26
Frigidaire, 180
Frigiking, 25, 26, 364
Fulton Co. (The), 100, 244
Fusion, 35, 503, 504, 505, 506, 507
Gilliam Manufacturing Co. (The), 56
Greb Co. (The), 122
GTO, 29, 30, 422, 437
HaDees, 7, 9, 10, 13, 124, 158
Handy Governor Corp, 10, 170
Helzen, 5, 100
Honda Civic DX, 29
Honeywell, 35
Hubbard, 6, 119
Hudson, 3, 13, 19, 25, 26, 67, 213, 214, 216, 246, 304, 305, 373
Hyundai Accent GL, 29
Impact, 32, 482
Imperial, 26, 342
Instant Heater, 10, 26
Interstate Electric Co., 62
Isuzu, 32, 478
J. Stevens Arms & Tool Co, 40
J.M. Shock Absorber Co., 4, 84
Jaguar, 32
John C. Hoof Co., 132
John W. Jepson, 4, 88
Johnson Co., 94
Jordan, 3, 73
K.P. Foot Rest Heater Co., 63
Kelch, 8, 148
Kelvinator, 24
Kia Rio, 29
Kingsley-Miller Co. (The), 9, 152
Kissel, 5, 6, 104, 120
Kokomo Electric Co., 106, 116
Kool Kooshion, 8, 16, 17, 20, 21, 147, 260, 274, 311
Kunkel, 173
Kunkel, 173
Kunkle Manufacturing Co., 108
LE SABRE, 20
Lehman Brothers, 1
Lexus, 33, 34, 35, 488, 497, 502, 511
Libby Owens Ford, 33

Liberty Foundries Co., 124, 157, 158, 169
Lincoln, 20, 23, 27, 28, 29, 30, 31, 33, 34, 218, 317, 318, 319, 331, 332, 333, 335, 384, 385, 386, 387, 400, 406, 426, 439, 462, 474, 487, 496, 501, 503, 504, 505, 506, 507, 509, 511, 512
Linendoll Corp., 175
Lion Products Co, 212
Lo-Merc Corp., 26
Long Life Vehicle, 32, 480
Lotus, 32, 478
Lucas, 31
Magic Air, 18, 292, 294, 295
Mandeville Steering Wheel Muff Co., 1, 50
MARK IV, 26
Mark V, 31, 462, 512
Marmon, 12
Mason Thermo, 4, 85
Mercedes-Benz, 32, 478
Mercury, 18, 19, 21, 23, 26, 27, 299, 380, 381, 382, 383, 476, 512
Merrimac Chemical Co., 37
Metal Stamping Co., 164
MG, 19, 27, 300, 404
Milwaukee Specialty Co., 85
Minlon, 31
Mitsubishi Lancer, 29
Mobilette Weather Matic, 26
Monogram, 9
MoPar, 19, 22, 26, 303, 351, 372, 509
Morgan, 27
Morrison, McIntosh & Co., 46
Motor Car Heater Co. (The), 48
Motor Meter, 3
Motorola, 17, 279, 509
Nash, 14, 16, 21, 24, 25, 26, 227, 228, 229, 230, 231, 329, 354, 355, 509, 512
National Carbon Co., 141
Navigator, 34, 496
Newhouse Automotive Industries, 21
Nissan Versa, 29

Noblitt-Sparks Industries, Inc., 157, 240
Novi Equipment Co., 15, 26, 242, 335
Old Pac, 1, 40
Oldsmobile, 11, 12, 14, 15, 17, 20, 23, 88, 89, 188, 204, 209, 232, 233, 234, 235, 245, 246, 275, 313, 314, 509, 510, 512
Packard, 1, 10, 11, 12, 15, 16, 18, 20, 21, 40, 41, 178, 179, 196, 197, 206, 249, 250, 251, 261, 291, 336, 509
Packard Overland, 1
Parkomat Manufacturing Co., 26
Peerless Radiator Co., 83
Pence, 51
Perfection Heater & Mfg. Co., 95
Pierce, 6, 12, 113
Plymouth, 13, 23, 215, 509, 511
PolarAire, 27, 393
Pontiac, 16, 20, 21, 29, 30, 223, 224, 273, 274, 328, 422, 437, 509
Post & Lester Co., 39
Prestone, 141, 241
Quad Duty Heating System, 17, 278
Quick Defrost Windshield, 31, 443
Radi-Air Heater, 175
Ranger, 32, 33, 486, 493
Red Cat Instant Heater, 10
Riviera, 29, 427
Robe Lock Co., 77
S. Smith & Sons, 8, 137
Saab, 30, 31, 32, 33, 34, 438, 440, 456, 491, 509
Scott Muffler Co. (The), 45
Sears, Roebuck & Co., 1
SelectAire, 27, 393
Smith Machine Co., 210
Smith Silent Wiper, 8
South Wind, 16, 18, 257, 258, 283, 284, 285, 509
ST, 33
Stadco See-Safe, 4
Stadeker Metal Specialty Co., 4, 92
Stanley Works, The, 30, 433
Stromberg Motor Devices Co., 5, 102
Studebaker, 12, 13, 14, 16, 17, 19, 206, 207, 220, 221, 222, 225, 236, 269, 270, 271, 276, 277, 278, 300, 509, 512
Subaru, 32, 34, 478, 498
Taurus, 32, 33, 476, 477, 484, 512
Temptrol, 4
Ternstedt, 111
Thermo Royal, 17, 278
Torino, 30, 442
Torrid-Hete, 170
Towne and Country, 26
Trico Marchal, 379
Trico Products Corp., 349, 350
Trico-Folberth Ltd., 361
Tropic-Aire Inc., 117, 184, 243
U-Auto-C Corp (The), 76
Universal Control, 30, 31, 442
Vasco, 2, 56
VDO Autolufter, 509
Venti Heat, 16, 273
Visionall, 8, 144
Volkswagen, 32, 403
Vornado, 26
Weather Eye, 24, 227
Whippet, 6, 118
Wizard, 26
XP 300, 20, 22, 308
Zenith Manufacturing Corp, 176

## MISCELLANEOUS

Bowden cable, 18, 19, 25, 26, 27, 32, 33, 305, 434
cable adjustment types 1981, 471
cool cushion, 35, 267, 311
dead air devices, 13
desert water bag, 21, 339
door cooler screen, 352
door curtain, 69
drafting room image, 32, 479
evaporative cooler, 20, 21, 268, 311, 330, 353, 358, 374, 391, 394
fan, 3, 5, 6, 7, 8, 10, 12, 15, 16, 17, 20, 21, 29, 32, 34, 80, 117, 127, 143, 167, 176, 189, 208, 231, 244, 281, 338, 421, 480, 499, 500

Federal Motor Vehicle Safety Standard, 32
FMVSS, 32
foot muffs, 8, 146
Forecast 1992, 485
heat pump, 492
Heat-A-Seat, 337
heated rear view mirror, 31, 462
hood cover, 56
insulated, 33, 204
insulation, 13, 225
isinglass, 2
Kool Kooshion, 8, 16, 17, 20, 21, 147, 260, 274, 311
molded thermoset plastic housing, 390
Neutral Net, 34
ozone layer, 32
pedal slot closures, 5
pillows, 8, 13, 146, 213
plastic replaces metal bracket, 457
Postal Long Life Vehicles, 32, 480
pull-pull cable, 33
radiator shutter, 3, 5, 7, 10, 67, 164
refrigerated glove box, 34
side window demisters, 18
slide lever controls, 481
storm aprons, 45
sunshade, 6, 15, 35, 253, 274, 502
thermostat, 4, 7, 8, 10, 14, 17, 18, 19, 85, 94, 142, 155, 177
tire chains, 8, 147, 278
turnbuckle adjusters, 434
umbrella, 1, 40
vacuum motor, 381
vacuum storage tank, 17, 281
venetian blinds, 17, 19, 21, 274, 300, 338
visor, 6
wind tunnel, 14, 237

## PATENTS
hot water heater with electric fan, 117
Intermittent Windshield Wiper, 431
R-12 Refrigerant, 180
Rubber bladed fan, 15
Spring Lock Coupling, 459
Windshield Wiper, 42

Wiper Shaker, 501

## PEOPLE
Amin, Jay, 33
Amman, Jim 214
Anderson, John W., 22, 510
Anderson, Mary, 2, 42
Ayres, Paul, 16
Beatty, Jack, 3, 64
Caesar, Orville S, 6
Cole, Nick, 27
Connor, R.W., 301, 510
Davis, Michael W. R., 47
Daykin, Ted, 18, 26
DeVaux, 11, 181
Dickirson, Gene, 479
Donnelly, Jim, 4
Elson, Jerold, 11
Elton, Bob, 304
Etheridge, G.T., 24
Fermoyle, Ken, 22
Fetch, E.T. (Tom), 1
Fischer, Philip, 21, 22, 339, 437
Gardner, Greg, 9, 150
Gore, W.E, 301
Gray, D., 8, 149
Gunnell, John, 509
Henne, Albert L, 180
Houston, C.D., 16
Jandrey, Richard & Regina, 221
Jepson, John W., 4, 88
Johnson, Ryan, 3, 4, 9, 94, 156, 375
Joyce, Harold, 14, 19, 151, 510
Kearns, R.K., 29
Kimes, Beverly Rae, 15, 509
Kohl, E.E, 20, 314
Krarup, Marius C., 1
Kughn, Richard, 6, 16, 113, 215, 253, 262, 263, 264
McNary, Robert R., 180
Meurer, Ed, 189, 297
Midgley, Thomas, 180
Morningstar, Jim and Barb, 6, 118
Oishei, John R., 4
Pastoria, Andy, 32
Rowand, Debbie, 307
Rowand, Mike, 307

Schild, Jim, 9, 509
Schnaidt, Wayne, 479
Sisk, Ken, 479
Snitzer, Harvey and Julie, 19, 300
Waugaman, Candy, 51
Welch, Lou, 23
Width, Doug, 11, 188
Wood, Norm, 33

## REAR WINDOW
defogger, 29, 455, 502
electric grid rear window defroster, 30, 437
Ford system 1981, 470
lowering, 11, 13, 24
venetian blinds, 274, 300
wiper, 288, 299

## ROBE
cord, 487
fabric, 6, 8, 13, 16
lap, 1, 8, 13, 19, 44, 49, 56, 96, 146, 155, 213, 299
lock, 31, 34, 77, 459
rail, 5, 96, 112

## SEATS
cool cushion, 35, 267, 311
cooled, 274, 496
Heat-A-Seat, 337
heated and cooled, 496
heated Saab 1972, 440
Kool Kooshion, 8, 16, 17, 20, 21, 147, 260, 274, 311
ventilated, 33, 491
ventilated Saab 1997, 491

## STEERING WHEEL
electrically heated, 2, 20, 61, 440
muffs, 8, 50, 146

## VENTILATION
air intake under rear seat, 13
Astro ventilation, 432
Buick system 1954, 334
Chrysler Flow-Through 1967, 433
cowl-top, 4, 5, 10, 11, 12, 14, 15, 29, 89, 98, 178, 190, 204, 208, 217, 223, 238, 296
dual scoops, 8, 12, 18, 178, 208, 216, 266, 300, 327, 336, 385, 488
eave, 6
eaves, 8, 119, 145
Fisher Body No-Draft Ventilation, 11, 12, 190, 195
Fisher VV, 5, 6, 8, 111
flow thru, 427
flow thru ventilation, 427
Ford system 1975, 453
Pinto and Bobcat system 1975, 454
power, 473
rainproof, 193
rear facing cowl top inlet, 12, 204
rear window, 13, 17, 18, 19, 20, 29, 30, 211, 220, 221, 274, 288, 299, 309, 374, 437, 455, 470
side cowl, 14, 16, 178
vent kit, 363
ventilator, 6, 12, 14, 98, 245
ventipanes, 8, 11, 12, 15, 29, 145, 190, 196, 199, 245, 253, 427
windshield, 14, 196, 198, 238

## WEARING APPAREL
coats, 1, 43
electrically heated gloves, 2
gloves, 1, 2, 36, 39, 46, 61
goggles, 52, 87
hats, 1, 36
leather, 1, 43, 502

## WINDSHIELD
celluloid, 2, 5
coating, 2, 34
defogger, 29, 455, 502
defroster, 9, 13, 14, 15, 16, 17, 18, 20, 29, 30, 150, 151, 159, 166, 212, 218, 224, 225, 227, 229, 233, 234, 273, 277, 280, 305, 309, 341, 437, 470
deicer, 502
Fisher VV, 5, 6, 8, 111
French plate glass, 2
glycerin, 2
isinglass, 2
kerosene, 2
rock salt deicer, 6
roll up, 109
SLEETOFF, 7

swing out, 6, 14, 196, 198, 206, 238
Triplex, 2, 6, 119

## WINDSHIELD WIPER & WASHER

airfoil blades, 405
airfoil wiper blades, 27
antifreeze, 348
auxiliary arms, 13
booster, 12
Buick system 1954, 341
cable drive, 8, 137, 456
cable linkage, 8, 26, 383, 398
Cam-O-Matic, 349
Centrifugal, 4
Clear Advantage, 33
Corvair washer system 1965, 422
Corvette cover system 1968, 436
cowl-mounted, 12, 15, 195, 206, 207, 209
curved blades, 286
dual, 8, 12, 18, 178, 208, 216, 266, 300, 327, 336, 385, 488
Duckbill, 13
electric motor, 3, 4, 5, 8, 9, 18, 19, 20, 25, 27, 28, 33, 64, 133, 156, 310, 364, 372, 482, 489, 493, 504
electric washer, 407
electronic intermittent wiper control, 431
felt tipped blade, 5
Ferrari articulated arm system 1975, 456
fittings, 287
Flash of Genius movie, 29
Ford Fusion system 2011, 507
Ford system 1957, 377
four wipers, 20
four-link wiper mechanism, 34
fuel pump booster, 274
hand-operated, 2, 3, 4, 5, 8, 9, 72, 98, 100, 149
headlamp wipers, 30, 438
hydraulic motor, 27, 28
infrared sensor system, 33
intermittent windshield wiper control-electronic, 29, 431
intermittent windshield wiper control-mechanical, 407
jar, 13, 22, 29, 339
Lincoln system 1961, 27, 400
Lucas 14W two-speed motor, 31
multi spray, 373
one and a half blade system, 31, 456
parallel vertical, 6, 8, 10, 113, 144, 171
Power-Boosted Super Speed, 350
refillable blades, 4
single, 5, 11, 32, 195, 362, 478
storage tank, 17, 281
styled arms and blades Nissan 2009, 34, 498
styled arms and blades Subaru 2007, 498
technology chart Ford 1965-1970, 419
three wipers, 27, 404
vacuum booster fuel pump, 12
vacuum motor, 5, 11, 18, 19, 20, 25, 26, 27, 28, 30, 101, 144, 183, 304, 341, 368, 377, 381, 382
vacuum storage tank, 17, 281
vibrator blade Wiper Shaker, 34
washer, 4, 15, 17, 18, 19, 22, 25, 26, 27, 28, 29, 30, 246, 274, 281, 291, 303, 339, 348, 359, 372, 379, 399, 407, 422, 441
washer nozzle, 18, 29, 30, 291, 422, 441
wet arm, 26, 30, 373, 441

This page intentionally blank.

# ABOUT THE AUTHOR

Gene D. Dickirson, P.E. was born in Vincennes, Indiana and grew up in Lawrenceville, Illinois, Las Vegas, Nevada and Corona, California. He has an Associate in Science degree in Drafting Technology from Vincennes University and a Bachelor of Science degree in Mechanical Engineering from Lawrence Technological University. He began his career at Ford Motor Company in 1963 as a draftsman and retired in 1998 as a manager in the Product Engineering Office.
He is a Registered Professional Engineer in the State of Michigan. He has been awarded four U.S. Design Patents and had several technical papers published. He is the author of *GDT Speedster from dream to reality* and *The Sheldon Road Project*.